NEUROMETHODS

Series Editor
Wolfgang Walz
University of Saskatchewan
Saskatoon, Canada

For further volumes:
http://www.springer.com/series/7657

Neuromethods publishes cutting-edge methods and protocols in all areas of neuroscience, as well as translational neurological and mental research. Each volume in the series offers tested laboratory protocols step-by-step methods for reproducible lab experiments and addresses methodological controversies and pitfalls in order to aid neuroscientists in experimentation. *Neuromethods* focuses on traditional and emerging topics with wide-ranging implications to brain function, such as electrophysiology, neuroimaging, behavioral analysis, genomics, neurodegeneration, translational research, and clinical trials. *Neuromethods* provides investigators and trainees with highly useful compendiums of key strategies and approaches for successful research in animal and human brain function, including translational "bench to bedside" approaches to mental and neurological diseases.

Ultrastructural Neuroscience Methods

Edited by

Marie-Éve Tremblay

School of Medical Sciences, University of Victoria, Victoria, Canada

Humana Press

Editor
Marie-Éve Tremblay
School of Medical Sciences
University of Victoria
Victoria, Canada

ISSN 0893-2336 ISSN 1940-6045 (electronic)
Neuromethods
ISBN 978-1-0716-5455-2 ISBN 978-1-0716-5456-9 (eBook)
https://doi.org/10.1007/978-1-0716-5456-9

This Humana imprint is published by the registered company Springer Science+Business Media, LLC, part of Springer Nature.
The registered company address is: 1 New York Plaza, New York, NY 10004, U.S.A.

Preface

It is with great enthusiasm that I present this collection of chapters exploring the study of brain ultrastructure using various methodological approaches and from diverse perspectives.

The last few decades have seen a renaissance in our understanding of the structure–function relationship of the nervous system, including neuronal and non-neuronal cells, and their complex interactions with peripheral cells and tumors, among others, across many compartments. This transformative understanding was notably achieved through recent advances in transmission and scanning electron microscopy, enabling the investigation of brain ultrastructure at accelerated throughput in both two and three dimensions.

This collection of chapters combines well-established and emerging methods for studying brain ultrastructure, from biological sample preparation to transmission and scanning electron microscopy imaging and cutting-edge qualitative and quantitative analysis. The chapters provide essential considerations and protocols for the study of animal models, human cellular models, and human post-mortem brain samples across contexts of health and disease.

Overall, I am hopeful that these chapters will become key resources for biomedical, translational, and clinical researchers interested in delving into the mysteries of the brain at nanoscale resolution. I find these chapters highly informative and inspiring, paving the way for making significant discoveries that will push forward the limits of electron microscopy, in combination with complementary approaches such as fluorescence microscopy and spatial omics.

Chapter 1 details the principles and methodologies of freeze-fracture electron microscopy with particular emphasis on describing the procedures and interpreting images of freeze-fracture replicas generated from nervous system samples. Freeze-fracture electron microscopy provides an unparalleled view of the cytoarchitecture of cell membranes, notably forming synapses, and has become increasingly popular in recent years.

Chapter 2 explains how to combine light and electron microscopy to detect intracellular lipids within the nervous system. The intracellular accumulation of lipids is increasingly associated with neurological disease conditions, including both genetic and infectious ones. The chapter describes state-of-the-art protocols for preserving brain samples and generating correlative light and electron microscopy images in different contexts.

Chapter 3 focuses on immunocytochemical electron microscopy, a powerful technique for unraveling the subcellular localization of brain proteins and their spatial relationships with other receptors, ion channels, and various signaling pathways. The chapter describes the use of pre-embedding immunogold as an optimized method to define the distribution of glutamate and cannabinoid receptors and a complementary post-embedding approach.

Chapter 4 proposes a chromogenic immunostaining method as an alternative to the conventional horseradish peroxidase system. This method is based on an alkaline phosphatase system and is applicable to both light and electron microscopy examinations. As an example, staining for the neuronal marker NeuN is performed on mouse brain samples using an established pre-embedding approach.

Chapter 5 discusses mitochondria, dynamic cellular organelles playing unique and necessary roles across many contexts of health and disease. The chapter covers the ultrastructural phenotypes of mitochondria that can be studied using scanning and transmission electron microscopy, their functional relationships with many other organelles, and their alterations in the context of ischemic stroke pathology.

Chapter 6 illustrates an ultrastructural characterization of glial cells, including microglia labeled with the marker IBA1 and astrocytes labeled with GFAP, in unguided neural organoids with Alzheimer's disease pathology. A step-by-step protocol is described for the study of these glial cells, their organelles, and their relationships with other architectural elements of the organoids using scanning electron microscopy.

Chapter 7 details a guide for the analysis of microglial ultrastructure. These immune cells present various states in which they can play beneficial or detrimental roles, determining brain health throughout life. The chapter provides a step-by-step protocol for the study of microglial cell bodies, processes, organelles such as phagocytic inclusions, and interactions with other parenchymal elements, including synapses, axonal myelin, and other cell types.

Chapter 8 describes a quantitative analysis method allowing to discriminate between microglial states in electron microscopy images. In particular, this method differentiates between typical, intermediate, and dark microglia based on the electron density of these cells. Dark microglia present distinctive markers of cellular stress and appear to play a key role in brain remodeling. Intermediate microglia display intermediate features between typical and dark microglia.

Chapter 9 provides a comprehensive guide for the implementation of quantitative electron microscopy. The chapter includes step-by-step protocols for mouse perfusion, brain section immunostaining, resin embedding, and ultrathin sectioning. Pre-embedding immunoperoxidase and immunogold approaches are both covered. The chapter additionally comprises schematic representations showing the subcellular trafficking of receptors in different conditions.

Chapter 10 explains a well-established methodology for the investigation of breast cancer brain metastasis, a devastating progression associated with significant morbidity and mortality. This methodology includes the in vivo detection, as well as localization and monitoring of metastatic tumors, the histological assessment of transcriptional and protein-level changes in the brain tumor microenvironment, and the ultrastructural analysis of subcellular alterations in cultured cells.

Chapter 11 provides an automated and user-friendly software tool, SnapG, for studying the g-ratio. Calculated as the axon diameter divided by the total myelinated fiber diameter, the g-ratio is widely used to assess myelination in the nervous system. Changes in the g-ratio were found across neurological conditions throughout life. While manual tracing is the gold standard to measure g-ratios, this process is laborious, which is circumvented by SnapG to increase throughput.

Chapter 12 describes a unique workflow for the three-dimensional reconstruction of white-matter structures, including myelinated axons, from serial transmission electron microscopy sections of the mouse spinal cord. The innovative pipeline includes sample fixation, resin embedding, and imaging, as well as image processing, alignment, segmentation, three-dimensional reconstruction, and visualization, and applies to both contexts of physiology and pathology.

Chapter 13 illustrates the preparation steps needed to examine the human cerebral cortex in postmortem tissues. The provided protocol covers brain fixation, resin embedding, and ultrathin sectioning for visualization using transmission electron microscopy. It is specifically optimized for the preservation of myelin sheaths, as demonstrated at high resolution through the visualization of various critical features of myelin ultrastructure.

Victoria, Canada *Marie-Ève Tremblay*

Contents

Contributors

YOKO BEKKU • *Department of Neuroscience and Physiology, NYU Grossman School of Medicine, New York, NY, USA*

BIANCA CAROLINE BOBOTIS • *School of Medical Sciences, University of Victoria, Victoria, BC, Canada; Centre for Advanced Materials and Related Technology (CAMTEC), University of Victoria, Victoria, BC, Canada*

RAPHAËL BOURGEOIS • *School of Medical Sciences, University of Victoria, Victoria, BC, Canada; Department of Biology, Université Paris-Saclay, Paris, France*

STEPHEN CAI • *DART Microscopy Core, New York University Langone Medical Center, New York, NY, USA*

HECTOR J. CARUNCHO • *School of Medical Sciences, University of Victoria, Victoria, BC, Canada; Mental Health Research Cluster, University of Victoria, Victoria, BC, Canada*

IASONAS DERMITZAKIS • *Laboratory of Histology-Embryology, School of Medicine, Aristotle University of Thessaloniki, Thessaloniki, Greece*

EMILY FEDORA • *Department of Human Anatomy and Cell Science, University of Manitoba, Winnipeg, Manitoba, Canada*

SOFIA GARGANI • *Laboratory of Histology-Embryology, School of Medicine, Aristotle University of Thessaloniki, Thessaloniki, Greece*

INMACULADA GERRIKAGOITIA • *Department of Neurosciences, Faculty of Medicine and Nursing, University of the Basque Country (EHU), Leioa, Spain; Achucarro Basque Center for Neuroscience, Science Park of the EHU, Leioa, Spain*

ALEXANDRA GIANNAKI • *Laboratory of Histology-Embryology, School of Medicine, Aristotle University of Thessaloniki, Thessaloniki, Greece*

SONIA MARÍA GÓMEZ-URQUIJO • *Department of Neurosciences, Faculty of Medicine and Nursing, University of the Basque Country (EHU), Leioa, Spain; Achucarro Basque Center for Neuroscience, Science Park of the EHU, Leioa, Spain*

FERNANDO GONZÁLEZ IBÁÑEZ • *School of Medical Sciences, University of Victoria, Victoria, BC, Canada*

PEDRO GRANDES • *Department of Neurosciences, Faculty of Medicine and Nursing, University of the Basque Country (EHU), Leioa, Spain; Achucarro Basque Center for Neuroscience, Science Park of the EHU, Leioa, Spain*

ANITA GROOTEMAAT • *Electron Microscopy Centre Amsterdam, Department of Medical Biology, Amsterdam Institute for Immunology and Infectious Diseases (AI&I), Microscopy & Cytometry Core Facility, Amsterdam UMC, Amsterdam, Netherlands*

SOPHIA HAVAKI • *Molecular Carcinogenesis Group, Department of Histology and Embryology, Medical School, National and Kapodistrian University of Athens, Athens, Greece*

ELLY M. HOL • *Department of Translational Neuroscience, University Medical Center Utrecht Brain Center, Utrecht University, Utrecht, The Netherlands*

SABINE HOMBACH-KLONISCH • *Department of Human Anatomy and Cell Science, University of Manitoba, Winnipeg, Manitoba, Canada; Department of Pathology, University of Manitoba, Winnipeg, Manitoba, Canada*

DANYYL IPPOLITOV • *Department of Human Anatomy and Cell Science, University of Manitoba, Winnipeg, Manitoba, Canada*

LISA E. KALYNCHUK • *School of Medical Sciences, University of Victoria, Victoria, BC, Canada*

CHRISTINA KATSIGIANNI • *Laboratory of Histology-Embryology, School of Medicine, Aristotle University of Thessaloniki, Thessaloniki, Greece*

FRANCES KESTEL • *Department of Neuroscience and Physiology, NYU Grossman School of Medicine, New York, NY, USA*

MOHAMMADPARSA KHAKPOUR • *School of Medical Sciences, University of Victoria, Victoria, BC, Canada*

THOMAS KLONISCH • *Department of Human Anatomy and Cell Science, University of Manitoba, Winnipeg, Manitoba, Canada; Department of Pathology, University of Manitoba, Winnipeg, Manitoba, Canada; Departments of Medical Microbiology and Infectious Diseases, University of Manitoba, Winnipeg, Manitoba, Canada; CancerCare Manitoba, Winnipeg, Canada*

MAARTEN H. P. KOLE • *Axonal Signaling Group, Netherlands Institute for Neurosciences (NIN), Royal Netherlands Academy for Arts and Sciences (KNAW), Amsterdam, The Netherlands; Cell biology, Neurobiology and Biophysics, Department of Biology, Faculty of Science, Utrecht University, Utrecht, The Netherlands*

VICTOR Z. LAU • *School of Medical Sciences, University of Victoria, Victoria, BC, Canada; Institute on Aging and Lifelong Health (IALH), University of Victoria, Victoria, BC, Canada; Centre for Advanced Materials and Related Technology (CAMTEC), University of Victoria, Victoria, BC, Canada; Neuroscience Graduate Program, University of Victoria, Victoria, BC, Canada*

ANDRÉE LESSARD • *Plateforme de microscopie électronique, Faculté de Médecine et des Sciences de la Santé, Université de Sherbrooke, Sherbrooke, QC, Canada*

ALICE F. LIANG • *DART Microscopy Core, New York University Langone Medical Center, New York, NY, USA*

BASTIAAN LODDER • *Axonal Signaling Group, Netherlands Institute for Neurosciences (NIN), Royal Netherlands Academy for Arts and Sciences (KNAW), Amsterdam, The Netherlands*

MARIA-ELENI MANTHOU • *Laboratory of Histology-Embryology, School of Medicine, Aristotle University of Thessaloniki, Thessaloniki, Greece*

SOULTANA MEDITSKOU • *Laboratory of Histology-Embryology, School of Medicine, Aristotle University of Thessaloniki, Thessaloniki, Greece*

EMILY K. MILLER • *School of Medical Sciences, University of Victoria, Victoria, BC, Canada*

PATRICK C. NAHIRNEY • *School of Medical Sciences, University of Victoria, Victoria, BC, Canada*

SANNE VAN DER NIET • *Electron Microscopy Centre Amsterdam, Department of Medical Biology, Amsterdam Institute for Immunology and Infectious Diseases (AI&I), Microscopy & Cytometry Core Facility, Amsterdam UMC, Amsterdam, Netherlands*

IOANNIS PANTELIDIS • *Laboratory of Histology-Embryology, School of Medicine, Aristotle University of Thessaloniki, Thessaloniki, Greece*

KATHERINE PICARD • *School of Medical Sciences, University of Victoria, Victoria, BC, Canada*
NAGORE PUENTE • *Department of Neurosciences, Faculty of Medicine and Nursing, University of the Basque Country (EHU), Leioa, Spain; Achucarro Basque Center for Neuroscience, Science Park of the EHU, Leioa, Spain*
ERIC REITS • *Microscopy and Cytometry Core Facility, Department of Medical Biology,, Amsterdam UMC, Amsterdam, Netherlands*
ITZIAR BONILLA-DEL RÍO • *Department of Neurosciences, Faculty of Medicine and Nursing, University of the Basque Country (EHU), Leioa, Spain; Achucarro Basque Center for Neuroscience, Science Park of the EHU, Leioa, Spain*
JOSEPH SALL • *DART Microscopy Core, New York University Langone Medical Center, New York, NY, USA*
JAMES L. SALZER • *Department of Neuroscience and Physiology, NYU Grossman School of Medicine, New York, NY, USA; Institute for Translational Neuroscience, New York University Langone Medical Center, New York, NY, USA*
CARLA L. SANCHEZ-LAFUENTE • *School of Medical Sciences, University of Victoria, Victoria, BC, Canada; Department of Microbiology and Immunology, Western University, London, ON, Canada*
EVA ŠIMONČIČOVÁ • *School of Medical Sciences, University of Victoria, Victoria, BC, Canada*
JULIA TCW • *Department of Pharmacology, Physiology & Biophysics, Boston University, Chobanian & Avedisian School of Medicine, Boston, MA, USA; Bioinformatics Program, Faculty of Computing & Data Sciences, Boston University, Boston, MA, USA*
PASCHALIS THEOTOKIS • *Laboratory of Histology-Embryology, School of Medicine, Aristotle University of Thessaloniki, Thessaloniki, Greece*
MARIANELA EVELYN TRAETTA • *School of Medical Sciences, University of Victoria, Victoria, BC, Canada*
MARIE-ÈVE TREMBLAY • *School of Medical Sciences, University of Victoria, Victoria, BC, Canada; Centre for Advanced Materials and Related Technology (CAMTEC), University of Victoria, Victoria, BC, Canada; Biochemistry and Molecular Biology Department, The University of British Columbia, Vancouver, BC, Canada; Institute of Aging and Lifelong Health (IALH), and Center for Advanced Materials and Related Technologies (CAMTEC), University of Victoria, Victoria, BC, Canada*
IOANNA TREMI • *Molecular Carcinogenesis Group, Department of Histology and Embryology, Medical School, National and Kapodistrian University of Athens, Athens, Greece*
MARLOES VERKERKE • *Department of Translational Neuroscience, University Medical Center Utrecht Brain Center, Utrecht University, Utrecht, The Netherlands*
NICOLE N. VAN DER WEL • *Electron Microscopy Centre Amsterdam, Department of Medical Biology, Amsterdam Institute for Immunology and Infectious Diseases (AI&I), Microscopy & Cytometry Core Facility, Amsterdam UMC, Amsterdam, Netherlands*
THOMAS XIN • *DART Microscopy Core, New York University Langone Medical Center, New York, NY, USA*
JARED VANDER ZWAAG • *School of Medical Sciences, University of Victoria, Victoria, BC, Canada; Centre for Advanced Materials and Related Technology (CAMTEC), University of Victoria, Victoria, BC, Canada; Neuroscience Graduate Program, University of Victoria, Victoria, BC, Canada*

Chapter 1

Freeze-Fracture Electron Microscopy Methods for Evaluation of Nerve Cells and Synapses

Carla L. Sanchez-Lafuente, Lisa E. Kalynchuk and Hector J. Caruncho

Abstract

Freeze-fracture electron microscopy offers an unparalleled view of the cytoarchitecture of cell membranes, which has been used in numerous studies and observations in nervous system cells and synapses. While these technologies were quite popular in the 1980s and 90s, they became less used thereafter but in the past 5 years there has been an increase in studies using these methodologies. This chapter summarizes the principles and methodologies of freeze-fracture electron microscopy with a special emphasis on describing procedures and interpretation of images of freeze-fracture replicas from nervous system samples.

Keywords Freeze-fracture electron microscopy, Intra-membrane particles, Fracture-label, Label-fracture, Fracture-flip

1 Introduction

Freeze-fracture techniques were initially developed in the 1960s, and involve the vacuum deposition of a replica, typically made of platinum and carbon, onto the fracture plane generated in a frozen specimen. Normally, the replica is examined under a transmission electron microscope, although there are also methods that allow studying the replica's topography using scanning electron microscopy techniques. While these techniques are not much used nowadays, they represent an unusual view of cell membranes' architecture that offers interesting insights into membrane events and proteins seen as intramembrane particles in freeze-fracture replicas.

These techniques have demonstrated their ability to reveal the macromolecular architecture of membranes, as well as the number, precise distribution, and biochemical identity of a wide variety of transmembrane molecules. This is because, in biological specimens, the fracture lines often pass through the lipid bilayer's midpoint, dividing it into two monolayers. In this region, the connections are weaker, primarily of a hydrophobic nature, and in the case of frozen

Marie-Ève Tremblay (ed.), *Ultrastructural Neuroscience Methods*, Neuromethods, vol. 238,
https://doi.org/10.1007/978-1-0716-5456-9_1,

specimens, there are also Van der Waals forces at play. Replicating the fractured surfaces allows the observation of numerous particles, known as intramembrane particles, which are believed to represent integral membrane proteins.

The purpose of this chapter is first to describe in detail freeze-fracture methodologies, and thereafter briefly analyze images and provide an interpretation of freeze-fracture replicas from nervous tissue. A general set of references or recommended reading list is given at the end of the chapter but not referred to in the text.

Freeze-Fracture Devices

In this section, we describe the various components of a freeze-fracture device (we are taking as an example the Balzers 400-D freeze-fracture machine (see Fig. 1), but the general statements can be applied to other freeze-fracture devices).

A freeze-fracture apparatus consists of the following components: a vacuum system, composed of a rotary pump and a diffuser; a freezing system that utilizes liquid nitrogen as the cryogenic element; and a shadowing system, which includes a platinum electrode and a carbon electrode. In this section, we are going to present step-by-step instructions on how to use the Balzers 400-D by describing its parts and operating instructions.

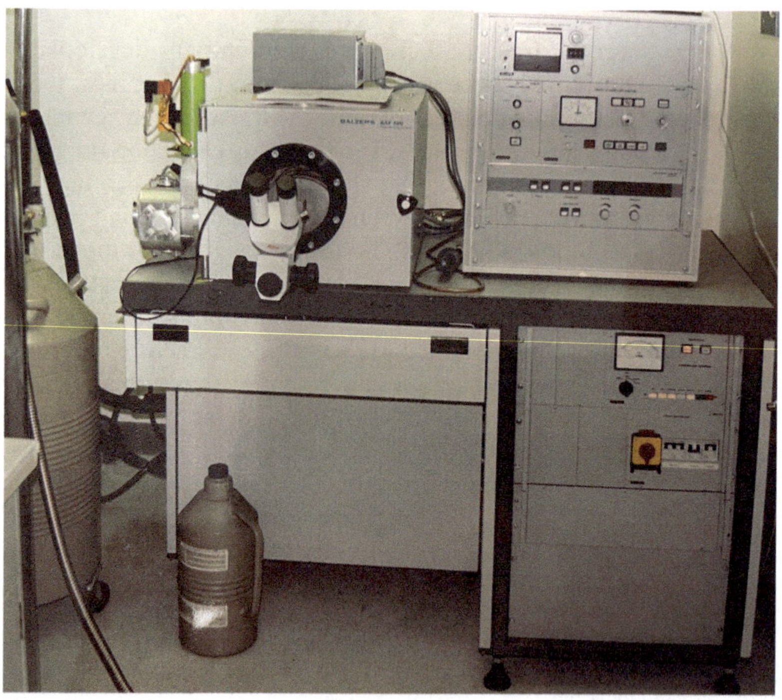

Fig. 1 Photograph of a Balzers 400-D freeze-fracturing device

1.1 System Startup and Electrode Cleaning

To start the system, it is first necessary to connect compressed air to allow the opening and closing of the high vacuum valves and water for the cooling circuit of the diffusion pump. You should also fill the liquid nitrogen tank, subject it to a pressure of 0.7 bar, and open valve FET004 to connect the nitrogen tank to the Balzers 400-D. Next, turn on the main switch of the ALC 101 system, which allows sample exchange without breaking the vacuum in the cryo-fracture chamber. Finally, turn on the switch for the vacuum control unit DPA 102 and set it to the STANDBY position until the diffusion pump reaches its operating temperature (which takes about 30 minutes). During this time, you can clean the carbon and platinum electrodes that may have become dirty in your previous cryo-fracture.

To clean the electrodes, start by venting the chamber, which is done by selecting the VENT position on the DPA 102 unit. Once the chamber is open, switch it back to STANDBY. To clean the electrodes, remove them from the chamber (always check that there are no electrical residues and, of course, wear gloves). Remove the hood and carefully clean the metal parts with a piece of "scotch-brite" soaked in alcohol. Then, check that the carbon and platinum rods are in a condition to be reused. If not, proceed to replace them. Finally, reattach the electrode hood and return it to its position in the cryo-fracture chamber, which has been closed during this time.

Once this is done, switch the DPA 102 to the PUMP position. The system will automatically proceed to create a vacuum in the chamber. Once a pressure of 0.01 mbar is reached, the high vacuum valve will open, allowing the diffusion pump to operate in the chamber. In approximately 30 min, you will have an adequate vacuum to proceed with the cryo-fracture process.

1.2 Performing Freeze-Fracture

At this point, samples should be ready for cryo-fracture (we will discuss how to prepare these samples later), and the Balzers 400-D should have an adequate vacuum level (around 2×10^{-6} mbar).

Sample Insertion

With the ALC 101 system, we can insert the sample into the chamber's sample holder without breaking its vacuum conditions. To do this, we activate the VENT switch on the ALC 101, which ventilates the intermediate chamber (like inserting grids in an electron microscope). We then close the VENT switch, place our sample with the transfer rod, and activate the PUMP switch. This action creates a vacuum in the intermediate chamber (approximately 20 s). Afterward, we close the PUMP switch and activate SV 1, which opens the valve connecting the intermediate chamber to the main chamber. We place the sample on the sample holder, which has been pre-cooled to −130°C, and remove the transfer rod to the intermediate chamber. Then, we disconnect the SV 1 switch, which closes the valve. We remove the transfer rod from the intermediate chamber, close it, and activate the PUMP switch for 20 s to maintain the vacuum in the intermediate chamber when not in use.

Fracture and Shadowing

The system should be under an adequate vacuum (2×10^{-6} mbar), and the sample and the blade should be at the operating temperature. Under the microscope and using the microtome controlled by the BMB 101 system, lower the blade and start cutting (fracturing) the specimen until you obtain an appropriate fracture surface. Before making the final cut/fracture, turn on the quartz crystal monitor QSG 301, which controls the thickness of the replica. Also, turn on the EVM 030 unit, which controls platinum and carbon shadowing. Make the final cut/fracture while keeping the blade over the sample, while we start with the shadowing (if you keep it in place for several minutes and raise the sample's temperature to about −100°C, an etching of the sample occurs because the ice in it sublimates and condenses onto the blade, which is at a much lower temperature of −196°C).

To create shading, we select electrode #1 (platinum) on the EVM 030 unit, which will be used at a voltage of 1.6 KV and a current of 60 mA. We then turn on the HIGH VOLTAGE switch, and the system will automatically heat the electrode until it begins to evaporate. This can be observed through the glass as a "rain" that gradually deposits onto the specimen. Once it starts, we remove the blade from above the sample so that the platinum deposits on it. When the thickness of the platinum on the specimen is around 2 nm (as indicated by the QSG 301 glass monitor), we turn off the HIGH VOLTAGE switch. We immediately select electrode #2 (carbon) and turn the HIGH VOLTAGE switch back on (which will set the carbon electrode to 1.9 KV and 90 mA), resulting in the evaporation of carbon onto the replica. We then verify that the carbon shading thickness reaches 20 nm (using the QSG 301 monitor) and turn off the HIGH VOLTAGE switch. NOTE: The platinum electrode is generally positioned at a 45° angle above the sample, while the carbon electrode is placed vertically above the sample. We then remove the sample from the chamber in the same way it was introduced (see Sect. 1.2.2).

1.3 Shutdown of the System

At least 2 hours should pass after performing a cryo-fracture before turning off the apparatus. If new cryo-fractures are planned for the same day or the next day, the system should be left running. Close the valve on the liquid nitrogen tank. Turn off the switches for the units QSG 301, the high vacuum gauge (selecting BACK), the DPA 102, and the ALC 101. The system should stop making noise, and all lights should be off.

2 Biological Samples Preparation Protocol for Freeze-Fracture

If you have a system that allows ultra-rapid freezing using liquid helium or propane, small tissue samples can be frozen directly without any prior steps. In this case, it is a form of physical fixation, and the rapid freezing causes water to solidify in an amorphous state instead of forming ice crystals, which could rupture cells. If you don't have access to an ultra-rapid freezing system and to prevent the problem of ice crystal formation mentioned earlier, a cryoprotectant solution is typically used. This involves sequentially immersing the tissue in aqueous solutions of glycerol at 10%, 20%, and 30% before freezing the sample. However, the use of cryoprotectants can introduce some artifacts (e.g., vesiculation of the endoplasmic reticulum and cell vacuolization). To mitigate this, a slight chemical fixation is typically performed first using 4% paraformaldehyde in phosphate buffer at pH 7.4. In these cases, freezing is often done by using isopentane cooled with liquid nitrogen or by directly freezing in semisolid (slush) nitrogen.

Before freezing the sample (and after fixing and embedding it in glycerol), it should be placed on sample holders, which are usually made of copper coated with a gold layer. To do this, a small drop of OCT is deposited on the sample holder, and the sample is placed on top of it. Freezing should be fast, and the sample should not be kept very close to the freezing substance before freezing. Simply, take the sample holder with tweezers and quickly immerse it in the freezing solution for about 4–5 s. After removing it, quickly transfer it to liquid nitrogen.Although it's advisable to clean the edges of the sample holder with a piece of filter paper, this should be done very quickly (no more than two seconds). Once the samples are frozen, they can only be handled with tweezers that have been pre-cooled in liquid nitrogen. The sample holders are then placed in the liquid nitrogen on the sample holder table, and this is introduced into the platform in the cryo-fracture chamber (see the section above on inserting the sample into the chamber).

As we will see later, great care must be taken during the sample freezing process because many potential artifacts occur during this stage. Once the freeze-fracture has been performed as explained in Sect. 1.2., the replicas have adhered to the fractured tissue, and the sample holders are removed from the freeze-fracture chamber. They are gently immersed in a solution of distilled water, causing the replica (with adhered tissue) to float on the water. In the case of immunocytochemistry techniques combined with cryo-fracture (see Sect. 5), the replicas are washed in distilled water several times before moving on to immunolabeling or being collected on grids coated with a formvar membrane. However, for conventional cryo-fracture techniques, the replicas are transferred to a solution of sodium hypochlorite (e.g., bleach). The replicas should be in the hypochlorite solution for at least 2 h before being transferred back to distilled water (at least half an hour) and collected on grids coated with a formvar membrane.

NOTE: The transfer of replicas from one solution to another is done with small metal washers, which hold a film of water (due to surface tension, like making soap bubbles, for example). The washer is passed under the replica, and upon lifting it, a film of water remains with the replica. It is then gently placed in the next solution. Another method is to collect the replicas on glass surfaces (e.g., the wall of a Pasteur pipette) and deposit them in the next solution. The grids with replicas are allowed to dry and are ready to be observed with a transmission electron microscope.

3 Freeze-Fracture Shadowing Techniques

3.1 Complementary Replicas Technique

While in the case of normal freeze-fracture, one face of the membrane is lost after fracture, the technique of complementary replicas (or double replicas) allows both faces of the same membrane to be observed. In this technique, the sample is frozen between two supports, which are placed on a sample holder that acts as a hinge so that when opened, a fracture occurs, and replication is performed on both sides of the fractured surface, without losing either face. In the case of complementary replicas, the particles on each of the faces should correspond exactly with the cavities on the opposite face. Nevertheless, due to the shading covering the original material or various contamination issues, it is very difficult to achieve an exact correspondence between the two replicas.

3.2 Rotational Shadowing Technique

Rotational shadowing techniques have been developed to overcome some of the problems with the relief contrast of the surface encountered in classical single-directional shadowing techniques. In this case, after the fracture, the sample holder begins to rotate rapidly, causing shading in all directions rather than just one. This way, it was possible to distinguish several subunits in some particles of chloroplast membranes that were not observable with conventional methods.

3.3 Deep Etching Technique

Another interesting technical variant is known as deep etching. The technique is like freeze-etching, but in this case, the ice on the fractured surface is allowed to sublimate for an extended period (usually about an hour). When using this technique, ultra-rapid freezing methods should be employed because glycerol solutions cannot be used as they do not sublimate and can lead to artifacts. These techniques are typically used for the detailed study of cytoskeletal components exposed after fractures through the cell, and after prolonged etching, they are shaded with platinum and carbon.

4 Interpretation of Obtained Images and Nomenclature

The most important characteristic of images of membranes obtained through cryo-fracture is their unconventional nature: the observed image corresponds to a given membrane or an area of this membrane and is not affected by neighboring membranes or the regions surrounding the membrane. Additionally, the presence of intramembrane particles (possibly corresponding to transmembrane proteins) and graininess are unique features observed in fractured surfaces of biological membranes.

As early as the 1960s, it was demonstrated that the fracture plane in biological membranes occurs between the two lipid layers, so cryo-fracture provides a view of the apolar interior of the membranes. Initially, the nomenclature to refer to the faces of the membrane and the two sublimation-exposed surfaces was: A, B, C, D, or convex and concave, or descriptive names (internal and external fracture faces, internal and external surfaces). However, to standardize the terminology, a new system based on cellular structure was later adopted. In this system, the face of the membrane closest to the cellular "plasma" (cytoplasm, nucleoplasm, mitochondrial matrix, stroma) was named the protoplasmic face or face P, and the face of the membrane closest to extracellular or exoplasmic spaces was named the exoplasmic face or face E. Likewise, the surfaces of the membranes were denoted as Ps and Es, respectively (Fig. 2).

For a proper interpretation of replica images, it should be noted the direction and angle of shading. Photographs should be presented in such a way that the shading direction (white shadows due to the non-deposition of platinum) goes from the lower end of the

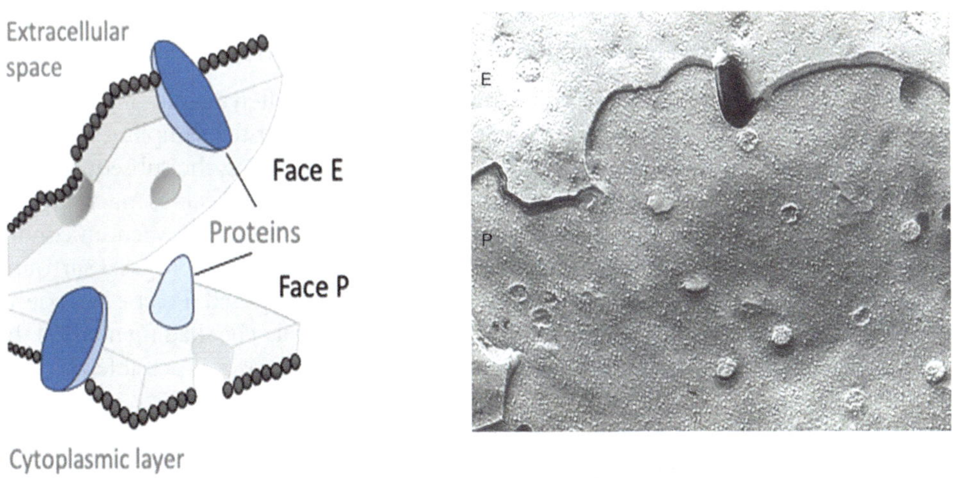

Fig. 2 Representation of faces E and P on the left and electron microscopy image of freeze-fractured tissue on the right (Cell nucleus showing the P-face of the internal nuclear membrane and the E-face of the external nuclear membrane). (Taken from Caruncho et al. [1])

photograph to the upper end. Incorrect orientation can lead to an image where the particles appear as cavities and vice versa. The interpretation of some particle details may be interesting when obtaining stereo images. Once the tissue is fractured, immediate replication takes place, which is when cryo-fracture truly occurs. In case the ice on the fractured surface is allowed to sublimate (see Sect. 1.2.3), it is referred to as freeze-etching, which allows the observation of the true surfaces of the membranes.

5 Artifacts

The protocol for obtaining replicas is laborious and errors or changes can occur, resulting in artifacts. The main causes of artifacts that can be induced during the sampling and pre-treatment of specimens are as follows:

1. Problems due to gravitational forces when centrifugation methods are used.
2. Mechanical stress in sections or during freezing.
3. Problems with hydrostatic pressure.
4. Changes in solute concentration.
5. Changes in the gaseous composition of the environment (e.g., anoxia).
6. Changes in specimen temperature before freezing.
7. Infiltration with cryoprotectants when chemical fixation was not used.
8. Problems with glutaraldehyde fixation.
9. Problems with osmium tetroxide fixation.
10. Treatment with non-penetrating cryoprotectants.

All these causes can lead to artifacts, often manifesting as membrane fusion or particle rearrangement. Even if the sampling and freezing were adequate, artifacts can still occur during the subsequent steps of fracture and replication. To minimize these defects as much as possible, the following considerations must be considered: vacuum conditions should be excellent with a minimum of condensable gases surrounding the specimen when the fracture plane is exposed. Precise and consistent specimen temperature control is essential. The shading method should generate as little heat and contaminating gases as possible.

6 Immunocytochemical Techniques Applicable to Cryo-Fracture

Over the last decade, various cryo-fracture techniques have been developed that can be combined with immunocytochemical methods. These methods have certain limitations but are of considerable interest for studying the topography of membrane receptors (e.g., neurotransmitter receptors).

We will briefly describe the four main techniques of this type, along with a routine protocol for their use. These techniques are Fracture and Labeling, Labeling and Fracture, Fracture and Flip, and SDS-Cryo-fracture.

6.1 Fracture and Labeling (Fracture-Label)

In this technique, the fracture is performed under nitrogen at atmospheric pressure. Once the samples are fractured, they are de-glycerinated, labeled using conventional immunocytochemical methods, and embedded for ultra-thin sectioning or taken to a cryo-fracture chamber to replicate the fractured surfaces. This allows the identification of integral membrane proteins and establishes a distribution pattern on the protoplasmic side of the fractured membrane, determines the partitioning and topology of lipids and proteins on the protoplasmic and exoplasmic sides of intracellular and plasma membranes, studies the intracellular trafficking of specific membrane components, and investigates the topology and compartmentalization of intracellular membranes, as well as marking some receptors exposed through nucleoplasm and cytoplasm after fracture.

6.2 Labeling and Fracture (Label-Fracture)

This technique is only applicable to cell suspensions or monolayer cultures. It involves labeling the external cell surface and then performing the cryo-fracture. The shaded specimens are rinsed with distilled water (sodium hypochlorite solutions are not used, so the cytoplasmic faces of the membranes remain attached to the replica). This allows us to obtain replica images that show the exoplasmic faces underneath, displaying electron-dense labeling (usually colloidal gold).

Cell Suspension Labeling and Fracturing Protocol

- The cell culture medium is taken, and cells are centrifuged at a relatively low speed (about 700–800 rpm), causing cell precipitation. The cells are immediately resuspended in a 4% paraformaldehyde solution in phosphate buffer (pH 7.4) recently prepared. The sample is kept in the fixing solution at 4°C for half an hour.
- The cell suspension is centrifuged again, the paraformaldehyde is discarded, and the cells are resuspended in phosphate buffer. They are kept in the phosphate buffer with agitation for a minimum of half an hour (it is advisable to change the phosphate buffer every 10 min, for which the cell suspension must be centrifuged again, the used phosphate buffer removed, and the new one added).
- We are now ready to perform immunolabeling, with steps like those performed, for example, in pre-inclusion immunocytochemistry. The main difference is that in this case, when we need to transfer the cell suspension from one solution to the next, we must centrifuge the cells and then resuspend them in the next

solution. Additionally, in this case, we can only use colloidal gold as the marker (good results are obtained using secondary antibodies or protein A conjugated with 10 nm colloidal gold).

- After completing immunolabeling, the cells are washed several times in phosphate buffer before being embedded in increasing concentrations of glycerol (10%, 20%, and 30%). This is done after centrifuging the cell suspension once again.
- Once the glycerol incubation at 30% is finished, the cells are centrifuged once more and then resuspended in a small volume of 30% glycerol so that we have a high cell concentration in a droplet of the suspension.
- A small droplet of the suspension is placed on flat sample holders for double replicas, and another sample holder is placed on top, like a sandwich.
- The sandwich is then frozen, for example, in semi-solid nitrogen, following the method described earlier, and the samples are ready for cryo-fracture.
- After cryo-fracturing, the replicas are washed several times with distilled water and collected on grids coated with a formvar membrane.

Monolayer Culture Labeling and Fracturing Protocol

- The simplest approach is to cultivate cells in Petri dishes and place small 4 mm diameter glass circles on them before coating them with poly-lysine. Cells growing on these circles will be used for immunolabeling.
- Once the culture is established, cells can be fixed by removing the culture medium and adding fresh 4% paraformaldehyde solution in PBS. Maintain the cultures in the fixative at 4°C for approximately half an hour, then replace the fixative with PBS.
- We then proceed with immunolabeling using a routine immunocytochemistry technique, but we will only obtain labels for epitopes located in the extracellular domain of the membrane. We will use a secondary antibody conjugated with 10 nm colloidal gold, or with 5 and 15 nm in the case of double labeling.
- After immunolabeling, we proceed with the inclusion of cultures in glycerol by simply changing the solution containing the culture to glycerol solutions of 10%, 20%, and 30%, 10 min each.
- After glycerol inclusion is completed, we create a sandwich with the Balzers culture sample holders and the small 4 mm diameter glass circles with cells.
- We freeze the sandwich and insert it into the Balzers 400-D for cryo-fracturing.

- Once replicas are obtained, they are washed repeatedly in distilled water before being collected on copper grids coated with a formvar membrane.

6.3 Fracture and Inversion (Fracture-Flip)

In this case, labeling is performed either before or after fracturing, but once the samples are frozen and fractured, they are shaded with a single layer of carbon (approximately 30 nm thick). The replicas are then repeatedly washed in distilled water and collected on a grid coated with a formvar membrane. This process should be carried out in such a way that the carbon replica is directly on the formvar, and the fractured surfaces are exposed for labeling if it was not done before fracturing. Once the replica and the adhered membranes have dried, they are placed back in the Balzers 400-D at room temperature, and shading is carried out again, this time using platinum and carbon. In this way, if the E and P faces were exposed during the fracture, after collecting the replica on the formvar membrane, the true external membrane surface is exposed in the case of the E face, and this surface is revealed with the second shading. This results in images similar to those from scanning electron microscopy (revealing the actual cell surface), with the advantage that fracture and inversion can provide better resolution.

6.4 SDS-Freeze-Fracture

As we have seen, the techniques mentioned so far only allow for the use of immunocytochemical techniques coupled with cryo-fracture in the case of monolayer cultures or suspended cells. Other techniques that could be coupled with tissue cryo-fracture were developed in the 1980s but presented multiple problems. However, a new technique, SDS-Freeze-fracture, has been developed, which allows the use of immunolabeling techniques on freeze-fractured tissue samples with relatively few issues. This technique involves the use of collagenase and a solution of sodium dodecyl sulfate (SDS) to digest part of the tissue adhered to the replica (like what bleach does in the case of regular freeze-fracture). In this case, the fractured membranes adhered to the replica are stable against the action of SDS and, therefore, do not fragment. After tissue digestion, it is washed several times in PBS, and immunolabeling is then carried out. Upon completion, the replicas are washed and collected on grids for observation using the electron microscope.

Briefly, the protocol for this technique is as follows:

- Tissue sections of approximately 50 μm (previously fixed) are prepared using a vibratome.
- After embedding the sections in glycerol, cryo-fracture is performed as previously described.
- Replicas with attached tissue are washed several times in PBS before undergoing agitation washing in a 5% solution of collagenase I or II at 37°C.

- Subsequently, the replicas are washed again in PBS and left to soak in a 3% SDS solution overnight with vigorous agitation.
- The next day, the replicas are washed once more in PBS, and immunolabeling is carried out.
- Upon completion of labeling (using, as always, a secondary antibody conjugated with colloidal gold), the replicas are washed several times in distilled water before being collected on grids.

7 Freeze-Fracture of Nervous System Cells and Synapses

This section helps to interpret images of freeze-fracture replicas from the nervous system through multiple photographs, including cell nucleus, cytoplasm, organelles, and plasma membrane of neurons,

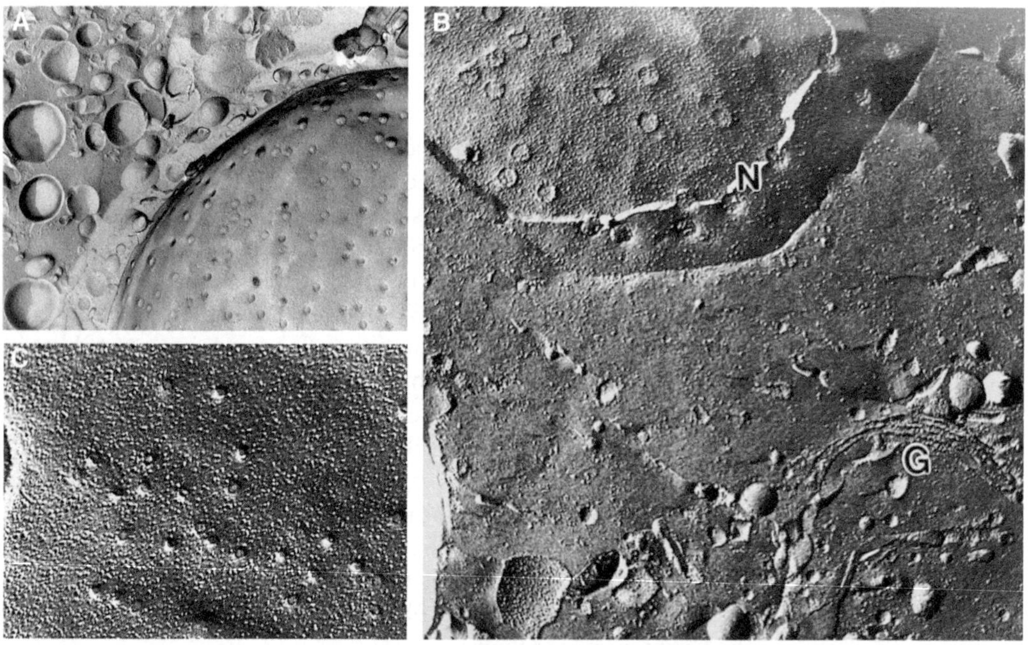

Fig. 3 (**A**) Freeze fracture image of the nucleus and cytoplasm of a hypothalamic neuron in the rainbow trout brain. Note the tridimensional fracture of the nuclear internal membrane showing multiple nuclear pores as well as fracture of organelle membranes (primarily secretory granules). (**B**) Cross-fracture of a cell in the inner meningeal layer of the goldfish, showing the external face of the outer nuclear membrane and the protoplasmic face of the inner nuclear membrane (left). Note the nuclear pores, and the numerous intramembrane particles that are more numerous in the protoplasmic face (N). The image also shows the cross-fracture of a Golgi apparatus dictyosome G) and the membranes of organelles (mitochondria and secretory vesicles). (**C**) Protoplasmic face fracture of the plasma membrane of a cell in the intermediate layer of the rainbow trout meninges, showing plasmalemma vesicles and numerous intramembrane particles. (Figures B and C are taken from Caruncho et al. [2])

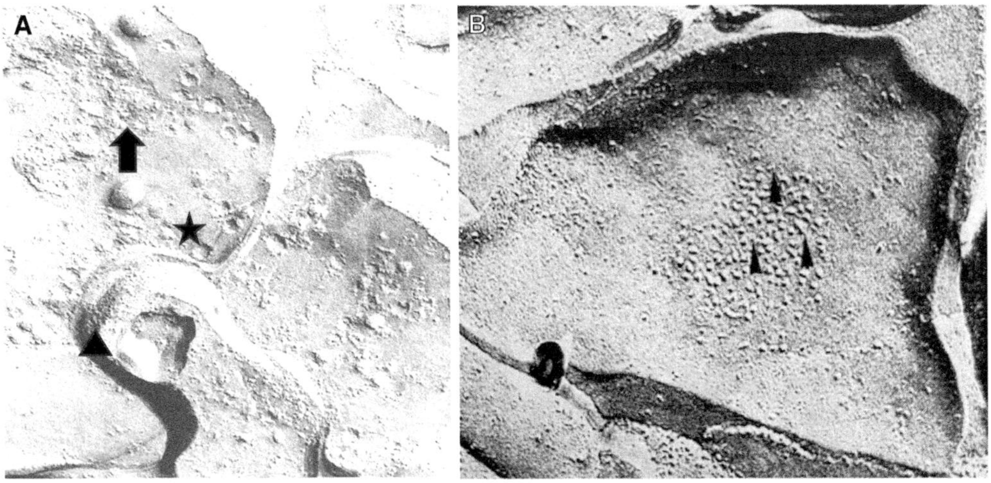

Fig. 4 Freeze-fracture of synaptic contacts in hypothalamic neurons in the rainbow trout. (**A**) Double synapse impinging onto one dendritic spine. The image shows two cup-like cross-fractures of presynaptic terminals where one can easily identify synaptic vesicles and a cross-fracture of a mitochondrion (arrow). Both terminals make synapses with a dendritic spine, and one can see synaptic vesicles lining inside the presynaptic terminal membrane (star) and the accumulation of intramembrane particles (arrowhead, representing neurotransmitter receptors) in the postsynaptic membrane. (**B**) Image of the external fracture face of a postsynaptic terminal plasma membrane showing an accumulation of intramembrane particles (clustering of neurotransmitter receptors, arrowheads) in the postsynaptic membrane active area. (Images taken from Caruncho et al. [3])

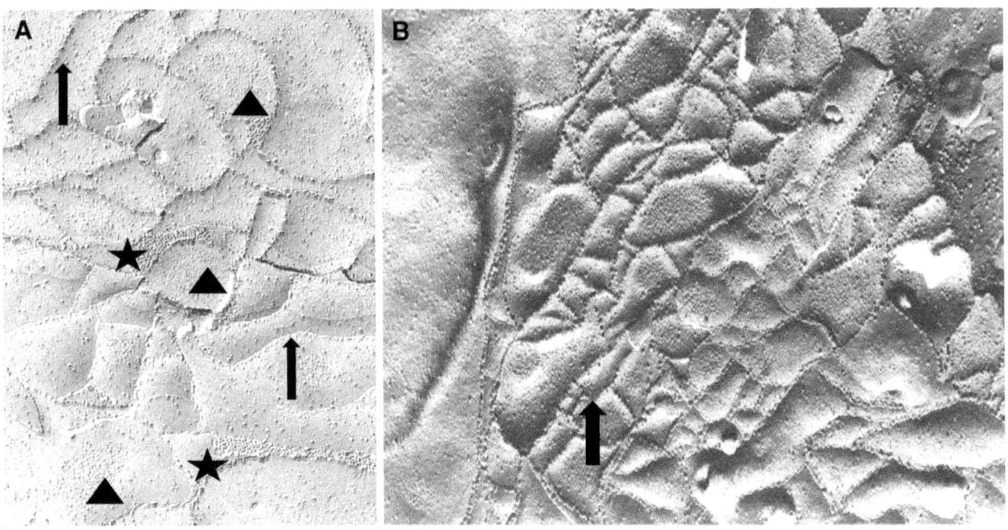

Fig. 5 Freeze-fracture of cells in the barrier layer of the goldfish meninges (**A** and **B**), showing rows of tight junctions, seen as lines of particles or furrows, that represent occludins and other tight junction proteins that basically are sewing the membranes of two adjacent cells and are the main component of the meningeal barrier (arrows). Gap junctions (stars) are seen as an accumulation of closely attached intramembrane particles that represent the connexons. Desmosomes are seen as loose accumulations of intramembrane particles (arrowheads). (Images taken from Caruncho et al. [3, 4])

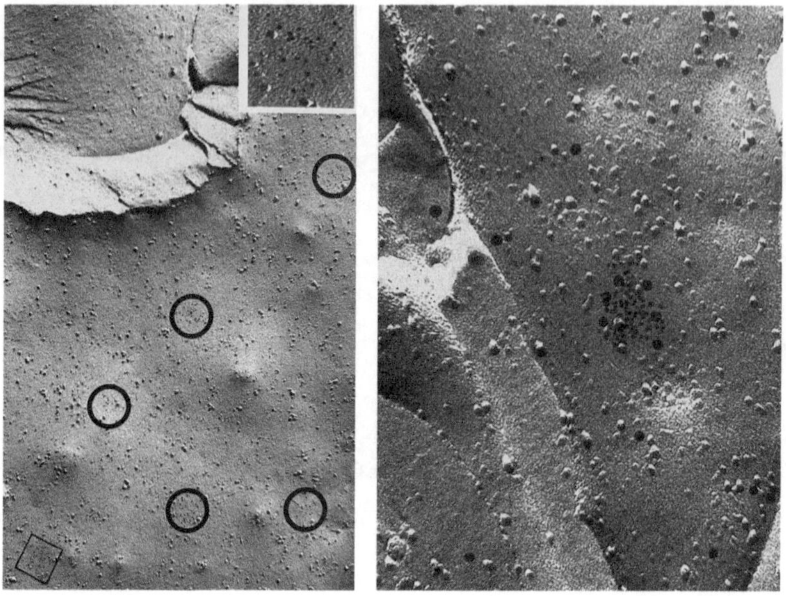

Fig. 6 Label-fracture images of GABAA receptors in the plasma membrane of cultured cerebellar granule cells from mouse pups. (**A**) GABAA receptors immunogold label is observed in clusters (circles) in label-fracture replicas of a cerebellar granule cell plasma membrane. (**B**) Double labeling of alpha (large gold particles) and beta (small gold particles) subunits of GABAA receptors in label-fracture replicas of cultured cerebellar granule cells. (Images taken from Caruncho et al. [5])

synaptic contacts, intercellular junctions, and label-fracture of cultured cells (Figs. 3–6):

8 Conclusions

While freeze-fracture technologies were widely used in the 1980s and 90s, they are less popular today (costs and being complicated technologies may be related to that), but after a gap of about a decade 2010–2020, they continue to be used as a very interesting avenue to explore the cytoarchitecture of cell membranes, which is of primary importance when studying the ultrastructure of the nervous system.

9 Additional Recommended Files for Reader

Recommended reviews on freeze-fracture of the nervous system (PubMed, arranged by alphabetical order):

A fiber matrix model for the restricting junction of the blood-brain barrier in a cephalopod mollusk: implications for capillary and epithelialpermeability.

Abbott NJ, Lane NJ, Bundgaard M. J Neurocytol. 1992;21(4):304-11. https://doi.org/10.1007/BF01224763.

Contributions of new impregnation methods and **freeze** etching to the problems of synaptic fine structure.

Akert K, Moor H, Pfenninger K, Sandri C. Prog Brain Res. 1969;31:223-40. https://doi.org/10.1016/S0079-6123(08)63241-0.

Ultrastructure of synapses in the waking state. A laboratory report on recent advances.

Akert K, Livingston RB, Moor H, Streit P. J Neural Transm. 1974;Suppl 11(0):1-11.

Ultrastructure and molecular organization of biological membranes.

Benedetti EL. Biol Cell. 1994;80(2-3):111-4. https://doi.org/10.1111/j.1768-322x.1994.tb00917.x.

Multiscale modeling of claudin-based assemblies: A magnifying glass for novel structures of biological interfaces.

Berselli A, Benfenati F, Maragliano L, Alberini G. Comput Struct Biotechnol J. 2022;20:5984-6010. https://doi.org/10.1016/j.csbj.2022.10.038. eCollection 2022.

Electron microscopy of cutaneous nerves and receptors.

Breathnach AS. J Invest Dermatol. 1977;69(1):8-26. https://doi.org/10.1111/1523-1747.ep12497857.

Three-dimensional morphological analysis of nerve cells by scanning electron microscopy. A **review**.

Castejón OJ. Scanning Microsc. 1991;5(2):461-75; discussion 475-6.

Three-dimensional morphology of cerebellar protoplasmic islands and proteoglycan content of mossy fiber glomerulus: a scanning and transmission electron microscope study.

Castejón OJ, Castejón HV. Scanning Microsc. 1991;5(2):477-92

Scanning electron microscopy of vertebrate cerebellar cortex.

Castejón OJ. Scanning Microsc. 1988;2(1):569-97.

Vesicle hypothesis of the release of quanta of acetylcholine.

Ceccarelli B, Hurlbut WP. Physiol Rev. 1980;60(2):396-441. https://doi.org/10.1152/physrev.1980.60.2.396.

On the cellular communication of amphibian embryogenesis.

Chuang TM. Prog Clin Biol Res. 1986;226:35-44.

Live-cell imaging of aquaporin-4 diffusion and interactions in orthogonal arrays of particles.

Crane JM, Tajima M, Verkman AS. Neuroscience. 2010;168(4):892-902. https://doi.org/10.1016/j.neuroscience.2009.08.034.

A **review** on neurosecretory granules: their contents and mechanisms of release.

Dreifuss JJ. Ann N Y Acad Sci. 1975;248:184-201. https://doi.org/10.1111/j.1749-6632.1975.tb34185.x.

"Non-synaptic" mechanisms in seizures and epileptogenesis.
Dudek FE, Yasumura T, Rash JE. Cell Biol Int. 1998;22(11-12):793-805. https://doi.org/10.1006/cbir.1999.0397.
On the mechanism of acetylcholine release.
Dunant Y. Prog Neurobiol. 1986;26(1):55-92. https://doi.org/10.1016/0301-0082(86)90020-1.
Phosphoinositides: Nanoscale Effects on Neuronal Membranes.
Eguchi K. Neuroscientist. 2025;31(6):576-590. https://doi.org/10.1177/10738584251337664.
Review of evidence for loss of motor nerve terminal calcium channels in Lambert-Eaton myasthenic syndrome.
Engel AG. Ann N Y Acad Sci. 1991;635:246-58. https://doi.org/10.1111/j.1749-6632.1991.tb36496.x.
Motor nerve terminal calcium channels in Lambert-Eaton myasthenic syndrome. Morphologic evidence for depletion and that the depletion is mediated by autoantibodies.
Engel AG, Nagel A, Fukuoka T, Fukunaga H, Osame M, Lang B, Newsom-Davis J, Vincent A, Wray DW, Peers C. Ann N Y Acad Sci. 1989;560:278-90. https://doi.org/10.1111/j.1749-6632.1989.tb24105.x.
The ultrastructure of membranes in sympathetic ganglia.
Forsman CA, Elfvin LG. Scanning Microsc. 1987;1(1):191-205.
Intra-synapse-type and inter-synapse-type relationships between synaptic size and AMPAR expression.
Fukazawa Y, Shigemoto R. Curr Opin Neurobiol. 2012;22(3):446-52. https://doi.org/10.1016/j.conb.2012.01.006.
Ependymal and choroidal cells in culture: characterization and functional differentiation.
Gabrion JB, Herbuté S, Bouillé C, Maurel D, Kuchler-Bopp S, Laabich A, Delaunoy JP. Microsc Res Tech. 1998;41(2):124-57. https://doi.org/10.1002/(SICI)1097-0029(19980415)41:2<124::AID-JEMT3>3.0.CO;2-U.
Compared the effects of extracellular K+ ions and soman, a neurotoxic, on cerebral astrocyte morphology. An in vitro study.
Grange-Messent V, Raison D, Bouchaud C. J Submicrosc Cytol Pathol. 1996;28(2):151-9.
Dual effect of potassium on transmitter exocytosis.
Grohovaz F, Fesce R, Haimann C. Cell Biol Int Rep. 1989;13(12):1085-95. https://doi.org/10.1016/0309-1651(89)90022-2.
Fine structural basis of pulmonary surfactant.
Groniowski JA. Int Rev Exp Pathol. 1983;25:183-238.
Preparing biological samples for stereomicroscopy by the quick-**freeze**, deep-etch, rotary-replication technique.
Heuser J. Methods Cell Biol. 1981;22:97-122. https://doi.org/10.1016/s0091-679x(08)61872-5.

Pathology of the transmissible spongiform encephalopathies with special emphasis on ultrastructure.
Jeffrey M, Goodbrand IA, Goodsir CM. Micron. 1995;26(3):277-98. https://doi.org/10.1016/0968-4328(95)00004-n.
Morphological features of central synapses, with emphasis on the presynaptic terminal.
Jones DG. Life Sci. 1977 ;21(4):477-92. https://doi.org/10.1016/0024-3205(77)90086-8.
Are there gap junctions between chief (glomus, type I) cells in the carotid body chemoreceptor? A **review**.
Kondo H. Microsc Res Tech. 2002;59(3):227-33. https://doi.org/10.1002/jemt.10196.
Several mutations in mice affect the cerebellum.
Landis DM, Landis SC. Adv Neurol. 1978;21:85-105.
Membrane and cytoplasmic structure at synaptic junctions in the mammalian central **nervous system**.
Landis DM. J Electron Microsc Tech. 1988;10(2):129-51. https://doi.org/10.1002/jemt.1060100203.
Membrane structure in mammalian astrocytes: a **review** of **freeze-fracture** studies on adult, developing, reactive and cultured astrocytes.
Landis DM, Reese TS. J Exp Biol. 1981;95:35-48. https://doi.org/10.1242/jeb.95.1.35.
Morphology of glial blood-brain barriers.
Lane NJ. Ann N Y Acad Sci. 1991;633:348-62. https://doi.org/10.1111/j.1749-6632.1991.tb15626.x.
High-resolution quantitative visualization of glutamate and GABA receptors at central synapses.
Masugi-Tokita M, Shigemoto R. Curr Opin Neurobiol. 2007;17(3):387-93. https://doi.org/10.1016/j.conb.2007.04.012.
Exocytosis and membrane recycling.
Meldolesi J, Ceccarelli B. Philos Trans R Soc Lond B Biol Sci. 1981;296(1080):55-65. https://doi.org/10.1098/rstb.1981.0171.
Morphology of cryofixed myelin sheath.
Meller K. Int Rev Cytol. 1998;184:81-108. https://doi.org/10.1016/s0074-7696(08)62179-8.
Ultrastructural aspects of olfactory signaling.
Menco BP. Chem Senses. 1997;22(3):295-311. https://doi.org/10.1093/chemse/22.3.295.
Qualitative and quantitative **freeze-fracture** studies on olfactory and respiratory epithelial surfaces of frog, ox, rat, and dog. IV. Ciliogenesis and ciliary necklaces (including high-voltage observations).
Menco M. Cell Tissue Res. 1980;212(1):1-16. https://doi.org/10.1007/BF00234028.
AAEE minimonograph #28: injury to peripheral motor nerves.

Miller RG. Muscle Nerve. 1987;10(8):698-710. https://doi.org/10.1002/mus.880100805.

The development of the human blood-brain and blood-CSF barriers.

Møllgård K, Saunders NR. Neuropathol Appl Neurobiol. 1986;12(4):337-58. https://doi.org/10.1111/j.1365-2990.1986.tb00146.x

Human cerebral microvessel endothelial cell culture as a model **system** to study the blood-brain interface in ischemic/hypoxic conditions.

Nagy Z, Vastag M, Kolev K, Bori Z, Karáidi I, Skopál J. Cell Mol Neurobiol. 2005;25(1):201-10. https://doi.org/10.1007/s10571-004-1384-9.

[Spatial distribution of thiamine pyrophosphatase and arylsulfatase activities in the area of the Golgi apparatus of the adenohypophyseal secretory cells in the white rat].

Nevorotin AI, Khokhlov SE. Tsitologiia. 1983;25(9):1073-6.

Lambert-Eaton myasthenic syndrome.

Newsom-Davis J. Springer Semin Immunopathol. 1985;8(1-2):129-40. https://doi.org/10.1007/BF00197251.

The cell surface membrane: biochemical aspects and biophysical probes.

Oseroff AR, Robbins PW, Burger MM. Annu Rev Biochem. 1973;42:647-82. https://doi.org/10.1146/annurev.bi.42.070173.003243.

Structural and functional correlates of synaptic transmission in the vertebrate neuromuscular junction.

Rash JE, Walrond JP, Morita M. J Electron Microsc Tech. 1988;10(2):153-85. https://doi.org/10.1002/jemt.1060100204.

Ultrastructure, histological distribution, and **freeze-fracture** immunocytochemistry of gap junctions in rat brain and spinal cord.

Rash JE, Yasumura T, Dudek FE. Cell Biol Int. 1998;22(11-12):731-49. https://doi.org/10.1006/cbir.1998.0392.

A pharmacological approach to the structure of sodium channels in myelinated axons.

Ritchie JM. Annu Rev Neurosci. 1979;2:341-62. https://doi.org/10.1146/annurev.ne.02.030179.002013.

Axolemmal abnormalities in myelin mutants.

Rosenbluth J. Ann N Y Acad Sci. 1990;605:194-214. https://doi.org/10.1111/j.1749-6632.1990.tb42393.x.

Electrophysiology and morphology of myelinated nerve fibers. V. Intramembranous particle distribution in nerve fiber membranes.

Rosenbluth J. Experientia. 1983;39(9):953-63. https://doi.org/10.1007/BF01989760.

Intracellular transport and exocytosis of the secretory granules.

Senda T. Kaibogaku Zasshi. 1997;72(1):1-7.

Gap junctions revealed by **freeze-fracture** electron microscopy.

Shivers RR, McVicar LK. Microsc Res Tech. 1995;31(5):437-45. https://doi.org/10.1002/jemt.1070310512.

The diversity of function and structure of cellular membranes.

Sjöstrand FS. Subcell Biochem. 1983;9:335-93. https://doi.org/10.1007/978-1-4613-3533-7_5.

Regulation of gap junctional conductance.

Spray DC, White RL, Mazet F, Bennett MV.Am J Physiol. 1985;248(6 Pt 2):H753-64. https://doi.org/10.1152/ajpheart.1985.248.6.H753.

Freeze-fracture replication of mammalian peripheral nerve--a **review**.

Stolinski C, Breathnach AS. J Neurol Sci. 1982;57(1):1-28. https://doi.org/10.1016/0022-510x(82)90107-1.

Is the MEPP due to the release of one vesicle or to the simultaneous release of several vesicles at one active zone?

Tremblay JP, Laurie RE, Colonnier M.Brain Res. 1983;287(3):299-314. https://doi.org/10.1016/0165-0173(83)90009-7.

Small vesicle bouton synapses on the distal half of the lateral dendrite of the goldfish Mauthner cell: **freeze-fracture** and thin section study.

Tuttle R, Masuko S, Nakajima Y.J Comp Neurol. 1987;265(2):254-74. https://doi.org/10.1002/cne.902650209.

Nicotinic acetylcholine receptor and the structural basis of neuromuscular transmission: insights from Torpedo postsynaptic membranes.

Unwin N. Q Rev Biophys. 2013;46(4):283-322. https://doi.org/10.1017/S0033583513000061.

Structural aspects of brain barriers, with special reference to the permeability of the cerebral endothelium and choroidal epithelium.

van Deurs B. Int Rev Cytol. 1980;65:117-91. https://doi.org/10.1016/s0074-7696(08)61960-9.

Structure, biophysics, and circuit function of a "giant" cortical presynaptic terminal.

Vandael D, Jonas P. Science. 2024;383(6687):eadg6757. https://doi.org/10.1126/science.adg6757. Epub 2024 Mar 8.

Structure of membranes and role of lipids therein.

Vandenheuvel FA. Adv Lipid Res. 1971;9:161-248.

Live-cell imaging of aquaporin-4 supramolecular assembly and diffusion.

Verkman AS, Rossi A, Crane JM. Methods Enzymol. 2012;504:341-54. https://doi.org/10.1016/B978-0-12-391857-4.00017-3.

Ontogenesis of the axolemma and axoglial relationships in myelinated fibers: electrophysiological and **freeze-fracture** correlates of membrane plasticity.

Waxman SG, Black JA, Foster RE. Int Rev Neurobiol. 1983;24:433-84. https://doi.org/10.1016/s0074-7742(08)60226-3.

Structure and functions of aquaporin-4-based orthogonal arrays of particles.

Wolburg H, Wolburg-Buchholz K, Fallier-Becker P, Noell S, Mack AF.Int Rev Cell Mol Biol. 2011;287:1-41. https://doi.org/10.1016/B978-0-12-386043-9.00001-3.

The disturbed blood-brain barrier in human glioblastoma.

Wolburg H, Noell S, Fallier-Becker P, Mack AF, Wolburg-Buchholz K. Mol Aspects Med. 2012;33(5-6):579-89. https://doi.org/10.1016/j.mam.2012.02.003.

Orthogonal arrays of intramembranous particles: a **review** with special reference to astrocytes.

Wolburg H. J Hirnforsch. 1995;36(2):239-58

Conflicts of Interest

The authors have no conflicts of interest to declare that are relevant to the content of this chapter.

References

1. Caruncho HJ, Rodriguez-Moldes I, Anadon R (1990) A freeze-fracture study of the soma membranes of monoaminergic neurons in the hypothalamic nucleus recessi posterioris of the rainbow trout (*Salmo gairdneri*, Rich.). J Hirnforsch 31:575–584
2. Caruncho HJ, Pinto da Silva P, Anadon R (1993) The morphology of teleost meningocytes as revealed by freeze-fracture. J Submicrosc Cytol Pathol 25:397–406
3. Caruncho HJ, Rodriguez-Moldes I, Lamas J, Anadon R (1990) Freeze-fracture study of synaptic contacts in the neuropil of the tuberal hypothalamus of the rainbow trout (*Salmo gairdneri*, Rich.). J Hirnforsch 31:689–696
4. Caruncho HJ, Pinto da Silva P (1994) Alterations in the intermediate layer of goldfish meninges during adaptation to darkness. J Anat 184:355–362
5. Caruncho HJ, Puia G, Slobodyansky E, Pinto da Silva P, Costa E (1993) Freeze-fracture immunocytochemical study of the expression of native and recombinant GABAA receptors. Brain Res 603:234–242

Check for updates

Chapter 2

How to Combine Light and Electron Microscopy to Detect Lipids

Anita Grootemaat, Sanne van der Niet, Eric Reits and Nicole N. van der Wel

Abstract

The levels of intracellular lipid accumulations can be upregulated in various genetic and infectious diseases, with lipids being present in lipid droplets as well as compartments with different morphologies. For detection of lipid accumulation, both electron microscopy (EM) and fluorescence microscopy (FM) can be used, and the combination of these techniques is even more powerful. It allows localizing specific markers for lipids by FM, followed by imaging these lipid-rich regions by EM, showing the organelles and structures at high resolution. As lipid accumulations in disease can be present in various shapes and sizes, confirmation with lipid markers and FM is useful. However, during the process of standard plastic embedding of EM samples, FM signals are often lost. Tokuyasu or immuno-EM methods with gently fixed, gelatin-embedded materials allow for the detection of FM markers like Nile Red and high-resolution EM imaging. Here, we describe different protocols to fixate brain tissues and generate CLEM images and explain when these protocols could be and have been applied.

Keywords Lipids, Tokuyasu method, Transmission electron microscopy (TEM), Fluorescence microscopy (FM), Correlative light and electron microscopy (CLEM)

1 Introduction

Besides imaging at high resolution, a big advantage of electron microscopy (EM) is the imaging of all organelles and components of cells and tissues without staining these structures separately. Most organelles are identifiable based on their ultrastructure, and there is no need for staining with antibodies. Still, even after decades of analysis of tissues and cells, some structures are difficult to recognize. For example, lipids can be present in various shapes and morphologies, specifically during diseases like pathogen infections or genetic disorders. We have used correlative light and electron microscopy

Anita Grootemaat and Sanne van der Niet. Both authors contributed equally

Marie-Ève Tremblay (ed.), *Ultrastructural Neuroscience Methods*, Neuromethods, vol. 238,
https://doi.org/10.1007/978-1-0716-5456-9_2,

(CLEM) to visualize lipid accumulation in infection models and tissues of patients with lysosomal storage disease and shown that lipid accumulations can be present in multiple forms (Fig. 1).

Originally, embedding cells or tissue in gelatin and cryo-sectioning is used for immuno-gold labeling using antibodies and 10 nm gold developed by Tokuyasu [4], a protocol that was improved to double labeling with two sizes of gold particles [5]. This method can be applied in many ways and tissues and is well established with

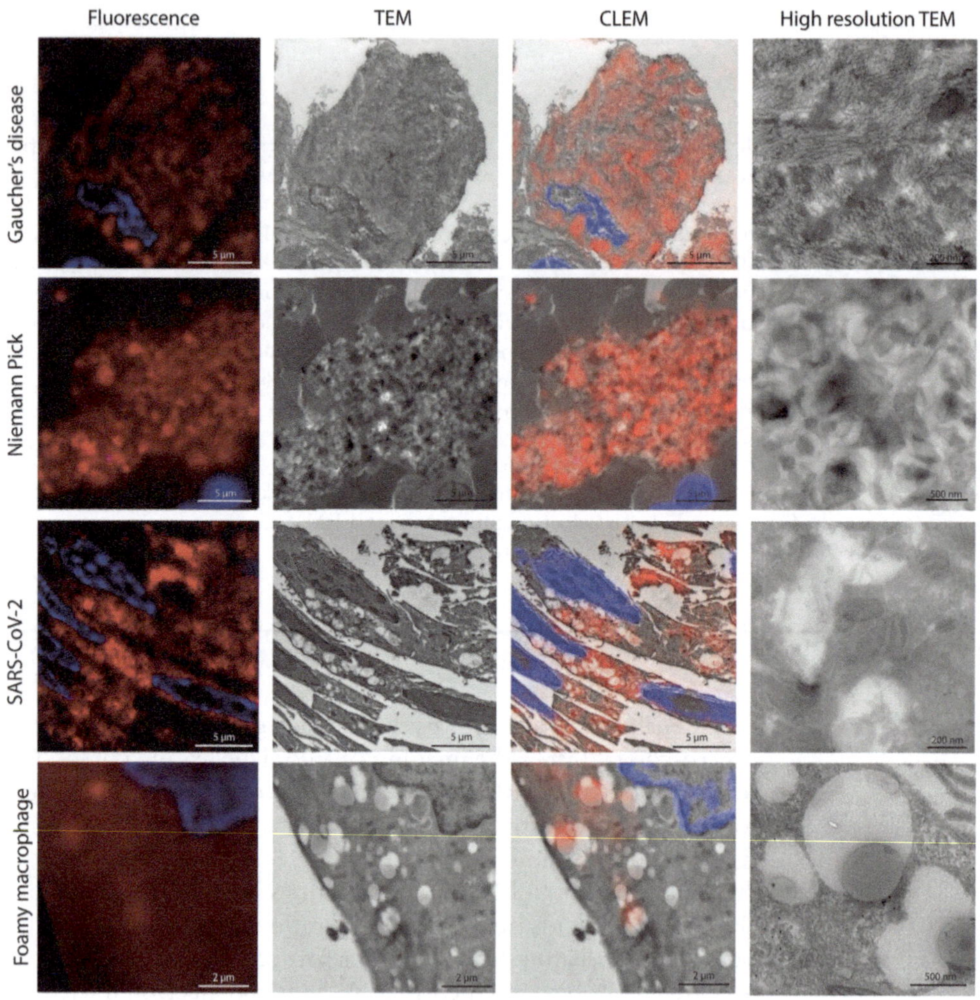

Fig. 1 CLEM to detect lipids in multiple disease models. Left column: widefield fluorescence image of Gaucher's disease spleen tissue, Niemann Pick patient spleen tissue (preliminary data, Aten and van der Wel labs), SARS-CoV-2-infected Vero cells (Grootemaat et al. [1, 2]), and foamy macrophages treated with *Mycobacterium tuberculosis* antacid 1-TbAd (Bedard et al. [3]). Lipids were stained with Nile red (red) and DNA with Hoechst (blue). Second column: Corresponding Transmission Electron Microscope (TEM) images. Third column: CLEM images. Overlays were produced using ICY software with the ec-CLEM plugin. Right column: High-resolution TEM images of lipid structures. The scale bar represents 200 nm, 500 nm, 2 μm, and 5 μm

numerous applications published during the last 50 years, including the identification of microglia cells in the brain [6, 7] or organelles like lysosomes [3, 6, 8, 9].

Besides the immuno-gold application, we and others have used cryo-immuno sections for fluorescence microscopy (FM), as these sections are very suited for visualization using FM [10–14]. Since the sections can be cut 100–500 nm thick, the background or out-of-focus signal is low, and high-resolution FM can be performed. Also, penetration of antibodies is optimal, as every cell is sectioned in 20–100 sections (depending on section thickness and cell size); therefore, no harmful incubation with detergents is needed. Another important advantage is that the identical section can be re-imaged using both FM and EM, generating so-called CLEM images. CLEM allows a combination of the ultrastructure visible by EM with any type of fluorescent marker. These FM markers can be expressed in cells (like GFP, mCherry, or any other fluorescent reporter protein); fluorescent dyes, immuno-labeled with antibodies and a fluorescent secondary antibody to detect a protein of interest; or novel FM markers like DNA-paint [14, 15]. Unfortunately, the combination of FM tools is often not compatible with standard epon/plastic embedding. The ethanol dehydration or the incubation at 60°C destroys, for example, the Nile Red and Hoechst signals. The more gentle gelatin-embedding technique from Tokuyasu is very flexible and can be applied in as many forms as we have tools and questions. Here, we show how to apply these protocols for the detection of lipid in brain tissues, showcase an example, and discuss the protocols needed to perform correlated light and electron microscopy.

2 Protocol 1 Fixation and Gelatin Embedding of Brain Tissue

Fixation of brain tissue can be a challenge, and the proper protocol needs to be chosen in advance. When protein detection with antibodies is needed, it is advisable to test if the antibody–antigen interaction is blocked by glutaraldehyde (GA) fixation. If so, the tissue is fixed in 4% paraformaldehyde (PFA) only. When no antibody labeling is needed or the antibody–antigen combination can resist GA fixation, a combination of PFA and GA is advised, as this will dramatically improve the ultrastructure in EM. Both 4%PFA/0.4%GA in PHEM buffer and the commercial 3%PFA 3%GA in phosphate buffer can be used. The advantage of PHEM-based fixative is the preservation of the cytoskeleton, but this fixative needs to be prepared freshly. For optimal fixation, fresh fixative is always best, but commercial fixatives can be used for 6 months after opening and storage at 4°C. Autofluorescence caused by the GA is no issue here, as we work with semithin sections. This protocol is based on the immuno-EM method [4, 16].

2.1 Method Fixation and Preparation of Brain Tissue

Materials

- Pieces of tissue-biopsy.
- Fixative:
  - PFA (see recipe),
  - PFA/GA (see recipe).
  - 3%PFA 3%GA in phosphate buffer, pH 7.4 (for example, EMS; 16538-07).
- Storage solution (see recipe).
- Razor blades, alcohol cleaned.
- Plastic vials with screw caps.
- Glass Pasteur pipette.
- High-precision tweezers.
- PBS + 0.02 M Glycine (PBSG).
- 2.3 M sucrose in 0.1 M phosphate buffer.

1. Add biopsy/tissue using tweezers (without exerting pressure) as fast as possible after dissection to fixative (PFA/GA or PFA in PHEM or 3%PFA/3%GA). Fresh brain tissue is difficult to cut into small blocks, so this is performed at Step 5, after fixation. Be careful; these fixatives are corrosive to the skin, nose, and eyes, so work in the hood with gloves. Perfusion fixation of tissue is preferable for achieving good morphology. It is normally used for small animals such as mice and rats.
2. Incubate tissue in fixative at room temperature for at least 2 days (for PFA) or 24 h (for GA/PFA).
3. When material is not directly used, transfer tissue to storage solution with 0.5% PFA and store at 4°C.
4. When preparing samples for (CL)EM, remove fixatives by rinsing tissue in PBS three times (or more with larger blocks), followed by 2 times in PBS + 0.02 M Glycine (PBSG).
5. Keep the tissue immersed in a small drop of PBSG at all times, dissect out the required structure using a clean razor blade, and cut 1–3 mm blocks.
6. Carefully transfer the tissue blocks to a preheated (37°C) glass vial filled with 2% gelatin and incubate at 37°C for 30 min.
7. Remove 2% and replace with 6% gelatin and incubate at 37°C for 30 min.
8. Remove 6% and replace with 12% gelatin and incubate at 37°C for 30 min. Gelatin at 12% sets quickly; keep everything as much as possible at 37°C!

9. Remove 12% and replace with fresh 12% gelatin and incubate at 37°C for 30 min.
10. Remove tissue blocks from gelatin and place under a dissecting microscope. Using forceps, immediately drag tissue away from the drop of gelatin, allowing the gelatin to set, forming a small gelatin-covered block.
11. Quickly transfer blocks into glass vials with a cap containing 2.3 M sucrose solution and gently rotate or shake overnight at 4°C.
12. Collect blocks from sucrose and place them on aluminum pins. Make sure that there is enough sucrose to hold the block on the pin, and blot away the excess sucrose with filter paper, plunge freeze the pin in liquid nitrogen, and store in liquid nitrogen until sectioning.

2.2 Method Sectioning Brain Tissue

Materials

- Tissue gelatin embedded in sucrose from liquid nitrogen stock.
- Diamond cryo-EM knife.
- Ultra-microtome with cryo-box.

Before sectioning brain tissues, first the front of the tissue block is trimmed until the sample is flat. Check using toluidine blue staining what will be the region of interest and trim a rectangle at −80°C using a glass knife or diamond trim knife. For EM analysis, a 150–200 µm by 400–600 µm block is created; for FM, this can be 400 by 800 µ m, but it will be more difficult to generate good sections for EM at this size, and more trimming might be required. In general, the smaller the block, the easier it is to generate good sections. The section thickness for FM can be 100–500 nm and cut at −80°C till −120°C, depending on the material. Larger but also thicker sections are preferably generated at −80 till −100°C. The section thickness for EM can be 60–100 nm thin and needs to be performed at −120°C. When sections stick to the knife, an ionizer might be used. We use a cryo-diamond knife and the Leica Ultracut UC7 with a cryo-box, which is cooled with liquid nitrogen.

2.3 Reagents

Paraformaldehyde (PFA) fixative,
5 ml 0.4 M PHEM buffer (see recipe)
10 ml 8% PFA stock (EM grade (commercially available))
5 ml H_2O

- Prepare fresh immediately before use.
- The final concentration for fixation is 0.1 M PHEM buffer and 4% PFA.

Paraformaldehyde/glutaraldehyde (PFA/GA) fixative,
5 ml 0.4 M PHEM buffer (see recipe)
10 ml 8% PFA stock solution (EM grade (commercially available))
1 ml 8% GA (EM grade (commercially available))
4 ml H_2O

- Prepare fresh immediately before use.
- The final concentration for fixation is 0.1 M PHEM, 4% PFA, and 0.4% GA.

12 % Gelatin in 200 ml 0.1 M phosphate buffer
24 g gelatin (Sigma Aldrich G2500-500gr)
200 ml 0.1 M phosphate buffer
200 μl 20% sodium azide

- Add 24 g of gelatin to 200 ml of 0.1 M phosphate buffer while stirring.
- Cover the solution to make sure the phosphate buffer does not evaporate.
- Incubate at 60°C for ~4 h until the gelatin is dissolved. Stir every 30 min.
- Cool the solution to 37°C and add 200 μl 20% sodium azide.
- Aliquot the gelatin and store at 4°C.
- Test a small volume of the gelatin if it solidifies at room temperature.
- Dilute to 2, 6, and 12 % gelatin in phosphate buffer just before use.

PHEM buffer, 0.4 M for 1000 ml
240 mM PIPES (72.5 g PIPES)
100 mM HEPES (23.8 g HEPES)
8 mM MgCl2 (1.63 g $MgCl_2$)
40 mM EGTA (15.2 g EGTA)

- Make a solution of 600 ml 0.375 M NaOH.
- Add PIPES while slowly stirring the solution.
- Step by step, add EGTA to the solution while slowly stirring.
- Dissolve HEPES in 200 ml H_2O and add this to the solution.
- Add $MgCl_2$.
- Adjust pH to 6.9 using NaOH.
- Add water till the solution is 1000 ml.
- Divide the solution into portions of 50 ml Falcon tubes or 15 ml Falcon tubes.
- Store in the freezer until needed.

Storage solution
5 ml 0.4 M PHEM buffer (see recipe)
1.2 ml 8% PFA stock solution (EM grade (commercially available))
13.8 ml H_2O
The final concentrations are 0.1 M PHEM and 0.5% PFA.
PBS stock solution [10×] for 1000 ml
PBS tablets Gibco 18912-014
Or:
80 g NaCl
2 g KCl
14.4 g $Na_2HPO_4.2H_2O$
2.3 g $NaH_2PO_4.2H_2O$
Dilute 1:10 prior to use.
Two different applications for CLEM are described separately.

3 Protocol 2 Direct-CLEM Using Section on a Grid

When it is crucial to image the identical section both in FM and EM, a 100–150 nm section is generated at −120°C and placed on a (finder) grid and imaged with widefield fluorescence or super-resolution microscope. As these sections are thin, a confocal laser microscope is not essential but can be used when preferred. This technique allows for the best correlation afterwards, as finder grids make it easier to find back the identical section/cell or even organelle in both FM and EM, but any other type of grid can be used. One should be aware that acquiring low- and high-magnification images for both FM and EM is vital for finding the correct regions and making proper overlays. After a section has been imaged with EM, FM is no longer possible, or much more difficult, so appropriate amounts of FM images are crucial. After the FM images are acquired, the grids are carefully washed and stained as normally for immuno-gold-labeled grids and imaged using a Transmission Electron Microscope (TEM).

3.1 Examples

Identical bacteria with lipid and DNA stained [17].
Lipid accumulation in mesenchymal-like cancer cells [18].
Lipid accumulation in mice brain (preliminary data, Bakker and van der Wel labs) (Fig. 2).

3.2 Method Direct CLEM

Materials:

- Cryo-diamond knife.
- Finder golden grid (FF200F2-AU, EMS) or any type of copper grid.
- 2% toluidine blue.
- Methyl cellulose 2% (Sigma; M6385-250G)—2.3 M Sucrose (Merck; 7653.500) mixed gently in a 1:1 combination.

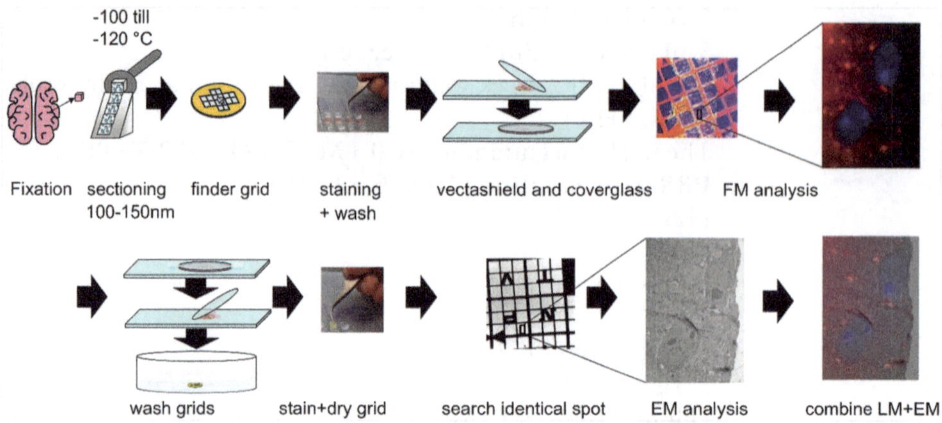

Fig. 2 Schematic representation of the direct-CLEM method for the imaging of a mouse brain tissue section. Section is placed on a finder grid, first stained with Nile Red for lipids and Hoechst for DNA, imaged using a widefield fluorescent microscope, washed and stained using Uranyl Acetate, and finally imaged using a TEM, see protocol 2

- PBS/Glycine 0.02 M (0.075 gram in 50 ml).
- Vectashield (Brunschwig Chemie B.V.; H-1000-10).
- 1 mg/ml Nile Red (9-diethylamino-5H-benzo[a]phenoxazine-5-one; Sigma-Aldrich).
- Hoechst 33342 (Thermo Fisher Scientific).

1. Pins with fixed blocks of tissue are transferred to a cryo-box ultramicrotome at −80°C.
2. Blocks are trimmed using a glass knife to select the region of interest based on light microscopical sections stained with toluidine blue.
3. Ultrathin sections (100–150 nm) are cut using a cryo-diamond knife at −100 till −120°C at a speed of 0.6–1 mm/sec.
4. Sections picked up with the MC-Sucrose combination in a metal loop.
5. Sections are transferred to grids by thawing the MC-Sucrose combination with sections and touching the formvar-coated grids when the MC-Sucrose is thawed.
6. Wash three times for 5 min in PBS 0.02 M glycine by incubating the grid on a droplet placed on parafilm.
7. When performing an immuno-labeling, dilute antibody in PBS with 0.1% BSA (when no antibody incubation is needed, wash grids 5 × 10 min in PBS and go to Step 12).
8. Incubate for 1 h in 10 μl of the antibody solution at room temperature.

9. Rinse in PBS with glycine 3×10 min.
10. Dilute fluorescent dye-labeled second fluorescent antibody in PBS with 0.1% BSA.
11. Incubate 25 μl for 20 min in the dark at room temperature.
12. After 15 min incubation, add 1 μl Nile Red and 0.5 μl Hoechst to the 25 μl droplet.
13. Rinse 3×5 min in PBS.
14. Place a droplet of Vectashield on a glass slide and transfer the grid inside the droplet with the section facing upwards.
15. Cover with a coverslip, make sure to place the coverslip slowly to prevent air bubbles being formed.
16. Direct imaging using FM is preferred, or store in the dark at 4°C, maximum 2 days. Please realize that while performing fluorescence microscopy, making overviews at high and low magnification makes correlation with EM images easier.
17. To remove the coverslip, place droplets of water on both sides of the coverslip and let the water mix with the Vectashield (this can take a few minutes).
18. When coverslip is loose, gently lift it so that the grid remains soaked in Vectashield and carefully transfer to small petri dish and wash the grid in water for 4 h at 37°C while replacing the water every 15 min.
19. Contrast grids with Uranyl acetate-methylcellulose solution as normal EM immune-gold-labeled sections [5, 16].

4 Protocol 3 Serial-CLEM Technique

When a sample is too big to analyze with EM, for example, when pathogens are present in a small portion of an organism or organ, CLEM can be used to search for the correct region of interest. As an example, a zebrafish embryo is used. The brain of a zebrafish embryo 5 days post-fertilization is 450 μm wide by 700 μm long by 320 μm high [19]. When the full brain is cut from top to bottom, with 100 nm thin sections, one should analyze 3200 sections. Nowadays, with automated 3DSEM techniques, Svara et al. [19] were able to image the whole brain. When those techniques are too time-consuming, or the research question is limited to, for example, the subcellular localization of a pathogen or bead, screening with FM is an elegant solution to find the region of interest. Then, 100–400 nm sections are placed on a glass slide and, when needed, stained with a fluorescent marker and imaged using FM. As soon as the infection [1, 2, 20] or bead [21] is detected, the next section will be cut at 80–100 nm and placed on a grid. Now, immuno-gold labeling can be performed to detect and label cell types or organelles, and subsequently, the section can be imaged using a TEM. An advantage is that the FM slides are stored and thus can be re-imaged for 1–2 days, when kept at 4°C in the dark. After 2 days, FM staining is often degraded, and detection is no longer possible. A disadvantage of serial CLEM is that the alignment can be a bit difficult, as in the next section, the morphological features of the cell and organelles might have changed [20].

4.1 Examples

In the brain, we identified FITC-dextran-filled blood vessels and determined how far the FITC-dextran penetrated into the brain tissue, out of the blood vessel [22].

In zebrafish embryos, we searched for fluorescently labeled mycobacteria [20] (Fig. 3).

In brain, lung and kidney of COVID patients, we searched for viral proteins and lipid accumulations [1, 2].

4.2 Materials

- 2% toluidine blue.
- Methyl cellulose 2% (Sigma; M6385-250G)–2,3 M Sucrose (Merck; 7653.500) mixed gently in a 1:1 ratio.
- PBS/Glycine 0.02 M (0.075 gram in 50 ml).
- Vectashield (Brunschwig Chemie B.V.; H-1000-10).
- 1 mg/ml Nile Red (9-diethylamino-5H-benzo[a]phenoxazine-5-one; Sigma-Aldrich).

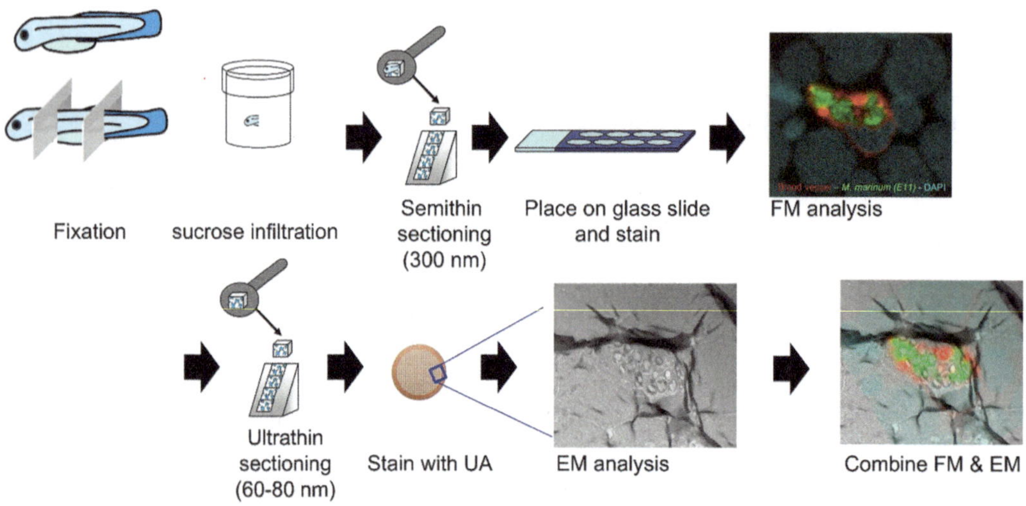

Fig. 3 CLEM method using serial sections of 9 days post-fertilization and infected with GFP-tagged *Mycobacterium marinum* as done in van Leeuwen et al. [20], and protocol 3. In short, after fixation, zebrafish embryos are cut into three pieces and incubated in 2.3 M sucrose, snap frozen, and sectioned at −80°C in semithin 300 nm sections. Sections were analyzed on glass slides to determine the region infected with bacteria. When an infected region is identified, an ultrathin section is cut at −120°C, which is placed on an EM grid and imaged using TEM

- Hoechst 33342 (Thermo Fisher Scientific).
- 8-well glass slides cleaned with ethanol.

4.3 Method

1. Pins with fixed blocks of tissue are transferred to the cryo-box Ultramicrotome at −80°C.
2. Blocks are trimmed using a glass knife to select the region of interest based on light microscopical sections stained with toluidine blue.
3. Semithin sections (300–500 nm) are cut using a cryo-diamond knife at −80 till −100°C at a speed of 0.6–1 mm/sec.
4. Sections are picked up with sucrose in a metal loop.
5. Sections are transferred to glass slides with 8 wells.
6. Staining and labeling of the section are identical to protocol 2.
7. When the FM images demonstrate that the region of interest is reached, generate ultrathin sections of 60–100 nm as in protocol 2.

5 Software for Generating CLEM Image

The images generated with FM and EM can be combined, and an overlay can be generated using various software tools. Both Adobe Photoshop (Fig. 2) and ICY software with ec-CLEM plugin (Figs. 1 and 3) can be used. When the background FM signal is present, deconvolution can be performed. We used Huygens deconvolution software in Fig. 4.

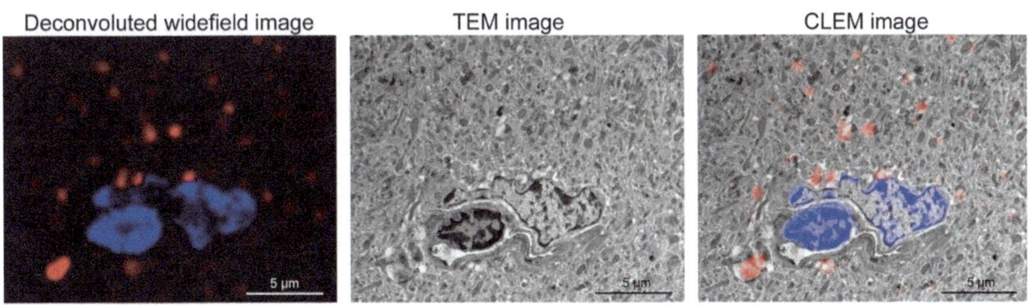

Fig. 4 CLEM on brain tissue to visualize lipid accumulations. Left: deconvoluted widefield image; lipids were stained using Nile red (red) and DNA using Hoechst (blue). The fluorescence image is deconvoluted using Huygens deconvolution software. Conservative deconvolution was used with three iterations. Middle: corresponding TEM image. Right: Correlated image. Overlay is performed using ICY software with the ec-CLEM plugin. The scale bar represents 5 μm. Brain tissues from van der Panne et al., submitted. Brains of C57BL/6JOlaHsd mice aged 13–23 weeks were perfusion-fixed with 3% PFA + 3% GA (Electron Microscopy Sciences, 16538-07), and the mid-cortical area between the anterior and posterior regions was cut into 1–2 mm cubes and stored in 3% PFA/3% GA and processed as in protocol 2

Acknowledgments

Shakira van der Panne and Erik Bakker for brain tissues. Jan Aten and Martijn van Lienden for Niemann Pick and Gaucher tissues. Daisy Picavet and Lisanne van Leeuwen for CLEM of zebrafish, and Ireen Schimmel for EM support.

Author Contributions AG and SvdN generated images and contributed to writing, ER contributed to writing, and NvdW wrote the manuscript.

Competing Interests The authors declare no competing interests.

References

1. Grootemaat AE, van der Niet S, Scholl ER, Roos E, Schurink B, Bugiani M et al. (2022) Lipid and nucleocapsid N-protein accumulation in COVID-19 patient lung and infected cells. Microbiol Spectr 10:1
2. Grootemaat AE, Wiersma N, van der Niet S, Schimmel IM, Florquin S, Reits EA et al. (2023) Nucleocapsid protein accumulates in renal tubular epithelium of a post-COVID-19 patient. Microbiol Spectr 11:6
3. Bedard M, van der Niet S, Bernard EM, Babunovic G, Cheng TY, Aylan B et al. (2023) A terpene nucleoside from M. tuberculosis induces lysosomal lipid storage in foamy macrophages. J Clin Invest 133:6
4. Tokuyasu KT (1973) A technique for ultracryotomy of cell suspensions and tissues. J Cell Biol 57:551–565
5. Geuze HJ, Slot JW, van der Ley PA, Scheffer RCT, Griffith JM (1981) Use of colloidal gold particles in double-labeling immunoelectron microscopy of ultrathin frozen tissue sections. J Cell Biol 89:653–665
6. Guo S, Cazarez-Marquez F, Jiao H, Foppen E, Korpel NL, Grootemaat AE et al. (2022) Specific silencing of microglial gene expression in the rat brain by nanoparticle-based small interfering RNA delivery. ACS Appl Mater Interfaces 14(4):5066–5079
7. Kloosterman DJ, Erbani J, Boon M, Farber M, Handgraaf SM, Ando-Kuri M et al. (2024) Macrophage-mediated myelin recycling fuels brain cancer malignancy. Cell 187(19):5336–5356.e30
8. van der Wel N, Hava D, Houben D, Fluitsma D, van Zon M, Pierson J et al. (2007) M. tuberculosis and M. leprae translocate from the phagolysosome to the cytosol in myeloid cells. Cell 129(7):1287–1298
9. van der Niet S, van Zon M, de Punder K, Grootemaat A, Rutten S, Moorlag S et al. (2021) IL-1R1-dependent signals improve control of cytosolic virulent mycobacteria in vivo. mSphere 6:3
10. van Elsland DM, Bos E, de Boer W, Overkleeft HS, Koster AJ, van Kasteren SI (2016) Detection of bioorthogonal groups by correlative light and electron microscopy allows imaging of degraded bacteria in phagocytes. Chem Sci 7(1):752–758
11. van Elsland DM, Pujals S, Bakkum T, Bos E, Oikonomeas-Koppasis N, Berlin I et al. (2018) Ultrastructural imaging of salmonella-host interactions using super-resolution correlative light-electron microscopy of bioorthogonal pathogens. ChemBioChem 19(16):1766–1770
12. Bakkum T, Heemskerk MT, Bos E, Groenewold M, Oikonomeas-Koppasis N, Walburg KV et al. (2020) Bioorthogonal correlative light-electron microscopy of mycobacterium tuberculosis in macrophages reveals the effect of antituberculosis drugs on subcellular bacterial distribution. ACS Cent Sci 6(11):1997–2007
13. Tuijtel MW, Mulder AA, Posthuma CC, van der Hoeven B, Koster AJ, Barcena M et al. (2017) Inducing fluorescence of uranyl acetate as a dual-purpose contrast agent for correlative light-electron microscopy with nanometre precision. Sci Rep 7(1):10442
14. Stein J, Ericsson M, Nofal M, Magni L, Aufmkolk S, McMillan RB et al (2025) Cryosectioning-enhanced super-resolution microscopy for single-protein imaging across cells and tissues. Proc Natl Acad Sci U S A 122(32):e2504578122
15. Wu J, Su B, Gu L, Zhang J, Shi Q, Hu D (2025) Immunoelectron microscopy: a comprehensive guide from sample preparation to high-resolution imaging. Discov Nano 20(1):154

16. Peters PJ, Bos E, Griekspoor A (2006) Cryo-immunogold electron microscopy. current protocols in cell biology. Curr Protoc Cell Biol 2006:4.7.1–4.7.19.
17. Scutigliani EM, Scholl ER, Grootemaat AE, Khanal S, Kochan JA, Krawczyk PM et al. (2018) Interfering with DNA decondensation as a strategy against mycobacteria. Front Microbiol 9:2034
18. Arnoldus T, van Vliet A, Bleijerveld OB, de Groot AFH, Piao Q, Blomberg N et al. (2025) Cytidine diphosphate diacylglycerol synthase 2 is a synthetic lethal target in mesenchymal-like cancers. Nat Genet 57(7):1659–71.
19. Svara F, Forster D, Kubo F, Januszewski M, Dal Maschio M, Schubert PJ et al. (2022) Automated synapse-level reconstruction of neural circuits in the larval zebrafish brain. Nat Methods 19(11):1357–66.
20. van Leeuwen LM, Boot M, Kuijl C, Picavet DI, van Stempvoort G, van der Pol SMA et al. (2018) Mycobacteria employ two different mechanisms to cross the blood-brain barrier. Cell Microbiol 20(9):e12858
21. van der Wijk AE, Lachkar N, de Vos J, Grootemaat AE, van der Wel NN, Hordijk PL et al. (2019) Extravasation of microspheres in a rat model of silent brain infarcts. Stroke 50(6):1590–1594
22. Bedussi B, van der Wel NN, de Vos J, van Veen H, Siebes M, VanBavel E et al. (2016) Paravascular channels, cisterns, and the subarachnoid space in the rat brain: a single compartment with preferential pathways. J Cereb Blood Flow Metab 37(4):1374–85.

Chapter 3

Immunoelectron Microscopy of Brain Receptors: Focus on G Protein-Coupled Receptors

Nagore Puente, Itziar Bonilla-Del Río, Sonia María Gómez-Urquijo, Inmaculada Gerrikagoitia and Pedro Grandes

Abstract

High-resolution immuno-electron microscopy has been successfully used for the localization of a myriad of brain proteins, like receptors for classical neurotransmitters and neuromodulatory signaling molecules. We herein describe the pre-embedding immunogold method for electron microscopy used in our laboratory over the years, first for the brain localization of the main excitatory neurotransmitter glutamate and its target receptors, and more recently for defining the detailed topography of the endocannabinoid system (ECS). When used properly, it is an excellent tool to estimate the precise protein distribution in subcellular compartments of the brain cells. We also give a description of a post-embedding immunogold method that can nicely complement the pre-embedding technique when proteins have to be defined unequivocally at specific sub-synaptic or tiny membrane compartments. We focus in this book chapter on the cannabinoid CB1 receptor, one of the main molecular components of the ECS, and related metabotropic glutamate receptors (mGluRs). A detailed understanding of the specific subcellular localization of the CB1 receptor and mGluRs is crucial because cellular processes underlie most physiological functions and pathological dysfunctions. Such knowledge identifies the precise sites at which the ECS and mGluRs modulate synaptic transmission and plasticity and defines their spatial relationship with other receptors, ion channels, neurotransmitter release mechanisms, and enzymatic pathways. Finally, it highlights the targets for cannabinoids and therapeutic cannabinoid-based medications with potential clinical benefits. The key takeaway is that conclusions regarding the localization of a given protein, particularly one expressed at low levels, require a combination of highly sensitive immunocytochemical methods and multiple antibodies of unquestionable specificity that recognize distinct protein sequences.

Keywords Ultrastructural immunocytochemistry, Pre-embedding immunogold labeling, Post-embedding immunogold labeling, Brain receptors, Cannabinoid receptors, Glutamate receptors, Endocannabinoid system, Subcellular localization, Synaptic transmission and plasticity, Molecular neuroanatomy

1 Introduction

The endocannabinoid system (ECS), which comprises cannabinoid receptors (CB1, CB2, and others), endogenous ligands, primarily the endocannabinoids 2-arachidonoylglycerol (2-AG) and

Marie-Ève Tremblay (ed.), *Ultrastructural Neuroscience Methods*, Neuromethods, vol. 238,
https://doi.org/10.1007/978-1-0716-5456-9_3,

N-arachidonoylethanolamine (anandamide, AEA), enzymes responsible for the synthesis and degradation of endocannabinoids, intracellular metabolic pathways, and transport systems [1–3], is extensively distributed throughout the central and peripheral nervous systems, as well as across the mammalian body [2–4]. This system plays a crucial role in modulating cellular physiology. Through the activation of CB1 receptors, which are considered the most abundant G-protein-coupled receptors (GPCRs) in the brain, the ECS contributes to brain function and dysfunction [2, 5–12].

In the 1990s, autoradiography of radioligand binding ([^{3}H] CP55,940) was employed to qualitatively and quantitatively assess cannabinoid receptor distribution in the brain [13]. Following the cloning of the CB1 receptor, *in situ* hybridization techniques were routinely used to study the distribution of CB1 mRNA in brain cell bodies [14]. Subsequently, polyclonal antibodies were developed targeting a specific N-terminal amino acid sequence of the CB1 receptor protein [15]. Shortly thereafter, a CB1 receptor antiserum was generated against a C-terminal amino acid sequence of the rat CB1 receptor [16].

The transition from regional distribution studies using autoradiography to cellular mapping with *in situ* hybridization, and eventually to CB1 receptor immunohistochemistry, marked a significant advancement in cannabinoid research by uncovering novel receptor localizations. These immunological tools, in combination with advanced histochemical techniques for light microscopy, enabled the visualization of patterns of CB1 receptor-like immunoreactivity in the brain [16, 17]. Thus, CB1 receptor-like immunoreactivity appears uniform in brain regions where the dominant neuronal populations widely express the receptor, exhibiting a widespread but regionally variable expression pattern. CB1 receptor density is particularly high in the olfactory bulb, cerebral cortex, basal ganglia (including the substantia nigra pars reticulata, globus pallidus, striatum, and entopeduncular nucleus), nucleus accumbens, hippocampal formation, and cerebellum. In contrast, there is a lower CB1 receptor density in the hypothalamus, amygdala, and brainstem [17]. This distribution is consistent with the critical role of the receptor in prenatal and postnatal development and in numerous brain functions, including food intake, cognition, behavior, motor control, motivation, reward, and mood regulation [12, 18, 19].

The precise subcellular distribution of CB1 receptors remained unresolved until the application of pre-embedding and post-embedding immunocytochemical techniques for electron microscopy. These methods proved invaluable for studying the exact localization of CB1 receptors in brain tissue and peripheral organs. Thus, they revealed that CB1 receptors are not uniformly distributed but display heterogeneity in brain regions where specific cell types or subcellular compartments exhibit higher CB1 levels [1]. In fact, the detrimental effects of cannabinoids on locomotion, perception, learning, and memory; the beneficial properties of cannabinoids; and the low toxicity and lethality of marijuana are closely correlated with the distinct distribution of CB1 receptors in specific cell types and subcellular compartments [20–24]. In this sense, although the density of CB1 receptors at inhibitory synaptic terminals varies across brain regions, it is consistently much higher at inhibitory than at excitatory synapses [25]. Notably, CB1 receptors are highly enriched in GABAergic synaptic terminals, particularly within cholecystokinin-positive and parvalbumin-negative basket cells in the cortex and hippocampus. In contrast, CB1 receptor expression is relatively sparse at excitatory glutamatergic synapses and remains low in astrocytes [14, 26–30]. At the subcellular level, CB1 receptors are predominantly localized to perisynaptic sites, positioned outside the active zone of presynaptic membranes [25, 26, 31].

It became evident that the levels of CB1 receptor expression do not necessarily correlate with their physiological significance [14]. This is because even low or very low levels of CB1 receptor expression, previously dismissed as background staining in certain brain cell types or subcellular compartments [26, 32], have been demonstrated to hold functional and behavioral relevance [6].

a. **Pre-embedding immunogold technique:**
   Antibodies conjugated with gold nanoparticles are directly applied to permeabilized tissues, which are subsequently embedded in resin for observation under an electron microscope.
   The pre-embedding immunogold method has been a cornerstone technique in our laboratory for many years, enabling precise localization of many receptor subtypes and channels in various tissues (Figs. 1 and 2). For example, we have successfully applied this method to study the distribution of metabotropic glutamate receptors (mGluRs), including mGluR2/3 and mGluR4 in the spinal cord, mGluR1b and mGluR4 in the cerebellum, mGluR2/3 and mGluR4 in the medial nucleus of the trapezoid body, and mGluR4 in the piriform cortex and superior colliculus. In addition, we have used this approach to investigate mGluR1a in cultured cerebellum, ionotropic glutamate receptor (GluR) subunits 1–4 in the MNTB, potassium channels in the medial nucleus of the trapezoid body (MNTB) and cerebellum, and potassium channels in cardiomyocyte mitochondria.
   We have also used this technique to investigate the subcellular localization of CB1 receptors and other components of the ECS in several brain regions. We identified the localization of N-acyl phosphatidylethanolamine phospholipase D (NAPE-PLD, AEA-synthesizing enzyme) and diacylglycerol lipase-alpha (DAGL-α, 2-AG-synthesizing enzyme) in the ventromedi-

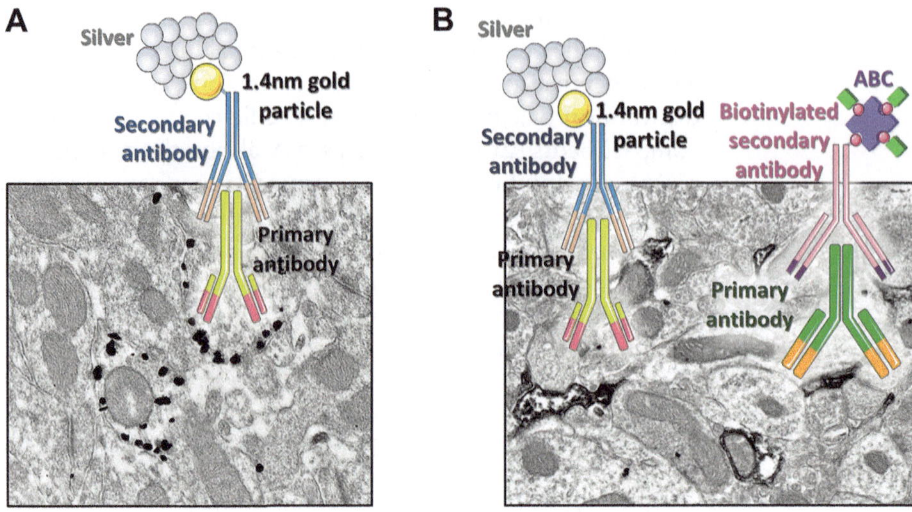

Fig. 1 Single (**A**) and double (**B**) pre-embedding immunogold techniques for TEM

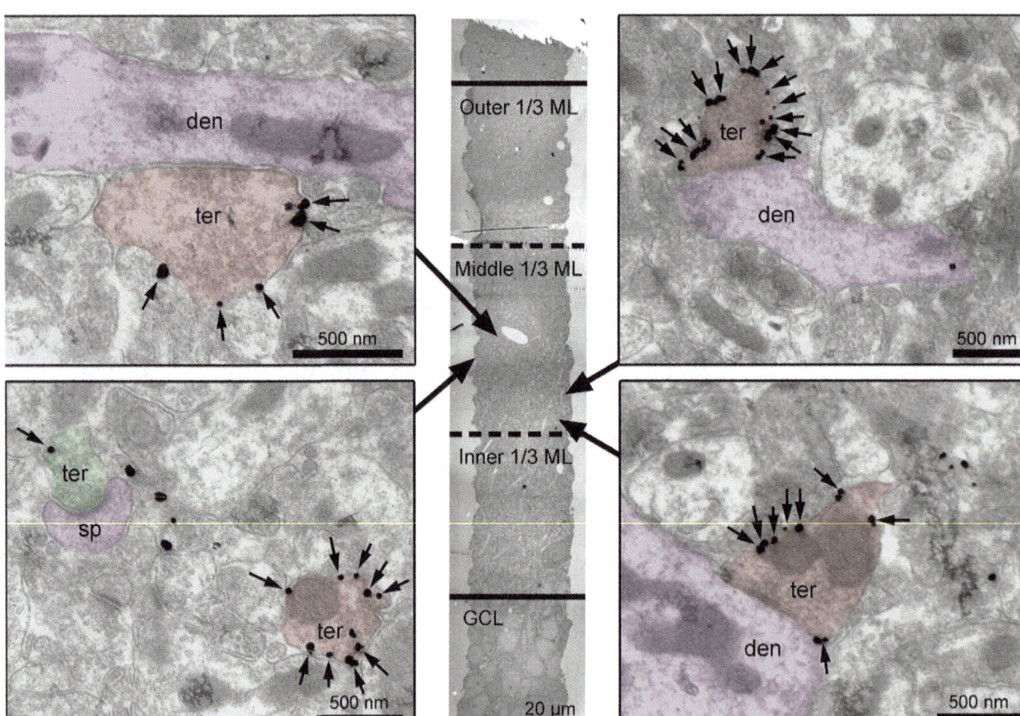

Fig. 2 CB1 localization in the middle one-third of the dentate molecular layer (ML) of the mouse hippocampus. Pre-embedding immunogold method for TEM. CB1 labeling, revealed by specific antibodies and silver-enhanced gold particles (arrows), is present on the presynaptic membranes of inhibitory (reddish) and excitatory (green) terminals (ter) forming symmetric and asymmetric synapses with dendrites and spines, respectively

al nucleus of the hypothalamus [33, 34] and the hippocampus [35]. We also studied CB1 receptors, mGluR5, transient receptor potential vanilloid-1 (TRPV1), and calcium channels in the bed nucleus of the stria terminalis (BNST) [36]; CB1 receptors in the olfactory bulb [37]; TRPV1 in the hippocampus [38, 39]; CB1 receptors in astrocytes of the hippocampus, cerebral cortex, hypothalamus, and nucleus accumbens [40–43]; and CB1 receptors in the hippocampus [21, 42, 44, 45], in the lateral habenula [46], and in the susbstantia nigra [47].

With this method, our laboratory was the first to demonstrate the localization of CB1 receptors at neuronal mitochondria [48–50], astroglial mitochondria [24, 42, 51], and muscle mitochondria [52, 53]. Furthermore, the use of a semi-quantitative analysis of CB1 receptor labeling in the CA1 stratum radiatum of the hippocampus revealed that approximately 56% of the total labeling (silver-intensified gold particles) is localized in GABAergic terminals, 12% in glutamatergic terminals, 6% in astrocytes, 15% in mitochondria, and 11% in other compartments [45]. Noticeably, antibodies against the glutamate aspartate transporter (GLAST) label nearly three times more astroglial area and about four times more astroglial membrane than antibodies against glial fibrillary acidic protein (GFAP), enabling the detection of approximately 12% of total CB1 receptor labeling within GLAST-positive astrocytes [54] (Fig. 3). The other subcellular CB1 localizations are not yet fully identified. Evidence indicates that CB1 receptors can also be found in microglia and oligodendrocytes [23, 55, 56], as well as within endosomes and at mitochondria–endoplasmic reticulum contact sites (MERCs) [57]. We also take advantage of this technique to investigate other proteins in the hippocampus, such as the subcellular localization of a mitochondria-targeted recombinant designer receptor exclusively activated by designer drugs (mitoDREADD-Gs) [58]. This engineered DREADD (a muscarinic receptor modified to couple to specific G-protein subtypes) is directed to mitochondria and is capable of activating endogenous Gs proteins.

However, the pre-embedding immunogold method has certain limitations. Mostly, the labeling does not precisely correlate with the exact localization of the antigen, is only semi-quantitative rather than fully quantitative, and is not suitable for multiple labeling to visualize several synaptic proteins simultaneously unless combined with other techniques, such as pre-embedding immunoperoxidase.

Our pioneer study of the ECS focused on the co-distribution of the CB1 receptor and mGluR1a in the MNTB. At that

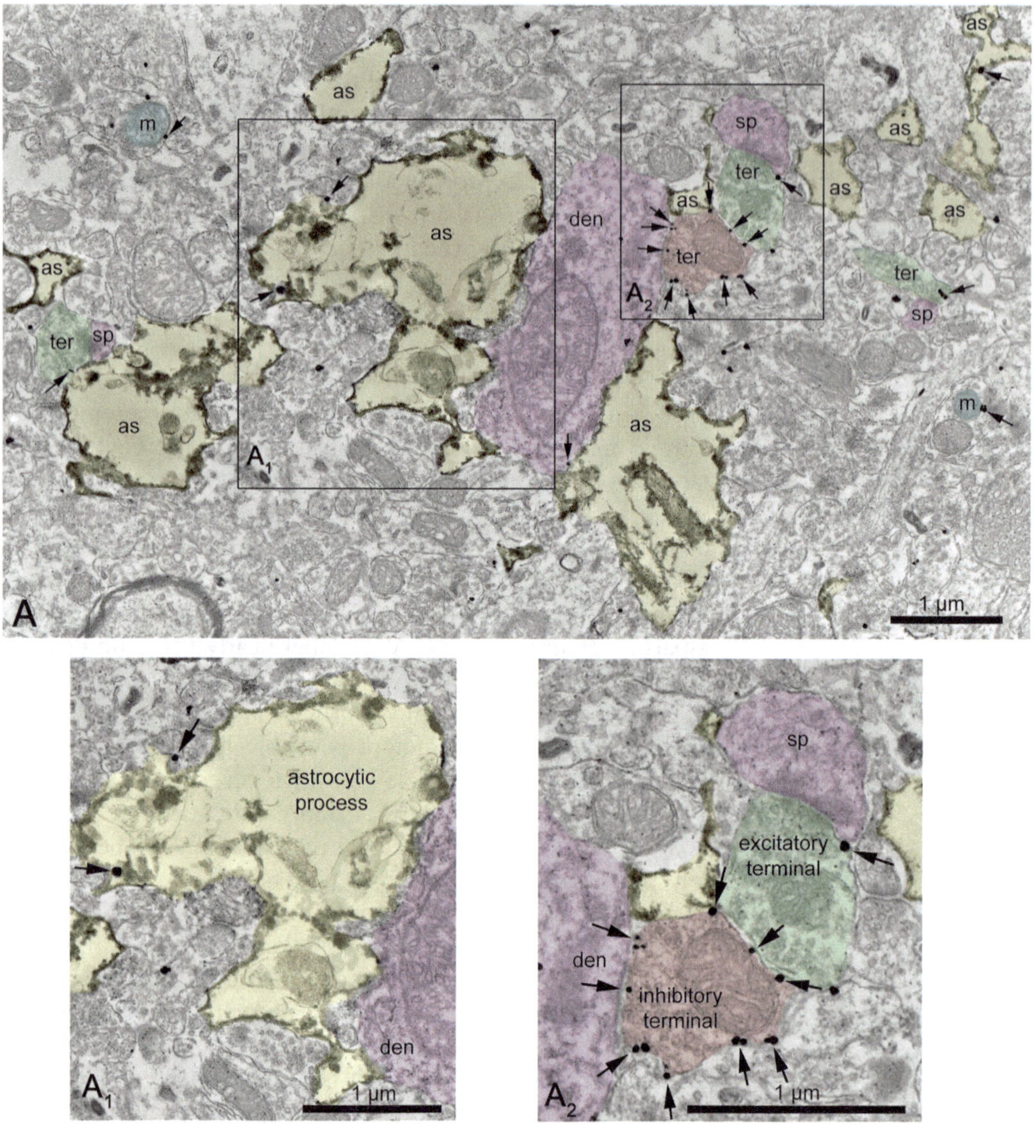

Fig. 3 Immunolocalization of CB1 receptors in the dentate molecular layer of the mouse hippocampus. Double pre-embedding immunogold and immunoperoxidase method for TEM. (**A**) CB1 particles (arrows) are localized on the membranes of excitatory (green) and inhibitory (reddish) synaptic terminals (ter), on mitochondria (m, blue), and on astrocytes identified by GLAST labeling (as, yellow). **A1** and **A2** show higher-magnification views of the boxed regions in panel A. Abbreviations: den, dendrite; sp, spine

time, we used a single pre-embedding immunogold method for the localization of CB1 and mGluR1a. Silver-intensified CB1-immunogold particles were decorating presynaptic membrane specializations at calyx of Held–principal cell synapses [59]. MGluR1a immunolabeling revealed the presence of the receptor distributed in membranes of principal cell bodies.

Notably, mGluR1a immunoparticles were markedly concentrated at perisynaptic sites of the numerous postsynaptic specializations of the calyciform synapses.

Later, we addressed the limitations of single immunolabeling by combining pre-embedding immunogold and immunoperoxidase methods to study co-localization. For example, we simultaneously examined CB1 receptors (immunogold), mGluR5 (immunogold or immunoperoxidase), and DAGL-α (immunoperoxidase) [60]. Similarly, we investigated the co-localization of CB1 receptors (immunogold) and dopamine D2 receptors (immunoperoxidase) in the prefrontal cortex [61]; CB1 receptors (immunogold) in conjunction with anterograde and retrograde tracers (immunoperoxidase) [62]; and CB1 receptors (immunogold) with vesicular glutamate transporter 1 (vGLUT1) (immunoperoxidase) in the BNST [63].

It is important to note that the chromogen 3,3′-diaminobenzidine (DAB), commonly used in the pre-embedding immunoperoxidase method, tends to diffuse from the initial oxidation sites. This diffusion can result in the unintended labeling of adjacent organelle membranes, such as those of mitochondria or the endoplasmic reticulum, or even neighboring cellular compartments. Moreover, when nickel-intensified DAB is used as an alternative in the pre-embedding immunoperoxidase method, it results in significant nonspecific CB1 receptor labeling of mitochondria [49]. Mitochondrial false positives may also arise from the presence of endogenously biotinylated proteins within the organelle [64].

b. **Post-embedding immunogold technique:**

These methods involve the detection of surface and intracellular antigens in ultrathin tissue sections that have already been embedded in resin. Pioneering studies have demonstrated that colloidal gold particles are among the most versatile markers for immunolabeling in post-embedding techniques used in electron microscopy [65–67]. Their ability to bind to antibodies, combined with their high electron density, enables rapid identification and quantification under an electron microscope. This facilitates precise spatial correlation between the gold particles and the target antigen. Consequently, this method offers several advantages over the pre-embedding technique. Antigens exposed on the surface of the section have an equal probability of being immunodetected, regardless of their subcellular localization. Additional benefits include the absence of immunoreagent diffusion within the tissue, the reliable labeling of antigens across serial ultrathin sections, and the ability to generate quantitative results. Moreover, this technique allows for simultaneous multi-labeling using gold particles of different sizes, enabling the visualization

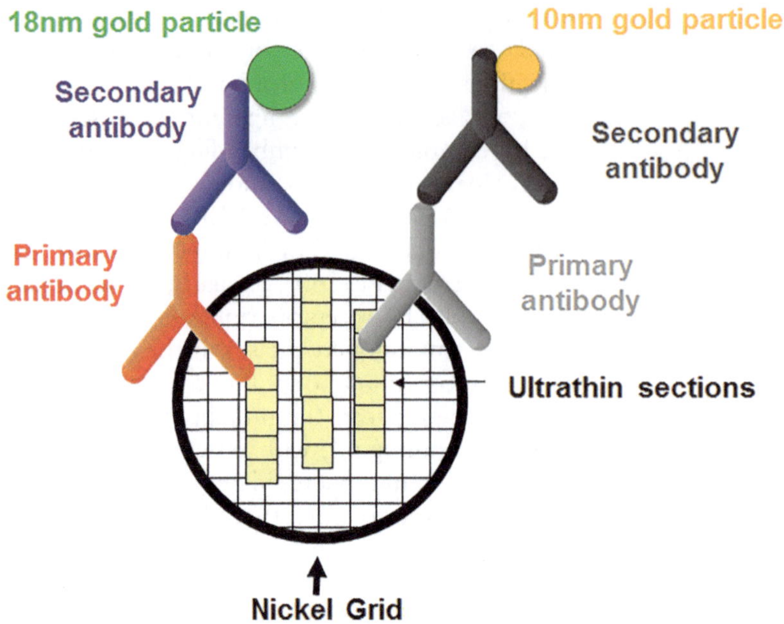

Fig. 4 Post-embedding immunogold method for TEM. Secondary antibodies are gold-labeled with different particle sizes

of multiple target proteins in a single sample with a conventional transmission electron microscope [68, 69] (Fig. 4).

However, this method also has several disadvantages. These include moderate sensitivity, as immunodetection is limited to antigens exposed on the surface of the section. In addition, one of the most significant challenges of this technique is that many antibodies fail to work properly due to the embedding procedure, which occurs during the embedding of the sample in resins. This phenomenon adversely affects antigen recognition, which is essential for immunocytochemistry. Other drawbacks include the high cost of the specialized equipment needed for resin embedding and the inability to correlate findings between light and electron microscopy [69].

Unlike pre-embedding methods, where samples are immunolabeled prior to embedding in resin, post-embedding techniques follow a different sequence: samples are first fixed, embedded in resin, sectioned, and mounted on grids before immunolabeling.

Tissue dehydration is performed at low temperatures, thereby minimizing antigen alteration. As a result, structural integrity and antigenicity are better preserved, enhancing the quality of the samples for immunogold labeling. Of note, sample quality can be compromised during conventional resin embedding due to the high temperatures required for polymerization (60°C) and the need for prior dehydration with organic solvents (e.g., acetone or alcohols).

These steps can also alter the antigens, leading to reduced antigenicity. To overcome these limitations, methacrylate-based plastics, such as Lowicryl K4M and HM20, are used for post-embedding techniques as alternatives to conventional high-temperature resin embedding. These resins, composed of cross-linked acrylate and methacrylate, undergo photo-polymerization when exposed to long-wavelength ultraviolet light (360 nm) at low temperatures (−35°C and −70°C, respectively).

We used this technique to investigate the localization of ionotropic glutamate receptor (GluR) subunits 1–4 in the MNTB [68, 70] and to explore the co-localization of mGluR1b and mGluR5 at hippocampal CA3 mossy fiber synapses [71].

2 Subcellular Localization of mGluR1a and mGluR5, and CB1 and mGluR2/3, with the Post-Embedding Immunogold Method

2-AG plays a pivotal role in various forms of synaptic plasticity through phasic endocannabinoid signaling. At excitatory synapses, neuronal membrane depolarization or the activation of GPCRs initiates the synthesis of 2-AG. Synaptic glutamate spillover stimulates postsynaptic group I mGluRs (mGluR1 and mGluR5) situated at perisynaptic membrane sites, leading to diacylglycerol (DAG) synthesis and the subsequent production of 2-AG at the plasma membrane. Once synthesized, 2-AG travels retrogradely to activate presynaptic CB1 receptors, which inhibit voltage-gated calcium channels and suppress neurotransmitter release. The functional and signaling interplay between mGluRs and CB1 receptors relies on a complex network of intracellular pathways that govern neurotransmitter release, synaptic plasticity, and neuronal excitability. These processes are orchestrated through the generation of second messengers, modulation of ion channels, and retrograde endocannabinoid signaling [72].

We analyzed the ultrastructural localization of mGluR1a and mGluR5 in the CA3 region of the hippocampus using the single post-embedding immunogold method for electron microscopy (Fig. 5).

Immunoparticles were distributed postsynaptically, primarily along the membranes of dendritic spines. These spines are innervated by mossy fibers, which are easily identifiable at the ultrastructural level due to their large size and dense content of synaptic vesicles. In consecutive images, the immunolabeling of both receptors was seen on the same dendritic spines identified in both sections.

Next, we applied the double-labeling post-embedding immunogold technique for electron microscopy to study the co-localization of mGluR1a and mGluR5 in more detail within the same tissue section of the CA3 hippocampal region (Fig. 6). Immunoparticles identifying

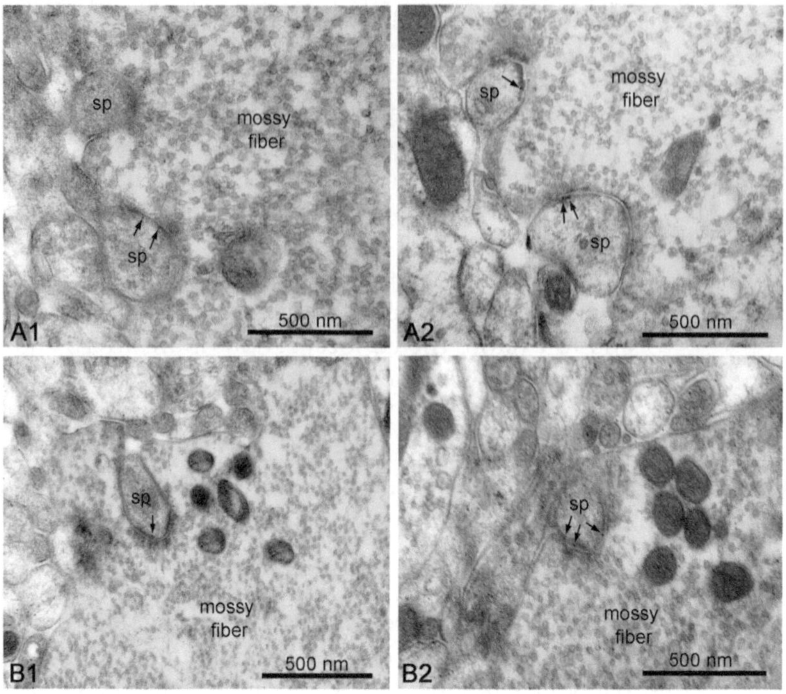

Fig. 5 Ultrastructural localization of mGluR1a (**A1**, **B1**) and mGluR5 (**A2**, **B2**) in two examples of consecutive sections of the stratum lucidum in the hippocampus CA3. Gold particles (10 nm diameter) identifying the mGluR1a or mGluR5 receptors (arrows) are visible on the membranes of dendritic spine excrescences (sp, pink), which receive synaptic input from mossy fibers (light purple). Both receptor subtypes are present on the same dendritic excrescence of a pyramidal neuron, readily identifiable by tracking the spine profile in the two sections (A1-A2; B1-B2). The giant mossy fiber terminals contain an abundance of loosely packed, round, clear synaptic vesicles, mitochondria, and form excitatory synaptic contacts with postsynaptic spine excrescences, hallmark features of these terminals

mGluR1a (smaller 10 nm particles, indicated by white arrowheads) and mGluR5 (larger 20 nm particles, indicated by black arrows) were both localized postsynaptically, specifically within dendritic elements and prominently on the spiny protrusions of dendrites from CA3 pyramidal neurons innervated by mossy fibers.

Notably, both receptors were present within the same dendritic spines, positioned at varying distances from the synaptic specialization. This observation confirmed the co-localization of mGluR1a and mGluR5 in dendritic spines of the CA3 hippocampal region.

We then performed a statistical analysis of the images obtained from the double-labeling experiment to quantify the percentage of spines immunolabeled for mGluR1a, mGluR5, or both receptor subtypes. Only spines receiving synaptic input from mossy fibers were considered (Fig. 7). Among the analyzed spines, 20.3% contained immunoparticles exclusively for mGluR1a, while 11.8% were

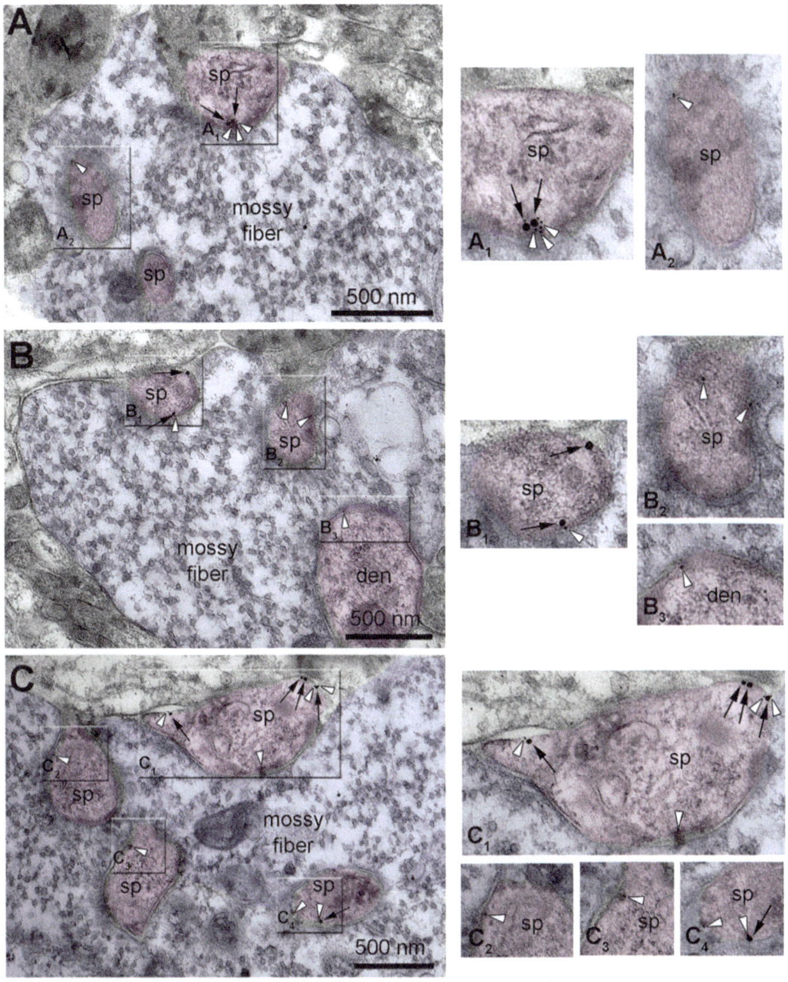

Fig. 6 Ultrastructural co-localization of mGluR1a and mGluR5 in the stratum lucidum of the hippocampus CA3. mGluR1a (10-nm particles, white arrowheads) and mGluR5 (20-nm particles, black arrows) are distributed along the membranes of dendritic spines (sp, pink) that receive synaptic input from mossy fibers (light purple). The boxed areas are shown at higher magnification on the right. Some spines co-localize both receptors (**A$_1$**, **B$_1$**, **C$_1$**, and **C$_4$**), whereas others express only mGluR1a (**A$_2$**, **B$_2$**, **C$_2$**, and **C$_3$**). In **B**, a dendrite (den) is immunopositive for mGluR1a (**B$_3$**), and in **A**, there is a spine immunonegative for both receptors

immunopositive solely for mGluR5. Notably, the majority of spines (41.3%) were simultaneously immunopositive for both receptors, whereas 23.9% showed no immunolabeling for either receptor.

We also analyzed the distribution of both receptors relative to the edge of the postsynaptic density. Most gold particles were located in the region closest to the synapse (Fig. 7). Specifically, 64.5% of mGluR1a particles and 56.8% of mGluR5 particles were found within the first 180 nm from the edge of the postsynaptic density.

These findings demonstrated that the majority of postsynaptic dendritic spines of CA3 pyramidal neurons contain mGluR1a and

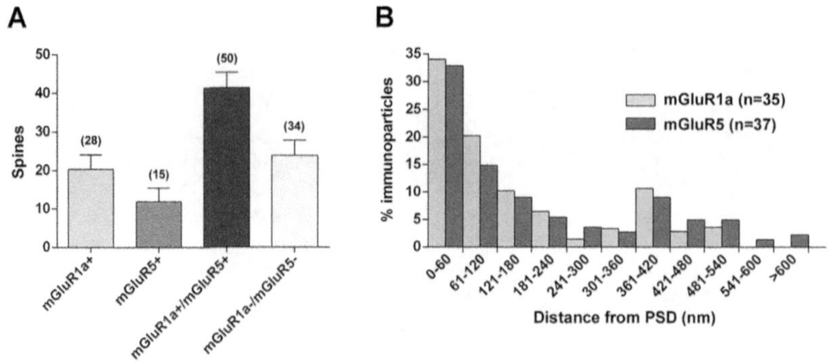

Fig. 7 Statistical analysis of the mGluR1a and mGluR5 distribution in the stratum lucidum of the hippocampus CA3. (**A**) Of the spines receiving input from mossy fibers, 20.3 ± 3.8% express only mGluR1a, 11.8 ± 3.6% express only mGluR5, 41.3 ± 4.0% are immunopositive for both receptors, and 23.9 ± 3.9% express neither. The total area analyzed was 443 μm², with the number of spines analyzed for each category indicated in parentheses. The total number of spines analyzed was 127. (**B**) The distribution of mGluR1a and mGluR5 immunoparticles relative to the postsynaptic density (PSD) shows a characteristic perisynaptic localization for both receptors. Approximately 34% of mGluR1a and 33% of mGluR5 immunoparticles are located within the first 60 nm from the edge of the PSD

mGluR5. Both receptors are positioned perisynaptically relative to the specialized excitatory synaptic contacts formed by mossy fibers.

We also observed co-localization of the CB1 receptor and mGluR2/3 in the middle third of the dentate molecular layer, which corresponds to the termination zone of the glutamatergic medial perforant path. CB1 immunoparticles (18 nm in diameter) were on excitatory axon terminals forming synapses with dendritic spines. These spines displayed 10 nm gold particles labeling mGluR2/3 on their membranes (Fig. 8). This labeling likely indicates the localization of mGluR3, as suggested by Tamaru and collaborators [73].

3 Materials

- Fixative solution: Heat 600 ml of distilled water in a microwave to 60°C.

  Add 11.5 g of Na_2HPO_4 (anhydrous), 40 g of paraformaldehyde, shake at 60°C. Add 2.62 g of $NaH_2PO_4 \cdot H2O$ and 2 ml of saturated picric acid, filter, and make up to 1 with distilled water. Add 4 ml of glutaraldehyde (25%) just before use.
- Phosphate-buffered saline (PBS 1×) (0.1 M, pH 7.4): To 800 ml of distilled water, add 8 g NaCl, 0.2 g KCl, 1.44 g Na_2HPO_4, and 0.24 g KH_2PO_4, and make up the final volume to 1 L with distilled water.

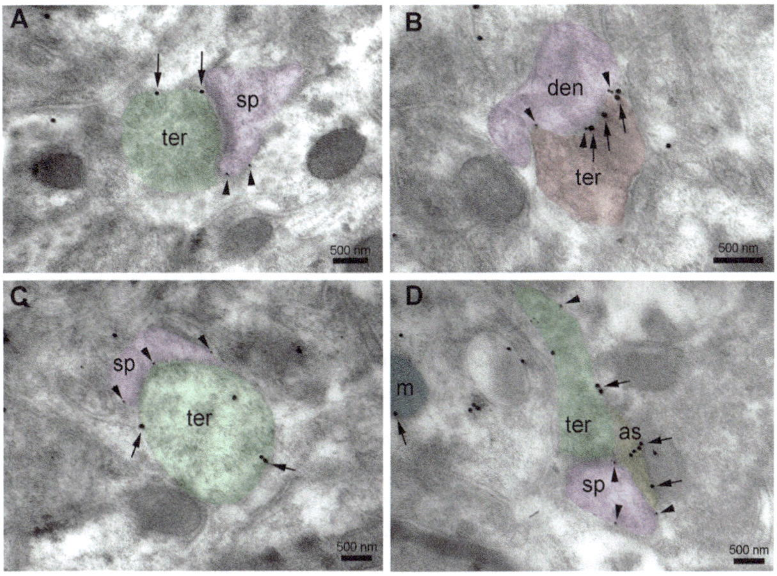

Fig. 8 Double immunolocalization of CB1 receptors and mGluR2/3 in the dentate molecular layer of the mouse hippocampus using the post-embedding immunogold method. (**A, B**) CB1 gold particles (18 nm, arrows) are distributed along the presynaptic membranes of both excitatory (ter, green) and inhibitory terminals (ter, reddish in B), where they co-localize with postsynaptic mGluR2/3 particles (10 nm, arrowheads) present on the postsynaptic spine (sp, pink) or dendritic membranes (den, pink in B). (**C**) Additional example of presynaptic CB1 localization (arrows) on a terminal (ter, green) forming an excitatory synapse with a spine (sp, pink) decorated with mGluR2/3 immunoparticles (arrowheads). (**D**) Beyond neuronal elements (ter, sp), CB1 (arrows) and mGluR2/3 immunoparticles (arrowheads) are also found on profiles consistent with astrocytic membranes (as, yellow). Also note a CB1 particle (arrow) localized to a mitochondrial section (m, blue)

- Phosphate buffer stock solution (0.2 M PB, pH 7.4): Dissolve 5.24 g $NaH_2PO_4{\cdot}H2O$ and 23.0 g Na_2HPO_4 in distilled water, and make up the solution to 1 L with distilled water.
- PB (0.1 M): dilute the 0.2 M PB stock solution 1:1 with distilled water.
  - Tris–hydrogen chloride buffered saline (TBS 1×, pH = 7.4): 3 M NaCl, 1 M Tris–HCl.
- Stock solution 10× TBS (1 L, pH = 7.4): 0.3 M NaCl, 0.1 M Tris-HCl,175 g NaCl,19.4 g Trizma base,132.2 g Trizma-HCl, make up solution to 1 L with distilled water.
- Tris-buffered saline with Triton X-100 (TBST, pH to 7.4): 3.03 g Trizma base, 4.5 g of NaCl, make up the volume to 500 ml with distilled water, and add 0.5 ml Triton X-100.
- Epon resin: 81.3 EPON 812, 53.0 g EPON HÄRTER DDSA, 35.7 MNA, 2.24 ml N-Benzyldimethylamine Fluka 13370, mix for at least 2 h before storing at −20°C in syringes.
- 2% aqueous uranyl acetate staining solution: 0.04 g uranyl acetate in 2 ml of distilled water, centrifuge at 11,600 × *g* for 5 min, use supernatant.

- Reynold's lead citrate staining solution: 2.66 g lead nitrate, 3.52 g Tri-sodium citrate dehydrate, 84 ml double-distilled water, 0.8 g 1 N NaOH in 20 ml double-distilled water, 16 ml NaOH solution to the lead citrate solution, filter the solution to remove undissolved material.

4 Methods

4.1 Brain Tissue Preservation

1. Anesthesia by intraperitoneal injection of ketamine/xylazine (80/10 mg/kg body weight).
2. Perfusion through the heart with phosphate-buffered saline (0.1 M PBS, pH 7.4, 20–25°C) for 20 s at RT, followed by an ice-cold fixative solution made up of 4% formaldehyde (freshly depolymerized from paraformaldehyde), 0.2% picric acid, and 0.1% glutaraldehyde in phosphate buffer (0.1 M PB, pH 7.4) for 10–15 min (*see* **Notes 1 and 2**).
3. Careful removal of the brain from the skull and post-fixation in the fixative solution for about 1 week at 4°C. Store the brains in 0.1 M PB diluted fixative (1:10) plus 0.025% sodium azide at 4°C until used.

4.2 Pre-Embedding Immunogold Method for Electron Microscopy

1. Cut 50 µm-thick sections on the vibratome and collect them in 0.1 M PB (pH 7.4) at RT (*see* **Note 3**).
2. Pre-incubate the sections in a blocking solution containing 10% bovine serum albumin (BSA), 0.1% sodium azide, and 0.02% saponin prepared in Tris–hydrogen chloride buffered saline 1× (TBS 1×), pH 7.4 for 30 min at RT.
3. Incubate with the primary antibodies in 10% BSA/TBS 1× containing 0.1% sodium azide and 0.004% saponin on a shaker for 2 days at 4°C.
4. Wash several times in 1% BSA/TBS 1×.
5. Incubate in a secondary 1.4 nm gold-labeled Immunoglobulin-G (Fab' fragment) in 1% BSA/TBS 1× with 0.004% saponin on a shaker for 3 h at RT.
6. Wash several times in 1% BSA/TBS 1× overnight on a shaker at 4°C.
7. Post-fix with 1% glutaraldehyde prepared in TBS 1× for 10 min at RT.
8. Wash several times in double-distilled water for 30 min.
9. Intensify the gold particles with an HQ Silver kit for about 12 min in the dark.
10. Wash several times in double-distilled water for 10 min (*see* **Note 4**).

11. Wash several times in 0.1 M PB (pH 7.4) for 30 min.
12. Osmicate (1% osmium tetroxide, in 0.1 M PB pH 7.4) for 20 min (*see* **Note 1**).
13. Wash several times in 0.1 M PB (pH 7.4) for 30 min.
14. Dehydrate in graded 50°, 70°, 96° alcohol (5 min/each) and 3×5 min in 100°.
15. Clear in propylene oxide 3×5 min.
16. Embed in a mixture of 1:1 propylene oxide and Epon resin 812 overnight on a shaker at RT.
17. Polymerize the resin in an oven at 60°C for 2 days.
18. Cut 1-μm semi-thin sections in an ultramicrotome.
19. Cut 50-nm ultra-thin sections and collect them on mesh nickel grids (*see* **Notes 5 and 6**).
20. Counterstain with 2.5% lead citrate for 20 min (*see* **Notes 7 and 8**).
21. Examine under the electron microscope.
22. Photograph the specimens by using a digital camera coupled to the electron microscope.

4.3 Double Pre-Embedding Immunogold and Immunoperoxidase Method for Electron Microscopy

1. Cut 50-μm-thick sections on the vibratome and collect them in 0.1 M PB (pH 7.4) at RT (*see* **Note 3**).
2. Pre-incubate in a blocking solution containing 10% BSA, 0.1% sodium azide, and 0.02% saponin prepared in TBS 1×, pH 7.4 for 30 min at RT.
3. Incubate with, for example, primary goat polyclonal antibodies and primary mouse monoclonal or rabbit polyclonal antibodies prepared in the blocking solution in 10% BSA/TBS 1× containing 0.1% sodium azide and 0.004% saponin on a shaker for 2 days at 4°C.
4. Wash several times in 1% BSA/TBS 1×.
5. Incubate with the corresponding biotinylated secondary antibody in 1% BSA/TBS 1× with 0.004% saponin on a shaker for 3 h at RT.
6. Wash several times in 1% BSA/TBS 1× overnight on a shaker at 4°C.
7. Incubate in a secondary 1.4-nm gold-labeled rabbit anti-goat Immunoglobulin-G (Fab' fragment) in 1% BSA/TBS 1× with 0.004% saponin on a shaker for 3 h at RT.
8. Wash several times in 1% BSA/TBS 1× for 30 min on a shaker at 4°C.
9. Incubate in avidin-biotin complex (ABC) (1:50) prepared in the washing solution for 1.5 h at RT.

10. Wash several times in 1% BSA/TBS 1× overnight on a shaker at 4°C.
11. Post-fix with 1% glutaraldehyde in TBS 1× for 10 min at RT.
12. Wash several times in double-distilled water for 30 min.
13. Intensify the gold particles with an HQ Silver kit for about 12 min in the dark.
14. Wash several times in double-distilled water for 10 min (*see* **Note 4**).
15. Wash several times in 0.1 M PB (pH 7.4) for 30 min.
16. Incubate in 0.05% DAB and 0.01% hydrogen peroxide prepared in 0.1 M PB for 5 min at RT.
17. Osmicate (1% osmium tetroxide, in 0.1 M PB pH 7.4) for 20 min.
18. Wash several times in 0.1 M PB (pH 7.4) for 30 min.
19. Dehydrate in graded 50°, 70°, 96° alcohol (5 min/each) and 3×5 min in 100° alcohol.
20. Clear in propylene oxide 3×5 min.
21. Embed in a mixture of 1:1 propylene oxide and Epon resin 812 overnight on a shaker at RT.
22. Polymerize the resin in an oven at 60°C for 2 days.
23. Cut 1-μm semi-thin sections in an ultramicrotome.
24. Cut 50-nm ultra-thin sections and collect them on mesh nickel grids (*see* **Notes 5 and 6**).
25. Stain with 2.5% lead citrate for 20 min (*see* **Notes 7 and 8**).
26. Examine under a Philips EM208S electron microscope. Photograph tissue preparations by using a digital Morada Camera from Olympus coupled to the electron microscope.

4.4 Single or Double Labeling with Post-Embedding Immunogold Method for Electron Microscopy

Primary antibodies are incubated directly on the grids containing the ultrathin sections. Incubations are carried out with drops of solution, typically 50 μl for most steps, 30 μl for primary antibodies, and 20 μl for secondary antibodies.

The detailed steps are as follows:

Tissue Preparation

1. Anesthetize, perfuse, and store the brains as mentioned above (*see* **Notes 1 and 2**).
2. Rinse in PB (4°C, overnight) small rectangular pieces of 0.5×0.5×1 mm from the region(s) of interest.
3. Cryoprotect them in glycerol (10%, 20%, and 30% in PB) and rapidly freeze in liquid propane in a cryofixation unit.

4. Freeze–substitute with methanol and 0.5% uranyl acetate, and embed in Lowicryl HM20 (*see* **Notes 9 and 10**).
5. Cut semi-thin sections of 1 μm thickness using an ultramicrotome.
6. Collect 70–90-nm ultra-thin sections on Formvar-coated single-slot nickel grids (*see* **Notes 11–16**).
7. Wash specimens with Tris–buffered saline with Triton X–100 (TBST) containing 0.1% $NaBH_4$ and 50 mM glycine for 10 min.
8. Rinse 3 × 1 min in TBST.
9. Pre-incubate in blocking solution: 10% (w/v) BSA in TBST for at least 10 min.
10. Incubate tissue sections overnight with individual or a mixture of affinity-purified primary antisera in TBST with 2% BSA. Use moistened filter paper to maintain a humid environment and minimize evaporation (*see* **Note 17**).
11. Perform three consecutive 1-min washes in TBST, followed by a 10-min wash.
12. Repeat the three 1-min washes in TBST.
13. Pre-incubate with the blocking solution: 10% (w/v) BSA and 0.5% PEG in TBST for 10 min.
14. Incubate with the corresponding or with a mixture of secondary antibodies or F(ab')2 IgG fragments coupled to different sizes of colloidal gold particles (10, 18, 20 nm) diluted 1:20 in TBST containing 2% BSA for 2 h.
15. Perform three consecutive 1-min rinses in double–distilled water, followed by a 10-min wash.
16. Dry the grids with filter paper.
17. Counterstain with 2% uranyl acetate for 12 min and 2.5% lead citrate for 2 min (*see* **Note 18**).
18. Perform three consecutive 1-min rinses in double-distilled water.
19. Dry with filter paper and allow the grids to sit for a few hours before observation under a transmission electron microscope.
20. Electron micrographs (magnification 18,000–36,000×) are captured from the grids carrying the ultrathin sections. A positive labeling is defined as the presence of at least one immunoparticle located between the plasma membrane and within a maximum distance of 30 nm from it. The area analyzed is measured using Image-J software (version 1.43u, NIH, USA). The percentages of positive profiles for each protein studied were calculated, and the data are presented as mean ±

SEM, analyzed using statistical software (GraphPad Prism 4, GraphPad Software Inc, San Diego, USA) (*see* **Notes 19 and 20**).

21. The distribution of immunoparticles relative to the postsynaptic density is assessed by analyzing the frequency of immunoparticles in 60-nm segments starting from the edge of the postsynaptic density. The distance of the gold particles from the edge of the postsynaptic specialization is measured using Image-J software, and the data are presented using GraphPad Prism 4 (*see* **Note 21**).

5 Notes

5.1 Pre-Embedding Immunogold Method

1. If the fixation with glutaraldehyde or osmium is insufficient or uneven, the tissue sample may be more degraded, and the preservation of subcellular structures (such as membranes or organelles) may be poor or inadequate.
2. To ensure proper preservation of the ultrastructure, the glutaraldehyde concentration can be adjusted depending on the specific target of observation. However, if the goal involves protein labeling, it is essential to test different concentrations to identify one that allows the antibody to detect the antigen effectively while still achieving optimal ultrastructural preservation.
3. Once the 50-μm-thick sections are cut with a vibratome, it is important to minimize the time before starting the immunocytochemical procedure, as the ultrastructure deteriorates over time.
4. The initial washes following silver intensification should be performed with double-distilled water before transitioning to phosphate buffer. However, prolonged or excessive washing with double-distilled water can cause the tissue to shrink due to changes in osmolarity, leading to membrane rupture and impaired ultrastructure. This can be mitigated by minimizing the time spent in water. It is crucial, however, to use double-distilled water for the initial washes, as mixing HQ with the salts in phosphate buffer can result in precipitation or residue formation, which affects the immunostaining. Therefore, while it is essential to begin with double-distilled water washes, their duration should be kept short.
5. When images cannot be properly focused, it is often due to ultrathin sections being slightly too thick. For TEM, the optimal thickness ranges from 50 to 100 nm. Thicker sections result in a loss of contrast and make it more difficult

to interpret fine details, while overly thin sections are more prone to damage. To achieve the highest image clarity, we usually prepare 50-nm sections. If cutting issues arise, such as vibrations in the ultramicrotome causing defects like the "shutter effect," we reduce the cutting speed, which can sometimes mitigate the problem. This "shutter effect" can also occur due to uneven or suboptimal resin polymerization. In such cases, we often rotate the capsule slightly to avoid cutting through the harder areas of the resin.

6. Damage from ultrathin sectioning: If the ultramicrotome cutting is not optimal or the blade has a nick, the sample may exhibit knife marks.
7. Lead citrate counterstaining: Lead reacts with CO_2 in the air, forming lead precipitates. To prevent this, we typically use a mask and filters when handling lead citrate droplets. Avoid shaking the bottle containing the lead citrate to prevent disturbing the deposits at the bottom or suspending them in the solution. Use parafilm to cover the container, ensuring it is not touched to avoid grease stains.
8. Sometimes, silver may appear scattered or splattered. This tends to happen when a significant time has passed since the sample was counterstained with lead citrate. To prevent this, we usually perform the counterstain the day before using the electron microscope, ensuring minimal delay.

5.2 Post-Embedding Immunogold Method

1. Freeze substitution is a method by which ice within tissues frozen at low temperatures can be replaced with substances possessing fixative properties, provided the sample is maintained at temperatures between –40°C and –78°C. After the tissue is frozen, the ice is gradually replaced by organic anhydrides and solvent molecules at low temperatures, followed by infiltration with a resin.

   The final outcome of this process depends on several factors, including the thickness of the sample, the freezing temperature, the warming time after substitution, and the inherent properties of the tissue. Once the tissue is frozen, the next step involves replacing the ice with a medium that remains liquid at temperatures below the sample's recrystallization point. This medium must also be capable of dissolving substances such as osmium tetroxide, which is used for tissue fixation and contrast enhancement.

2. Low-viscosity lowicryl resins (methacrylate and acrylic formulations) are usually used, which enable substitution and polymerization at low temperatures. Methacrylates are currently the most commonly used resins:

- Lowicryl K4M and KM11: Used at temperatures as low as −35°C and −60°C, respectively, and are polar (hydrophilic).
- Lowicryl HM20 and HM23: Used at temperatures as low as −70°C and −80°C and are nonpolar (hydrophobic).

These resins can polymerize either through direct irradiation with long-wavelength UV light (360 nm) or chemically, via the action of peroxides at 60°C. When using ultraviolet light, fixation with osmium tetroxide should be avoided, as it blocks the penetration of UV rays. However, the initiation of the polymerization reaction is temperature-independent, allowing the resin blocks to polymerize at the same temperature at which the substitution process was carried out. The use of these resins at low temperatures has opened a new field of study, allowing the observation of proteins under conditions of minimal denaturation.

3. Fragility of ultrathin sections: Ultrathin sections are highly fragile and may break or become lost during the immunocytochemical procedure. For this reason, we usually prepare 80-nm-thick sections.
4. When using Formvar-coated grids, the coating is extremely thin (typically around 10–30 nm) and both fragile and delicate. Before use, check the Formvar-coated grids under a light microscope to ensure they are free of pre-existing defects, such as perforations or irregularities.
5. Handling grids with sections is challenging and can add mechanical artifacts.
6. During handling with tweezers or while placing ultrathin sections, the Formvar-coated film can easily tear, resulting in holes in it.
7. During the transfer of the floating sections onto grids, the drying of liquid on the Formvar layer can create surface tension, leading to the collapse or tearing of the film. To prevent this, dry the grids slowly in a controlled environment (e.g., under moderate humidity conditions).
8. The staining and washing for sections mounted on grids can damage or dislodge the samples, especially if they are not firmly attached to the Formvar film. It is essential to ensure that the ultrathin sections are properly dried and securely adhered to the grids after being collected.
9. Performing immunostaining after sectioning exposes the sections to environmental pollutants, such as dust or particles, which may deposit on the grid and distort the image. It is necessary to be extremely careful.
10. Negative staining with agents such as uranyl acetate may only deposit on the surface of the sections, failing to penetrate

properly into deeper regions. This results in a loss of structural details and areas with uneven contrast.

11. The resolution of the post-embedding technique is determined by the diameter of the gold nanoparticle (typically 15 nm) and the length of the antibodies (approximately 8 nm). Regarding quantification, in most cases, each gold particle deposit can be considered the result of an independent antigen–antibody reaction. This allows for a linear relationship to be extrapolated between the observed gold particles and the number of antigens [74].

    For this assumption to hold, all antigens in the sample must have an equal likelihood of encountering the antibody. While this is theoretically achieved in ultrathin sections prepared for post-embedding techniques, in practice, the surface topology of the sample may distort antibody access to certain antigens. Nonetheless, a linear correlation between the number of antigens and the density of gold particles has been observed. This correlation diminishes at high antigen concentrations due to steric hindrance effects.

12. In our specific case, the use of the post-embedding immunocytochemistry technique facilitated the precise study of the spatial relationship between CB1 and mGluR2/3, and mGluR1a and mGluR5. Colloidal gold particles were employed to label the location of epitopes with a precision determined by the size of the antibody molecules. In our procedure, 10-nm gold particles were conjugated to the Fab′ fragment of secondary anti-rabbit antibodies to detect mGluR1a, mGluR5, and mGluR2/3. Meanwhile, 18-nm or 20-nm gold particles were conjugated to whole IgG anti-guinea pig antibodies to detect CB1 or mGluR5, respectively.

    Theoretically, the distance between the epitope and the center of the gold particle corresponds to the sum of the diameters of the primary antibody (IgG: ~8 nm), the secondary antibody (Fab′ fragment: ~4 nm or IgG: ~8 nm), and the radius of the gold particle (5 nm or 9–10 nm). Thus, the estimated distance between the recognized epitope and the center of the gold particle is approximately ~17 nm for mGluR1a, mGluR5 and mGluR2/3, and ~25–26 nm for CB1 and mGluR5.

13. When interpreting the localization, particularly perisynaptic (within the first 60 nm from the edge of the postsynaptic density) and extrasynaptic (up to 180 nm) of mGluR1a and mGluR5 at mossy fiber synapses, several considerations must be taken into account. A CA3 pyramidal neuron receives input from approximately 50 mossy fibers, which predominantly terminate on the dendritic excrescences/spines of these cells. Each mossy fiber terminal, in turn, has an average of 25 active zones (slightly fewer in adults), and each excrescence/spine

receives multiple synapses. Moreover, the average size of these active zones is about 0.1 μm^2, the mean distance between two active zones is 0.45 μm, and the postsynaptic density area in adult measures approximately 1.74 μm^2. The resolution of the technique (see above), coupled with the tissue processing that induces some distortion, results in variability in the localization of epitopes. Also, the three-dimensional arrangement of presynaptic and postsynaptic elements must be considered.

Assuming that excrescences/spines are roughly spherical in shape, the plane in which the synapse is sectioned can influence the apparent distribution (e.g., when sectioned tangentially "en face," nearby postsynaptic densities may appear to overlap). Along with the aforementioned factors—including the close proximity of synapses, tissue distortion, technique resolution, and the section thickness (~80 nm)—this may lead to artifacts where some gold particles appear to be in unexpected locations, such as the synaptic cleft. This occurs despite the use of antibodies targeting the intracellular C-terminal sequence of mGluR1a and mGluR5, which are inherently intracellular markers.

We observed the presence of mGluR1a gold particles in perisynaptic positions relative to the postsynaptic density of the dendritic excrescences of CA3 pyramidal cells, in a distribution similar to that of mGluR5. Interestingly, labeling was not previously detected in the CA3 *stratum lucidum*, the termination zone of the mossy fibers, with mGluR1a antibodies [75]. This discrepancy may be explained by our use of a specific antibody against mGluR1a combined with the sensitivity of the post-embedding technique employed in our study, which may have unmasked mGluR1a epitopes that remained concealed with other methods [69]. However, the localization of mGluR1a in approximately 20% of the dendritic excrescences of CA3 pyramidal cells, along with its co-localization with mGluR5 in approximately 40% of the same structures, suggests that methodological limitations inherent in combining immunolocalization and imaging techniques must be carefully considered when interpreting the presence or absence of a protein.

Conclusions

High-resolution immunoelectron microscopy remains one of the most powerful approaches for the precise localization of proteins in the brain, providing unparalleled insights into the subcellular distribution and spatial relationships of receptors, ion channels, and components of intracellular signaling pathways. Pre-embedding immunogold labeling enables reliable detection of proteins throughout neuronal and glial compartments, while preserving tissue ultrastructure. In contrast, post-embedding immunogold labeling provides superior spatial resolution for the quantitative analysis of protein distribution within synapses and other small membrane domains. Although each approach has distinct advantages and limitations, they are highly complementary. When appropriately optimized and applied, these techniques provide accurate ultrastructural information that is essential for understanding the molecular organization of neural circuits and the cellular mechanisms underlying normal brain function and disease.

Acknowledgments

Funding was provided by The Ministry of Science, Innovation and Universities, grant PID2024-158287OB-I00 funded by MICIU/AEI/ 10.13039/501100011033 and by "ERDF/EU"; RIA-PAd Network funded by Instituto de Salud Carlos III-ERDF/EU (RD24/0003/0027); and the Basque Government (grant IT1620-22).

Conflict of Interest The authors have no conflicts of interest to declare that are relevant to the content of this chapter.

References

1. Kano M, Ohno-Shosaku T, Hashimotodani Y, Uchigashima M, Watanabe M (2009) Endocannabinoid-mediated control of synaptic transmission. Physiol Rev 89(1):309–380
2. Katona I, Freund TF (2012) Multiple functions of endocannabinoid signaling in the brain. Annu Rev Neurosci 35(1):529–558
3. Lu HC, Mackie K (2021) Review of the endocannabinoid system. Biol Psychiatry Cogn Neurosci Neuroimaging 6(6):607–615
4. Piazza PV, Cota D, Marsicano G (2017) The CB1 receptor as the cornerstone of exostasis. Neuron 93(6):1252–1274
5. Monory K, Massa F, Egertová M, Eder M, Blaudzun H, Westenbroek R et al. (2006) The endocannabinoid system controls key epileptogenic circuits in the hippocampus. Neuron 51(4):455–466
6. Busquets-Garcia A, Bains J, Marsicano G (2018) CB1 receptor signaling in the brain: extracting specificity from ubiquity. Neuropsychopharmacology 43(1):4–20. Jan 1
7. Bernal-Chico A, Tepavcevic V, Manterola A, Utrilla C, Matute C, Mato S (2023) Endocannabinoid signaling in brain diseases: emerging relevance of glial cells. Glia 71(1):103–126
8. Costas-Insua C, Moreno E, Maroto IB, Ruiz-Calvo A, Bajo-Grañeras R, Martín-Gutiérrez D et al. (2021) Identification of BiP as a CB 1 receptor-interacting protein that fine-tunes cannabinoid signaling in the mouse brain. J Neurosci 41(38):7924–7941
9. Gómez-Cañas M, Rodríguez-Cueto C, Satta V, Hernández-Fisac I, Navarro E, Fernández-Ruiz J (2003) Endocannabinoid-binding receptors as drug targets. Methods Mol Biol 2576:67–94
10. Martinez Ramirez CE, Ruiz-Pérez G, Stollenwerk TM, Behlke C, Doherty A, Hillard CJ (2023) Endocannabinoid signaling in the central nervous system. Glia 71(1):5–35
11. Noriega-Prieto JA, Kofuji P, Araque A (2023) Endocannabinoid signaling in synaptic function. Glia 71(1):36–43
12. Fernández-Moncada I, Rodrigues RS, Fundazuri UB, Bellocchio L, Marsicano G (2024) Type-1 cannabinoid receptors and their ever-expanding roles in brain energy processes. J Neurochem 168(5):693–703
13. Herkenham M, Lynn AB, Little MD, Johnson MR, Melvin LS, de Costa BR et al. (1990) Cannabinoid receptor localization in brain. Proc Natl Acad Sci U S A 87(5):1932–1936
14. Marsicano G, Lutz B (1999) Expression of the cannabinoid receptor CB1 in distinct neuronal subpopulations in the adult mouse forebrain. Eur J Neurosci 11(12):4213–4225

15. Twitchell W, Brown S, Mackie K (1997) Cannabinoids Inhibit N- and P/Q-type calcium channels in cultured rat hippocampal neurons. J Neurophysiol 78(1):43–50
16. Egertová M, Giang DK, Cravatt BF, Elphick MR (1998) A new perspective on cannabinoid signalling: complimentary localization of fatty acid amide hydrolase and the CB1 receptor in rat brain. Proc Biol Sci 265(1410):2081–2085
17. Tsou K, Brown S, Sañudo-Peña MC, Mackie K, Walker JM (1998) Immunohistochemical distribution of cannabinoid CB1 receptors in the rat central nervous system. Neuroscience 83(2):393–411
18. Mechoulam R, Parker LA (2013) The endocannabinoid system and the brain. Annu Rev Psychol 64(1):21–47
19. Araque A, Castillo PE, Manzoni OJ, Tonini R (2017) Synaptic functions of endocannabinoid signaling in health and disease. Neuropharmacology 124:13–24
20. Manzanares J, Cabañero D, Puente N, García-Gutiérrez MS, Grandes P, Maldonado R (2018) Role of the endocannabinoid system in drug addiction. Biochem Pharmacol 157:108–121
21. Bonilla-Del Río I, Puente N, Mimenza A, Ramos A, Serrano M, Lekunberri L et al. (2021) Acute Δ9-tetrahydrocannabinol prompts rapid changes in cannabinoid CB 1 receptor immunolabeling and subcellular structure in CA1 hippocampus of young adult male mice. J Comp Neurol 529(9):2332–2346
22. Robledo-Menendez A, Vella M, Grandes P, Soria-Gomez E (2022) Cannabinoid control of hippocampal functions: the *where* matters. FEBS J 289(8):2162–2175
23. Terradillos I, Bonilla-Del Río I, Puente N, Serrano M, Mimenza A, Lekunberri L et al. (2023) Altered glial expression of the cannabinoid 1 receptor in the subiculum of a mouse model of Alzheimer's disease. Glia 71(4):866–879
24. Gómez-Sotres P, Skupio U, Dalla Tor T, Julio-Kalajzic F, Cannich A, Gisquet D et al. (2024) Olfactory bulb astrocytes link social transmission of stress to cognitive adaptation in male mice. Nat Commun 15(1):7103
25. Kawamura Y, Fukaya M, Maejima T, Yoshida T, Miura E, Watanabe M et al. (2006) The CB1 cannabinoid receptor is the major cannabinoid receptor at excitatory presynaptic sites in the hippocampus and cerebellum. J Neurosci 26(11):2991–3001
26. Katona I, Sperlágh B, Sík A, Käfalvi A, Vizi ES, Mackie K et al. (1999) Presynaptically located CB1 cannabinoid receptors regulate GABA release from axon terminals of specific hippocampal interneurons. J Neurosci 19(11):4544–4558
27. Marsicano G, Goodenough S, Monory K, Hermann H, Eder M, Cannich A et al. (2003) CB1 cannabinoid receptors and on-demand defense against excitotoxicity. Science 302(5642):84–88
28. Navarrete M, Araque A (2008) Endocannabinoids mediate neuron-astrocyte communication. Neuron 57(6):883–893
29. Stella N (2010) Cannabinoid and cannabinoid-like receptors in microglia, astrocytes, and astrocytomas. Glia 58(9):1017–1030
30. Oliveira da Cruz JF, Robin LM, Drago F, Marsicano G, Metna-Laurent M (2016) Astroglial type-1 cannabinoid receptor (CB1): a new player in the tripartite synapse. Neuroscience 323:35–42
31. Puente N, Bonilla-Del Río I, Achicallende S, Nahirney P, Grandes P (2019) High-resolution immunoelectron microscopy techniques for revealing distinct subcellular type 1 cannabinoid receptor domains in brain. Bio Protoc 9(2):e3145
32. Hájos N, Katona I, Naiem SS, Mackie K, Ledent C, Mody I et al. (2000) Cannabinoids inhibit hippocampal GABAergic transmission and network oscillations. Eur J Neurosci 12(9):3239–3249
33. Reguero L, Puente N, Elezgarai I, Mendizabal-Zubiaga J, Canduela MJ, Buceta I et al. (2011) GABAergic and cortical and subcortical glutamatergic axon terminals contain CB1 cannabinoid receptors in the ventromedial nucleus of the hypothalamus. PLoS ONE 6(10):e26167
34. Reguero L, Puente N, Elezgarai I, Ramos-Uriarte A, Gerrikagoitia I, Bueno-López JL et al. (2014) Subcellular localization of NAPE-PLD and DAGL-α in the ventromedial nucleus of the hypothalamus by a preembedding immunogold method. Histochem Cell Biol 141(5):543–550
35. Suárez J, Ortíz O, Puente N, Bermúdez-Silva FJ, Blanco E, Fernández-Llebrez P et al. (2011) Distribution of diacylglycerol lipase alpha, an endocannabinoid synthesizing enzyme, in the rat forebrain. Neuroscience 192:112–131
36. Puente N, Cui Y, Lassalle O, Lafourcade M, Georges F, Venance L et al. (2011) Polymodal activation of the endocannabinoid system in the extended amygdala. Nat Neurosci 14(12):1542–1547
37. Soria-Gómez E, Bellocchio L, Reguero L, Lepousez G, Martin C, Bendahmane M et al. (2014) The endocannabinoid system controls food intake via olfactory processes. Nat Neurosci 17(3):407–415
38. Puente N, Reguero L, Elezgarai I, Canduela MJ, Mendizabal-Zubiaga J, Ramos-Uriarte A et al. (2015) The transient receptor potential vanilloid-1 is localized at excitatory synapses in the mouse dentate gyrus. Brain Struct Funct 220(2):1187–1194

39. Canduela MJ, Mendizabal-Zubiaga J, Puente N, Reguero L, Elezgarai I, Ramos-Uriarte A et al. (2015) Visualization by high resolution immunoelectron microscopy of the transient receptor potential Vanilloid-1 at inhibitory synapses of the mouse dentate gyrus. PLOS ONE 10(3):e0119401
40. Han J, Kesner P, Metna-Laurent M, Duan T, Xu L, Georges F et al. (2012) Acute cannabinoids impair working memory through astroglial CB1 receptor modulation of hippocampal LTD. Cell 148(5):1039–1050
41. Bosier B, Bellocchio L, Metna-Laurent M, Soria-Gomez E, Matias I, Hebert-Chatelain E et al. (2013) Astroglial CB1 cannabinoid receptors regulate leptin signaling in mouse brain astrocytes. Mol Metab 2(4):393–404
42. Gutiérrez-Rodríguez A, Bonilla-Del Río I, Puente N, Gómez-Urquijo SM, Fontaine CJ, Egaña-Huguet J et al. (2018) Localization of the cannabinoid type-1 receptor in subcellular astrocyte compartments of mutant mouse hippocampus. Glia 66(7):1417–1431
43. Martín-Monteagudo C, Sánchez Romero J, Adams JM, Puente N, Grandes P, Marsicano G et al (2025) An astrocytic ensemble at 1 vHip-NAc synapses modulates cognitive impairments induced by chronic THC exposure. Nat Commun 17(1):471
44. Gutiérrez-Rodríguez A, Puente N, Elezgarai I, Ruehle S, Lutz B, Reguero L et al. (2017) Anatomical characterization of the cannabinoid CB1 receptor in cell-type–specific mutant mouse rescue models. J Comp Neurol 525(2):302–318
45. Bonilla-Del Río I, Puente N, Peñasco S, Rico I, Gutiérrez-Rodríguez A, Elezgarai I et al. (2019) Adolescent ethanol intake alters cannabinoid type-1 receptor localization in astrocytes of the adult mouse hippocampus. Addict Biol 24(2):182–192
46. Berger AL, Henricks AM, Lugo JM, Wright HR, Warrick CR, Sticht MA et al. (2018) The lateral habenula directs coping styles under conditions of stress via recruitment of the endocannabinoid system. Biol Psychiatry 84(8):611–623
47. Soria-Gomez E, Pagano Zottola AC, Mariani Y, Desprez T, Barresi M, Bonilla-del Río I et al. (2021) Subcellular specificity of cannabinoid effects in striatonigral circuits. Neuron 109(9):1513–1526.e11
48. Bénard G, Massa F, Puente N, Lourenço J, Bellocchio L, Soria-Gofmez E et al. (2012) Mitochondrial CB1 receptors regulate neuronal energy metabolism. Nat Neurosci 15(4):558–564
49. Hebert-Chatelain E, Reguero L, Puente N, Lutz B, Chaouloff F, Rossignol R et al. (2014) Cannabinoid control of brain bioenergetics: exploring the subcellular localization of the CB1 receptor. Mol Metab 3(4):495–504
50. Hebert-Chatelain E, Desprez T, Serrat R, Bellocchio L, Soria-Gomez E, Busquets-Garcia A et al. (2016) A cannabinoid link between mitochondria and memory. Nature 539(7630):555–559
51. Jimenez-Blasco D, Busquets-Garcia A, Hebert-Chatelain E, Serrat R, Vicente-Gutierrez C, Ioannidou C et al. (2020) Glucose metabolism links astroglial mitochondria to cannabinoid effects. Nature 583(7817):603–608
52. Arrabal S, Lucena MA, Canduela MJ, Ramos-Uriarte A, Rivera P, Serrano A et al. (2015) Pharmacological blockade of cannabinoid CB1 receptors in diet-induced obesity regulates mitochondrial dihydrolipoamide dehydrogenase in muscle. PLOS ONE 10(12):e0145244
53. Mendizabal-Zubiaga J, Melser S, Bénard G, Ramos A, Reguero L, Arrabal S et al (2016) Cannabinoid CB1 receptors are localized in striated muscle mitochondria and regulate mitochondrial respiration. Front Physiol 7:476
54. Achicallende S, Bonilla-Del Río I, Serrano M, Mimenza A, Lekunberri L, Anaut-Lusar I et al. (2022) GLAST versus GFAP as astroglial marker for the subcellular study of cannabinoid CB1 receptors in astrocytes. Histochem Cell Biol 158(6):561–569
55. de Meij J, Alfanek Z, Morel L, Decoeur F, Leyrolle Q, Picard K et al. (2021) Microglial Cannabinoid Type 1 receptor regulates brain inflammation in a sex-specific manner. Cannabis Cannabinoid Res 6(6):488–507
56. Manterola A, Chara JC, Aguado T, Palazuelos J, Matute C, Mato S (2022) Cannabinoid CB1 receptor expression in oligodendrocyte progenitors of the hippocampus revealed by the NG2-EYFP-knockin mouse. Front Neuroanat 16:1030060
57. Serrat R, Covelo A, Kouskoff V, Delcasso S, Ruiz-Calvo A, Chenouard N et al. (2021) Astroglial ER-mitochondria calcium transfer mediates endocannabinoid-dependent synaptic integration. Cell Rep 37(12):110133
58. Pagano Zottola AC, Martín-Jiménez R, Lavanco G, Hamel-Côté G, Ramon-Duaso C, Rodrigues RS et al. (2025) Potentiation of mitochondrial function by mitoDREADD-Gs reverses pharmacological and neurodegenerative cognitive impairment in mice. Nat Neurosci 28(9):1844–1857
59. Kushmerick C, Price GD, Taschenberger H, Puente N, Renden R, Wadiche JI et al. (2004) Retroinhibition of presynaptic Ca 2+ currents by endocannabinoids released via postsynaptic mGluR activation at a calyx synapse. J Neurosci 24(26):5955–5965

60. Lafourcade M, Elezgarai I, Mato S, Bakiri Y, Grandes P, Manzoni OJ (2007) Molecular components and functions of the endocannabinoid system in mouse prefrontal cortex. PLoS ONE 2(8):e709
61. Chiu CQ, Puente N, Grandes P, Castillo PE (2010) Dopaminergic modulation of endocannabinoid-mediated plasticity at GABAergic synapses in the prefrontal cortex. J Neurosci 30(21):7236–7248
62. Massi L, Elezgarai I, Puente N, Reguero L, Grandes P, Manzoni OJ et al. (2008) Cannabinoid receptors in the bed nucleus of the stria terminalis control cortical excitation of midbrain dopamine cells in vivo. J Neurosci 28(42):10496–10508
63. Puente N, Elezgarai I, Lafourcade M, Reguero L, Marsicano G, Georges F et al. (2010) Localization and function of the cannabinoid CB1 receptor in the anterolateral bed nucleus of the stria terminalis. PLoS ONE 5(1):e8869
64. Hollinshead M, Sanderson J, Vaux DJ (1997) Anti-biotin antibodies offer superior organelle-specific labeling of mitochondria over avidin or streptavidin. J Histochem Cytochem 45(8):1053–1057
65. Page Faulk W, Malcolm Taylor G (1971) Communication to the editors. Immunochemistry 8(11):1081–1083
66. Van den Pol AN (1989) Neuronal imaging with colloidal gold. J Microsc 155(1):27–59
67. Roth J (1996) The silver anniversary of gold: 25 years of the colloidal gold marker system for immunocytochemistry and histochemistry. Histochem Cell Biol 106(1):1–8
68. Hermida D, Elezgarai I, Puente N, Alonso V, Anabitarte N, Bilbao A et al. (2006) Developmental increase in postsynaptic alpha-amino-3-hydroxy-5-methyl-4 isoxazolepropionic acid receptor compartmentalization at the calyx of held synapse. J Comp Neurol 495(5):624–634
69. Amiry-Moghaddam M, Ottersen OP (2013) Immunogold cytochemistry in neuroscience. Nat Neurosci 16(7):798–804
70. Hermida D, Mateos JM, Elezgarai I, Puente N, Bilbao A, Bueno-López JL et al. (2010) Spatial compartmentalization of AMPA glutamate receptor subunits at the calyx of Held synapse. J Comp Neurol 518(2):163–174
71. Hunt DL, Puente N, Grandes P, Castillo PE (2013) Bidirectional NMDA receptor plasticity controls CA3 output and heterosynaptic metaplasticity. Nat Neurosci 16(8):1049–1059
72. Stachowicz K (2024) Interactions between metabotropic glutamate and CB1 receptors: implications for mood, cognition, and synaptic signaling based on data from mGluR and CB1R-targeting drugs. Pharmacol Rep 76(6):1286–1296
73. Tamaru Y, Nomura S, Mizuno N, Shigemoto R (2001) Distribution of metabotropic glutamate receptor mGluR3 in the mouse CNS: differential location relative to pre- and postsynaptic sites. Neuroscience 106(3):481–503
74. Ottersen OP (1989) Postembedding immunogold labelling of fixed glutamate: an electron microscopic analysis of the relationship between gold particle density and antigen concentration. J Chem Neuroanat 2(1):57–66
75. Shigemoto R, Kinoshita A, Wada E, Nomura S, Ohishi H, Takada M et al. (1997) Differential presynaptic localization of metabotropic glutamate receptor subtypes in the rat hippocampus. J Neurosci 17(19):7503–7522

Chapter 4

An Alternative Chromogenic Immunostaining Method for Light and Electron Microscopy: A Demonstration Using NeuN Staining in Mouse Brain Samples

Victor Z. Lau, Jared Vander Zwaag and Marie-Ève Tremblay

Abstract

Conventional immunostaining methods for electron microscopy (EM) generally use horseradish peroxidase (HRP) systems, which often involve additional exposure to toxic and carcinogenic compounds. Here, we show a proof-of-concept with a safe, affordable, and accessible chromogenic staining method to visualize insoluble 5-bromo-4-chloro-3-indolyl phosphate (BCIP-based) staining. In particularly, using a BCIP-based substrate, using immunostaining for neurons (NeuN) on mouse brain samples as an example, we demonstrate precipitates observable with both light microscopy and EM. We compare BCIP to other conventional chromogen systems and show equivalent staining quality to other conventional stains, including Vector® VIP and immuno-gold labeling. Our system also uses an alkaline phosphatase system for staining, thereby opening up further possibilities for multiplex, correlative light and EM immunostaining. Finally, our findings can be applied to immunostaining applicable in academic or industrial fields involving biology and biochemistry investigations.

Keywords Immunocytochemical electron microscopy, Pre-embedding staining, BCIP, Neurons, NeuN, Mouse brain

Abbreviations

BB	Blocking buffer
BCIP	5-bromo-4-chloro-3-indolyl phosphate
BS	Bovine serum albumin
BV	Blood vessel
CNS	Central nervous system
DAB	Diaminobenzidine
DDI	5,5′-dibromo-4,4′-dichloro-indigo
EM	Electron microscopy
*g*TBS	Glycine and Tris-buffered saline
H_2O_2	Hydrogen peroxide
HPMA	2-Hydroxypropyl methacrylate

Marie-Ève Tremblay (ed.), *Ultrastructural Neuroscience Methods*, Neuromethods, vol. 238,
https://doi.org/10.1007/978-1-0716-5456-9_4,

LM	Light microscopy
MSDS	Material safety data sheet
$NaBH_4$	Sodium borohydride
NeuN	Neuronal nuclei
NBT	Nitro blue tetrazolium
NDS	Normal donkey serum
NeuN	Fox3 protein (i.e., common neuronal nuclear antigen)
PB / PBS	Phosphate-buffer/Phosphate-buffered saline
SEM	Scanning electron microscopy
TB	Tris hydrochloride buffer
TEM	Transmission electron microscopy
X-gal	5-bromo-4-chloro-3-indolyl-β-d-galactoside

1 Introduction

Electron microscopy (EM), consisting of transmission EM (TEM) and scanning EM (SEM), is a powerful technique that allows visualizing structures at nanoscale resolution [1]. TEM involves transmitting electrons through a thin specimen in order to create an image, whereas SEM utilizes scanning a focused electron beam over the surface of a specimen; subsequently generated backscattered and/or secondary electrons are then detected to generate an image [1]. Both TEM and SEM have significantly contributed to advancing multiple scientific fields, including biology, materials science, and nanotechnology. In biology, visualization of nanoscale resolution (i.e., ultrastructural) features of cells and surrounding environments provides unique strengths. These strengths include collecting simultaneous data on cell or microenvironmental features not visible or resolved at lower-resolution microscopy and being able to assess numerous features in parallel (e.g., nucleus features and multiple organelle appearances, organelle stress and dysfunction, and cellular behaviors including phagocytosis and exocytosis) [1].

To increase specificity in investigating cellular types, organelles, and/or localized targets of interest, staining is often employed via immunoenzymatic chromogenic staining (Fig. 1). Briefly, a sample is incubated with a species-specific primary antibody targeting an antigen of interest and a subsequent secondary antibody specific for the species of the primary antibody. This secondary antibody is often biotinylated, which allows for strong binding with avidin or streptavidin proteins conjugated with a catalyzing enzyme (e.g., commonly horseradish peroxidase (HRP) or alkaline phosphatase (AP)). The secondary-biotin and conjugated avidin/streptavidin allow for mass amplification and local increased density of these enzymes to cleave soluble, chromogenic substrates through a commonly based oxidization reaction; after exposure to catalyzing enzymes, these soluble chromogens become insoluble deposits that are intended to be embedded within the sample. Together, this immunoenzymatic chromogenic staining method is intended to offer a higher signal-to-noise ratio of staining *versus* sites lacking the antigen of interest.

As EM results in images generated by differences in electron density, these images cannot differentiate staining by color and instead involve visible chromogenic deposits as localized electron dense precipitates (often dots). Conventionally, both 3,3′-diaminobenzidine (DAB) and Vector® VIP Substrate have been used to localize discrete chromogenic deposits for immunostaining in EM; both these chromogens are also visible in light microscopy (LM) [2, 3]. However, a small drawback is that these substrates are potentially carcinogenic, as indicated in their Material Safety Data Sheet (MSDS). Alternative immunostaining methods also include conjugating secondary antibodies with nanogold-based formulations, which may also demonstrate variance in specificity in EM despite being able to produce precipitates of greater size [4].

5-bromo-4-chloro-3-indolyl-β-d-galactoside (X-gal), which has been widely used for multiple reporter systems and subsequent staining purposes, involves a staining reaction that results in the production of 5,5′-dibromo-4,4′-dichloro-indigo (DDI) precipitates [5, 6]. A conventional combined nitro blue tetrazolium (NBT) and 5-bromo-4-chloro-3-indolyl phosphate (BCIP) system has also been repurposed for immunostaining, wherein BCIP is initially oxidized by AP; the resulting indoxyl intermediate is then oxidized by NBT, which forms a combination of insoluble products [7]. While NBT leads to non-specific staining outside the targeted antigen in EM [8], BCIP can notably form an insoluble DDI precipitate. These precipitates have been shown to also lead to both LM and EM staining, as evidenced by previous studies [6, 9]. Given this information, it likely stands that while initial substrates may differ between X-gal and BCIP-based systems, the final product is a shared, DDI precipitate deposited within targets of interest. Given that BCIP is a much safer and not categorized as carcinogenic per its MSDS, BCIP-based staining systems for both LM and EM provide an accessible, inexpensive, and safer high-potential target for exploring additional chromogenic immunostaining.

Finally, while X-gal-deposited staining has been shown to be visualized in EM, such staining can be dissolved out by conventional propylene oxide steps used when transitioning samples between water-based solvents and epoxy resin for EM imaging; this issue is solvable via replacing propylene oxide with 2-hydroxypropyl methacrylate (HPMA) [5]. Another option would be to only use ethanol and proceed straight to resin [10, 11]. Altogether, we introduce the usage of a novel immunostaining method that takes inspiration from a conventional immunoenzymatic chromogenic staining that is visible at LM, particularly with streptavidin-conjugated AP and BCIP (Fig. 1). By also using HPMA over propylene oxide to preserve

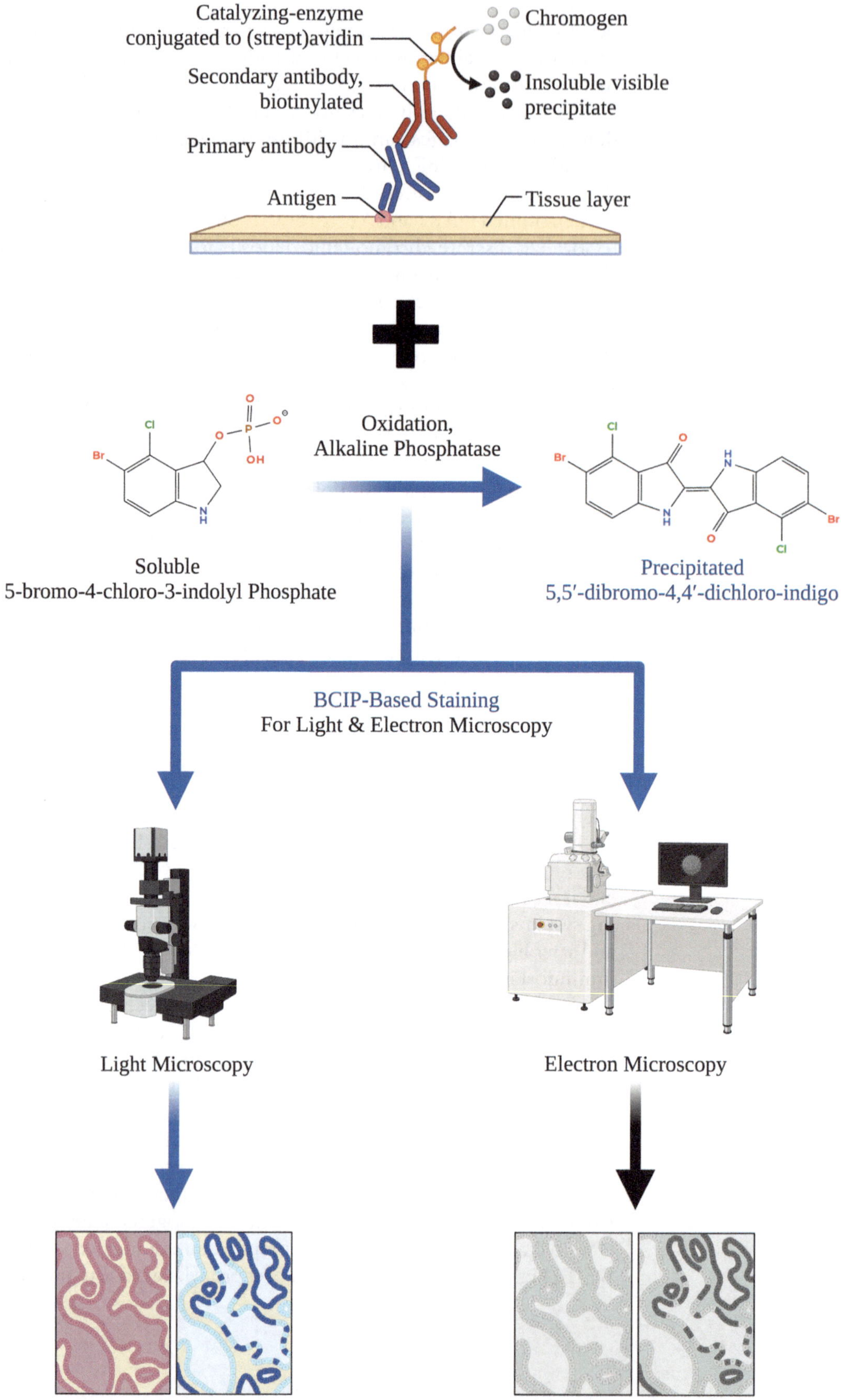
Catalyzing-enzyme conjugated to (strept)avidin
Chromogen
Secondary antibody, biotinylated
Insoluble visible precipitate
Primary antibody
Antigen
Tissue layer
Cl
Br
O
P
OH
N H
Oxidation,
Alkaline Phosphatase
Soluble
5-bromo-4-chloro-3-indolyl Phosphate
Precipitated
5,5′-dibromo-4,4′-dichloro-indigo
BCIP-Based Staining
For Light & Electron Microscopy
Light Microscopy
Electron Microscopy

BCIP-based staining in EM, we provide here a novel proof-of-concept application of this system using the marker neuronal nuclei (NeuN) for enriched staining specificity of neurons as previously revealed in multiple species [12, 13]. While demonstrated here on mouse brain samples, this system may be applicable to other tissue and cell culture samples, across species, and applicable for both academic and industrial usage.

2 Methods

2.1 Animal Preparation and Mouse Brain Tissue Cutting

In accordance with the guidelines of the Institutional Animal Ethics Committees, the Canadian Council on Animal Care, and the Animal Care Committee of Université Laval, 7-month-old male mice (C57BL/6J background) of a wild-type (WT) were used for experiments. Mice were housed under a 12-h light-dark cycle at 22–25°C with *ad libitum* access to food and water.

To optimize tissue quality for EM experiments, mice were anesthetized with an intraperitoneal injection of sodium pentobarbital (80 mg/kg), followed by transcardiac perfusion using 3.5% acrolein [diluted in phosphate buffer (PB): 100 mM at pH 7.4] and 4% paraformaldehyde solution [diluted in phosphate-buffered saline (PBS): 50 mM at pH 7.4]. Coronal brain sections were then sliced at 50-μ m thickness in ice-cold PBS using a vibratome (Leica VT1000S) and stored at −20°C in cryoprotectant [20% (v/v) glycerol and 20% (v/v) ethylene glycol in PBS] until used for further experimentation.

2.2 Chromogenic Immunostaining

50-μm brain sections containing the frontal cortex were used (Bregma 0–2.4 mm) [14]. Samples were washed with PBS to remove cryoprotectant, then incubated in 0.1% $NaBH_4$ (sodium borohydride) diluted in PBS for 30 min and washed with PBS. Afterward, sections were then incubated for 30 min in 0.3% H_2O_2 (hydrogen peroxide), diluted with

Fig. 1 Overview of BCIP-Based Staining for Light and Electron Microscopy. Here, a visual representation is given of how the novel immunostaining method works. Similar to conventional immunostaining methods with a chromogen, the sample is first incubated with a primary antibody to specifically target an antigen of interest. A biotinylated secondary antibody with specificity for the host species of the primary antibody is thereafter added, with subsequent addition of avidin or streptavidin conjugated with a catalyzing enzyme; this catalyzing enzyme is used to catalyze oxidation of a chromogen at the site of the antigen, creating an enriched signal-to-noise ratio showing staining localized to the antigen of interest. Here, alkaline phosphatase (AP) is used as a particularly selective enzyme to cleave 5-bromo-4-chloro-3-indolyl phosphate (BCIP) to produce enriched amounts of 5,5′-dibromo-4,4′-dichloro-indigo (DDI) precipitates at the antigen location. This novel method is applicable to both academic and industrial purposes and across different tissue and cell culture systems; it is also demonstrable across both light and electron microscopy. Figure made with Molview and BioRender.com

0.1 M glycine in Tris-buffered saline (gTBS): 50 mM at pH 7.4. After further gTBS washes, sections were incubated for 1 h with blocking buffer (BB) consisting of 5% bovine serum albumin (BSA), 5% normal donkey serum (NDS), and 0.05% Triton X-100. Afterward, in conditions involving 1:500 anti-rabbit NeuN (EMD Millipore, USA, cat # ABN78) in BB overnight, a primary control was used wherein sections were incubated in BB only instead of NeuN in BB.

The next day, sections were washed with gTBS. For conditions involving immuno-gold staining, sections were incubated with 1:100 anti-rabbit Nanogold®-IgG (Nanoprobes, USA, cat # 2003-1 ML) in BB for 3 h. Afterward, immuno-gold-condition sections were washed with double-distilled water and then 3% sodium acetate for 2 × 1 min each per sodium acetate wash. During washes, a HQ SILVER™ kit was prepared via mixing Components A, B, and C (Nanoprobes, USA, cat # 2012); after sodium acetate washes, immuno-gold-labeled sections were incubated with the silver kit for 12 min in the dark. After subsequent sodium acetate washes, the developed sections were processed for either LM or EM.

For the remaining conditions, sections were incubated with 1:300 anti-rabbit Biotin-SP AffiniPure™, Fc-specific (Jackson Immunoresearch, USA, cat # 111-066-046), for 2.5 h. After further washes with gTBS, sections were either incubated in BB (modified without Triton X-100) containing 1:100 streptavidin conjugated to HRP (Cedarlane sourced from Jackson Immunoresearch, Canada, cat # 016-030-084) or AP (Cedarlane sourced from Jackson Immunoresearch, Canada, cat # 016-050-084). Streptavidin-HRP was employed for DAB and VIP conditions, whereas streptavidin-AP was utilized for control and BCIP conditions. After streptavidin incubations, sections were washed with gTBS before the respective specific staining procedures. For the DAB staining conditions, section staining was revealed with 0.05% DAB (Millipore Sigma, cat # D5905-50TAB) and 0.015% H_2O_2 solution in TB (0.5 M, pH 8). For VIP conditions, a Vector® VIP substrate kit (Vector Laboratories, USA, cat # SK-4600) was used according to protocol: VIP Reagents 1–4 were mixed together and subsequently applied to sections. Regarding control and BCIP conditions, Reagents 2 and 3 were utilized from an NBT/BCIP kit (Vector Laboratories, cat # SK-5400) to omit NBT inclusion and staining. In particular, two drops from each of the supplied Reagent 2 and Reagent 3 bottles were added to 2.5 mL of TB (100 mM, pH 8.4), momentarily mixed, and then applied to sections to reveal potential staining. After chromogenic staining, regardless of conditions, sections were placed in PB.

2.3 Processing and Imaging for Light Microscopy

Stained sections for light microscopy experiments (n = 3 mice) were dehydrated in ascending ethanol concentrations of 5 min each (2 × 35%, 50%, 70%, 80%, 90%, 3 × 100%). Dehydrated sections were then washed three times, 5 min each, with Citrisolv (Fisher Scientific, Canada, cat# 04-355-121) for tissue clearing, and finally mounted with VectaMount Permanent Mounting Medium (Biolynx, Canada, cat #VECTH5000). To collect data for further quantitative microscopy, representative section staining was imaged via light microscopy using a Basler area scan camera (2.3 MP; acA1920–40uc) at 40x magnification resolution (1.51 μm/pixel). Mosaic images were then generated and stitched together via Manual Whole Slide Imaging (WSI) software (2019B, Microvisioneer).

NeuN+ cells were identified at the light microscopy level with enriched staining in the nucleus, but also with surrounding staining in the cytoplasm [12]. The majority of these NeuN+ cells were likely neurons, based on longer processes (n <3), indicating an axon and 1-2 dendrites.

2.4 Processing and Imaging for Electron Microscopy

To process the stained samples for SEM, protocols were slightly modified from previous procedures [15]. After immersing stained sections in a post-fixation solution of 0.6% glutaraldehyde (Electron Microscopy Sciences (EMS), USA, cat # 16320) in 4% paraformaldehyde in PB overnight, at room temperature, the stained sections (n = 3 mice) were processed for EM. Briefly, sections were washed in PB after glutaraldehyde post-fixation and incubated for 1 h in a solution containing an equal volume of 4% osmium tetroxide (EMS, cat # 19190) and 3% potassium ferrocyanide (Sigma-Aldrich, Canada, cat # P9387) in PB. After PB washes, sections were incubated in a filtered 1% thiocarbohydrazide solution for 20 min (diluted in double-distilled water; Sigma-Aldrich, Canada, cat # 223220). This was followed by further double-distilled water washes and then a second 30 min incubation with 2% aqueous osmium tetroxide (diluted in double-distilled water). After final double-distilled water washes, sections were dehydrated in ascending concentrations of ethanol for 5 min each (diluted in double-distilled water of 2 × 35%, 50%, 70%, 80%, 90%, 3 × 100%) and washed 3× 5 min each with HPMA (EMS, USA, cat # 16900).

Tissues were subsequently embedded in Durcupan resin (20 g component A, 20 g component B, 0.6 g component C, 0.4 g component D; Sigma Canada, Toronto, cat# 44610) overnight at room temperature. For flat-embedding, resin-infiltrated sections were placed the next day onto fluoropolymer films (ACLAR®, EMS, cat# 50425-25), painted with resin, and kept at 55°C in a convection oven for 4 days for sufficient resin polymerization. After resin polymerization, regions of interest were excised from resin-embedded sections and glued onto resin blocks for ultramicrotomy sectioning. Using a Leica ARTOS 3D ultramicrotome, 74 nm ultrathin sections were cut (2 levels, ~ 5–10 μm apart) and collected on silicon wafers for SEM. Imaging utilized a Zeiss Crossbeam 350 SEM using the SE2 detector and operating at a 1.4 kV voltage and 1.2 nA current [16]. Using a chip-mapping approach on Atlas 5 Engine (Fibics), images were first screened at 25 nm/pixel resolution and subsequently acquired at 5 nm/pixel resolution to qualitatively assess the ultrastructural localization of chromogenic staining.

At the ultrastructural level in EM, chromogenic staining was confirmed by electron dense precipitates distributed throughout the cytoplasm and nucleus (as consistent with literature reports of NeuN localization) [12]. Ultrastructural classification for each cell type in the brain is reviewed in-depth elsewhere [17, 18]. Briefly, neurons were distinguished by their inclusion of one or more electron-dense nucleoli, a large and round euchromatic nucleus, abundant nuclear pores, and also if such a nucleus was in continuous appearance with dendrites [17]. This level for neuronal classification was used to differentiate neurons from other cell types: oligodendrocytes were recognized by their heterogeneous heterochromatin, one-sided cytoplasm forming a rectangular or squarish shape, wide and short endoplasmic reticulum cisternae, and direct contacts with myelinated axons. Non-dark or "typical" astrocytes were distinguished via their pale nucleoplasm, thin heterochromatin rim, frequent appearance of intermediate filaments, and often irregularly- shaped cell body and processes forming acute angle [17]. Non-dark or "typical" microglia and infiltrating macrophages were identified via their smaller nucleus containing "cheetah-spotted" heterochromatin patterns, often-lengthened endoplasmic reticulum cisternae, common cellular inclusions (including lipofuscin and much more rarely in synaptic elements including dendritic spines, pre-synaptic axon terminals, and synaptic clefts), and association with pockets of extracellular space [17]. Dark microglia and infiltrating macrophages, which contain a darkened nucleoplasm *versus* typical "microglia," could be identified via cellular stress markers including dilated endoplasmic reticulum, altered mitochondrial ultrastructure, and frequent contacts with other structures including synaptic elements and blood vessels [15, 19]. Pericytes were recognized by their heterogeneous heterochromatin and containment within the basement membrane at blood vessels, and endothelial cells were identified by their appearance between the basement membrane and non-parenchymal space comprising blood vessels [18].

3 Results

To select a marker that would be reactive across numerous species, NeuN was chosen; for in-situ staining as done for our staining protocols, NeuN was selectively used for neuronal nuclei, with cytoplasm, and processes (contrasting reports for NeuN+ astrocytes being demonstrated for in-vitro cultures) [12, 13]. As demonstrated in Fig. 2, successful chromogenic staining was demonstrated for NeuN across DAB, VIP, and silver-enhanced gold conditions. Aside from assessing BCIP-based staining, a negative control was also employed by using the same steps as the BCIP-based staining, but without incubation of primary NeuN. While BCIP-based staining was not able to visualize processes as clearly as other chromogens at the light microscopy level, it was able to selectively stain for NeuN (Fig. 2). This contrasts with the control condition, which generally appeared with staining of non-specific BCIP oxidation. Here, in the control tissue, likely visible neurons are seen (indicated by oval shapes throughout the image); however, with no significant deposition or clearly enriched neuronal-staining contrast *versus* other staining conditions (Fig. 2).

After empirically staining with BCIP for LM, sections stained with the same methodology for results at LM (Fig. 2) were also verified and discerned for potential staining success in EM. Notably, significant BCIP-based precipitates were washed out with propylene oxide in preparing for EM; as detailed in the methods, HPMA was subsequently used as a propylene oxide alternative and was able to retain much more stronger staining visible even in EM (Fig. 3). Here, aside from the control condition, significant deposits of electron dense precipitates could be found in neurons (identified via ultrastructure [17]) in all chromogenic stains, including by the

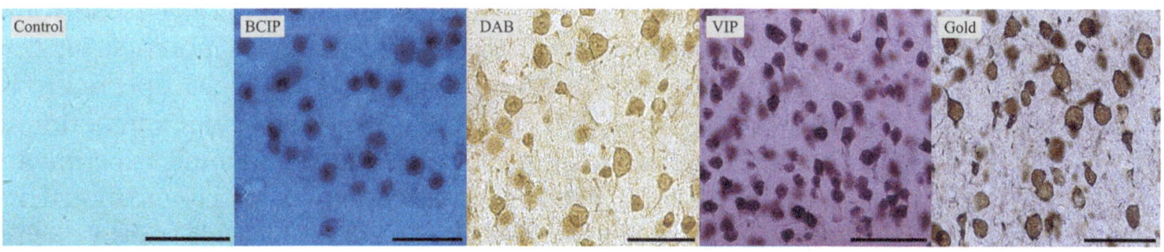

Fig. 2 BCIP-based staining is specific for NeuN and is comparable to other chromogenic stains at light microscopy. NeuN immunostaining conditions were conducted across the mouse frontal cortex. Particular chromogenic staining comprised control, BCIP, DAB, VIP, and silver-enhanced gold conditions as detailed in the methods; notable findings include positive neuronal staining in DAB, VIP, and silver-gold conditions, with the absence of clear neuronal staining *versus* tissue background in the control condition. Furthermore, the BCIP-based staining produced neuronal staining enhanced against the tissue background. Finally, findings observed here were also separately confirmed with electron microscopy (Fig. 3). All scale bars = 50 μm

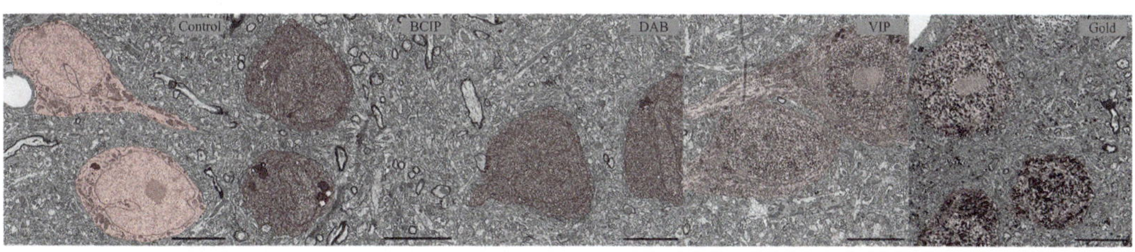

Fig. 3 BCIP-based staining for NeuN is comparable to other chromogenic stains in electron microscopy. To assess if the novel BCIP-based method could also be applicable to EM, sections processed using the same methodology were investigated. In all conditions except the control staining, significant staining for NeuN (as indicated by dot-like precipitates covering a neuronal nucleus and around the nucleus) was observed. This includes the novel BCIP-based methodology. Neuronal morphology stained by NeuN is colored red. All scale bars = 5 μm

novel, BCIP-based method. Although not shown in Fig. 3, occasional astrocytes could also be stained across conditions (except for DAB conditions, wherein no astrocytes were observed). This is consistent with the literature, wherein astrocytes can also be noted to be NeuN-positive [13]. The gold condition was less specific, with unintended positive staining in both endothelial and oligodendrocyte cells (Fig. 3). Finally, as BCIP-based staining produced dot-like precipitates, these precipitates visually appeared quite similar to those created by DAB oxidation. Given this data, the BCIP-staining method is unlikely to be compatible with double EM immunostaining panels with DAB.

4 Discussion

Here, by repurposing previously known chemical reactions and solutions applied for visualizing X-gal and EM [5], we provide a novel, safer, and affordable BCIP-based alternative to conventional chromogenic immunostaining systems. This method may act as a potential chromogen alternative to immunostaining applicable to both LM and EM, especially given that BCIP staining was more specific than gold-based staining. Other chromogens are likely expected to be still employed, due to their already-established usage and optimization with particular antigens across different scientific experiments; however, we hope that these findings can serve as a foundational first step to help other research teams find usage, provide improvements, and/or solve greater problems using the methods detailed here.

BCIP has also been shown to emit fluorescence with maximum intensity in the 750–800 nm range [7]. Our experiments were not able to successfully show significant fluorescent staining in tissues fixed with either 3.5% acrolein/4% paraformaldehyde in PB or 0.6% glutaraldehyde/4% paraformaldehyde. While both acrolein and glutaraldehyde generate autofluorescence that decreases the signal-to-noise ratio of immunofluorescence-based staining in tissues [20, 21], it is possible that our method could be extended toward assessing simultaneous light, fluorescent, and electron microscopy with a single reagent in tissues with lower levels of fixation (e.g., 4% paraformaldehyde only) or for cryo-EM applications.

Regarding solvents, it is likely that propylene oxide and HPMA both allow replacement of tissue from water to epoxy resin without significant ultrastructural damage; however, the chemical structure of propylene oxide is less polar relative to HPMA and so likely mechanistically explains why BCIP precipitate loss was significantly greater in propylene oxide conditions [5]. Additionally, as HPMA does not burn through plastic as propylene oxide does, HPMA is recommended as a safer substitute that requires less resources and involves fewer tissue transfers (i.e., allowing for reduced breaking of fragile, osmium-embedded samples) in processing steps before prepping semithin and ultrathin specimens for EM imaging. Finally, HPMA was used with all other chromogenic staining shown in Fig. 3. HPMA is therefore also likely a safer, easier, and cheaper alternative to propylene oxide to employ even for EM processing not requiring BCIP-based staining.

Acknowledgments

As part of the University of Victoria, we acknowledge and respect the lək' wə ŋn peoples on whose traditional territory the university stands and the Songhees, Esquimalt, and WSÁNEC peoples whose historical relationships with the land continue to this day.

Contributions VL conceived the experiment ideas, with support and critical improvements from MET. VL and JV participated in the experiments. VL wrote the manuscript and prepared figures. All the authors contributed to the article and approved the submitted version. All authors read and approved the final manuscript.

Funding V.L. was supported by graduate award scholarships from the Faculty of Graduate Studies, University of Victoria. J.V. is supported by a Canadian Institute for Health Research doctoral funding, alongside graduate award scholarships from the Faculty of Graduate Studies, University of Victoria. M.E.T. is a Tier I Canada Research Chair in Neurobiology of Healthy Cognitive Aging. Finally, EM images were generated via a Zeiss Crossbeam 350 scanning electron microscope acquired with funding from a Canada Foundation for Innovation John R. Evans Leaders Fund grant (#39965 Laboratory of ultrastructural insights into the neurobiology of aging and cognition). Funding was also obtained from the Canadian Institutes of Health Research (CIHR; Foundation Grant FDN341846) and a start-up grant received from the University of Victoria's School of Medical Sciences.

Data Availability All data presented in this study are available from the corresponding author upon reasonable request.

Additional Information: Ethical approval and consent to participate All animal experiments were performed ethically, according to the guidelines of the Institutional Animal Ethics Committees, the Canadian Council on Animal Care, and the Animal Care Committee of Université Laval (where the mice were bred, handled, and euthanized).

Competing Interests All authors declare that they have no competing interests.

References

1. Carrier M, Robert MÈ, González Ibáñez F, Desjardins M, Tremblay MÈ (2020) Imaging the neuroimmune dynamics across space and time. Front. Neurosci 14:903
2. Lanciego JL, Goede PH, Witter MP, Wouterlood FG (1997) Use of peroxidase substrate Vector VIP® for multiple staining in light microscopy. J Neurosci Methods 74:1–7
3. Zhou M, Grofova I (1995) The use of peroxidase substrate Vector® VIP in electron microscopic single and double antigen localization. J eurosci Methods 62:149–158
4. Tao-Cheng J-H, Crocker V, Moreira SL, Azzam R (2021) Optimization of protocols for pre-embedding immunogold electron microscopy of neurons in cell cultures and brains. Mol Brain 14:1–15
5. Masahira N et al. (2005) Improved preservation of X-gal reaction product for electron microscopy using hydroxypropyl methacrylate. Neurosci Lett 374:17–20
6. Itahana K, Itahana Y, Dimri GP (2013) Colorimetric detection of senescence-associated β galactosidase. Methods Mol Biol 965:143–156
7. Trinh LA et al. (2007) Fluorescent in situ hybridization employing the conventional NBT/BCIP chromogenic stain. Biotechniques 42:756–759
8. Novikoff AB, Shin W-Y, Drucker J (1961) Mitochondrial localization of oxidative enzymes: staining results with two tetrazolium salts. J Biophys Biochem Cytol 9:47–61
9. Childs BG, Bussian TJ, Baker DJ (2019) Cellular identification and quantification of senescence-associated β-galactosidase activity in vivo. Methods Mol Biol 1896:31–38
10. Spurr AR (1969) A low-viscosity epoxy resin embedding medium for electron microscopy. J Ultrastruct Res 26:31–43
11. Luft JH (1961) Improvements in epoxy resin embedding methods. J Biophys Biochem Cytol 9:409–414
12. Mullen RJ, Buck CR, Smith AM (1992) NeuN, a neuronal specific nuclear protein in vertebrates. Development 116:201–211
13. Darlington PJ, Goldman JS, Cui Q, Antel JP, Kennedy TE (2008) Widespread immunoreactivity for neuronal nuclei in cultured human and rodent astrocytes. J Neurochem 104:1201–1209
14. Van De Werd HJJM, Uylings HBM (2014) Comparison of (stereotactic) parcellations in mouse prefrontal cortex. Brain Struct Funct 219:433–459
15. St-Pierre MK, Carrier M, Lau V, Tremblay MÈ (2022) Investigating microglial ultrastructural alterations and intimate relationships with neuronal stress, dystrophy, and degeneration in mouse models of Alzheimer's disease. In: Jahani-Asl A (ed) Neuronal cell death: methods and protocols. Springer US, New York, NY, pp 29–58, 10.1007/978-1-0716-2409-8_3
16. St-Pierre M-K et al. (2022) Ultrastructural characterization of dark microglia during aging in a mouse model of Alzheimer's disease pathology and in human post-mortem brain samples. J Neuroinflammation 19: 235
17. Nahirney PC, Tremblay M-E (2021) Brain ultrastructure: putting the pieces together. Front Cell Dev Biol 9:629503
18. Nahirney PC, Reeson P, Brown CE (2016) Ultrastructural analysis of blood–brain barrier breakdown in the peri-infarct zone in young adult and aged mice. J Cereb Blood Flow Metab 36:413–425
19. Bisht K et al. (2016) Dark microglia: a new phenotype predominantly associated with pathological states. Glia 64:826
20. Luquin E, Pérez-Lorenzo E, Aymerich MS, Mengual E (2010) Two-color fluorescence labeling in acrolein-fixed brain tissue. J Histochem Cytochem 58:359–368
21. Tagliaferro P, Tandler CJ, Ramos AJ, Pecci Saavedra J, Brusco A (1997) Immunofluorescence and glutaraldehyde fixation. A new procedure based on the Schiff-quenching method. J Neurosci Methods 77:191–197

Chapter 5

Mapping Mitochondrial Architecture in the Brain: Methods and Insights

Emily K. Miller and Patrick C. Nahirney

Abstract

Mitochondria are dynamic cellular organelles that exhibit fascinating phenotypic differences across organisms, tissues, and cells. The variety of necessary and unique cellular roles they play in both health and disease has become evident with the use of high-resolution electron microscopy. This chapter introduces mitochondrial ultrastructural phenotypic differences visible by scanning electron microscopy (SEM) and transmission electron microscopy (TEM), the functional relationships they exhibit with other cellular organelles—such as the endoplasmic reticulum, nucleus, Golgi complex, and cytoskeletal elements—and how their normal structure and function are affected by ischemic stroke pathology.

Keywords Mitochondria, ATP, Energy, Transmission electron microscopy, Scanning electron microscopy, Histochemistry, Ultrastructure, Neuroscience, Neurons, Glia, Endoplasmic reticulum, Cytoskeleton, Nucleus, Golgi complex, Neuropathology, Stroke, Ischemia

1 Introduction

Mitochondria were first described in the late nineteenth century by Richard Altmann, who observed and named these intracellular structures "bioblasts." They were later renamed "mitochondria" in 1898 by Carl Benda, one of the early anatomists and pathologists to study cellular structure and intracellular components [1–3]. The word "mitochondria" is derived from the Greek *mito*, meaning "thread," and *chondria* or *chondros*, meaning "granule" or "grain."

Mitochondria break down macromolecules to form energy through aerobic mechanisms and are found in nearly all aerobic eukaryotic cells. But are mitochondria all the same? The simple answer is no. They serve different roles in different locations. Apart from the cell nucleus, mitochondria are the most conspicuous cytoplasmic organelle. Acting as small chemical generators—akin to self-sustaining batteries—they occur in different numbers, shapes, and sizes depending on cellular energy demands.

Marie-Ève Tremblay (ed.), *Ultrastructural Neuroscience Methods*, Neuromethods, vol. 238,
https://doi.org/10.1007/978-1-0716-5456-9_5,

This chapter illustrates a variety of microscopic techniques used to explore the unique features of brain mitochondria and their critical roles in neurons and glial cells. The importance of mitochondria is underscored by the prevalence of diseases associated with them, with epidemiologic evidence suggesting that at least 1 in 5000 individuals is affected [4]. Such diseases include disorders arising from mitochondrial gene mutations, including inherited neurologic conditions that affect the brain and can cause stroke-like symptoms, epilepsy, and dementia [4, 5]. Later in this chapter, we explore the morphological changes that mitochondria undergo following ischemic stroke in the brain.

2 Energy Production

Mitochondria are best known for their ability to power the cell through the production of adenosine triphosphate (ATP). They accomplish this through the citric acid cycle (CAC) in the mitochondrial matrix and through oxidative phosphorylation (OXPHOS) across the inner mitochondrial membrane (IMM). ATP synthesis results from a series of complex reactions in the cytoplasm and mitochondria and, beginning with a single glucose molecule, can be summarized as follows [6]:

- In the cytosol, a glucose molecule undergoes a series of nine to ten reactions that ultimately result in the formation of two pyruvate molecules. This biochemical pathway is termed glycolysis. Two ATP molecules initiate the pathway, and four ATP molecules are generated by the end, yielding a net of two ATP molecules. Additionally, two molecules of nicotinamide adenine dinucleotide (NAD+) each gain a hydrogen ion and an electron pair to become NADH.
- The pyruvate molecules then enter the mitochondrial matrix, crossing the outer mitochondrial membrane (OMM), intermembrane space (IMS), and IMM. In the matrix, pyruvate becomes acetyl-CoA with the help of the pyruvate dehydrogenase complex. Per pyruvate molecule, this oxidative decarboxylation reaction releases one carbon dioxide (CO_2), reduces one NAD+ to NADH, and adds one coenzyme A molecule.
- Acetyl-CoA then enters the CAC, where it joins oxaloacetate to form citric acid and releases coenzyme A. This is the first step of the CAC. The following seven steps are:
  1. Isomerization of citric acid to isocitrate.
  2. Oxidative decarboxylation of isocitrate to α-ketoglutarate, reducing NAD+ to NADH and releasing one H+ and CO_2.
  3. Conversion of α-ketoglutarate to succinyl-CoA, reducing NAD+ to NADH and releasing another H+ and CO_2.

4. Conversion of succinyl-CoA to succinate, producing one GTP (converted to ATP) and releasing coenzyme A.
5. Dehydrogenation of succinate to fumarate, reducing FAD to $FADH_2$.
6. Hydration of fumarate to malate.
7. Dehydrogenation of malate to oxaloacetate, reducing NAD+ to NADH and generating one H+.

- Overall, the CAC produces three NADH molecules, three H+ ions, one $FADH_2$, two CO_2 molecules, and one GTP per cycle. CO_2 is released as a waste product, GTP is converted to ATP, and the H+ ions and electrons carried by NADH and $FADH_2$ enter the electron transport chain (ETC) located in the IMM. Notably, each glucose molecule drives two CAC cycles—one per pyruvate.
- The ETC consists of five enzyme complexes located in the IMM. These complexes drive ATP formation through Complex V ATP synthase by maintaining an electrochemical gradient across the membrane. The transfer of electrons and pumping of protons into the IMS create a gradient that drives protons through ATP synthase to generate ATP. At Complex IV, oxygen serves as the final electron acceptor and, together with hydrogen ions, forms water [6].

These intricate biochemical reactions support ATP formation and contribute to pH buffering within the cell.

3 Histochemistry of Mitochondrial Enzymes

Enzyme histochemistry is commonly used to detect sites of energy production in cells. Mitochondria are rich in metabolic enzymes, including diaphorases and dehydrogenases [6–10]. Under normal physiologic conditions, diaphorases catalyze reactions involving NADH. In the presence of colorless nitro-blue tetrazolium (NBT) solution and excess NADH, they catalyze a coupled redox reaction that reduces NBT to a dark-purple, water-soluble formazan precipitate while oxidizing NADH to NAD+ [11, 12]. Dehydrogenases then reduce NAD+ back to NADH to continue the cycle. This staining method therefore identifies tissue regions with high mitochondrial enzymatic activity (Figs. 1 and 2).

4 Ultrastructure of Mitochondria and Quantification in Brain Regions

In TEM images, mitochondria appear as double-membrane-bound organelles with internal tubular or shelf-like cristae [13, 14], while in SEM images they appear spherical or ovoid with worm-like cristae

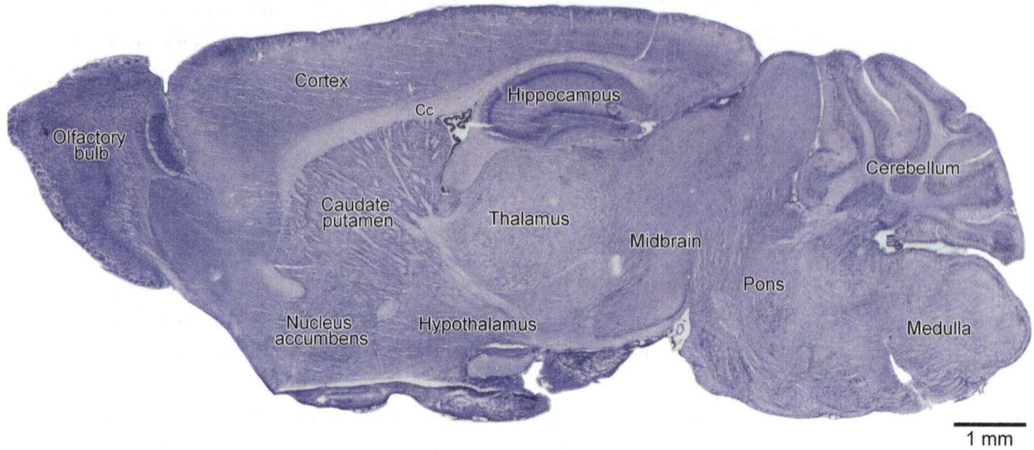

Fig. 1 NBT-NADH enzyme histochemical staining of mitochondria in the adult mouse brain. Low-magnification light microscopy (LM) image of a near mid-sagittal cryosection of the mouse brain reveals dark purple staining in areas rich in mitochondrial enzymatic activity. Mice were perfused with 5% sucrose and 1% paraformaldehyde in tris-phosphate buffered solution and equilibrated in 15% sucrose. Brains were snap-frozen in liquid nitrogen and sectioned at 30 μm thickness on a cryostat followed by staining with NBT-NADH solution. Cc, Corpus callosum

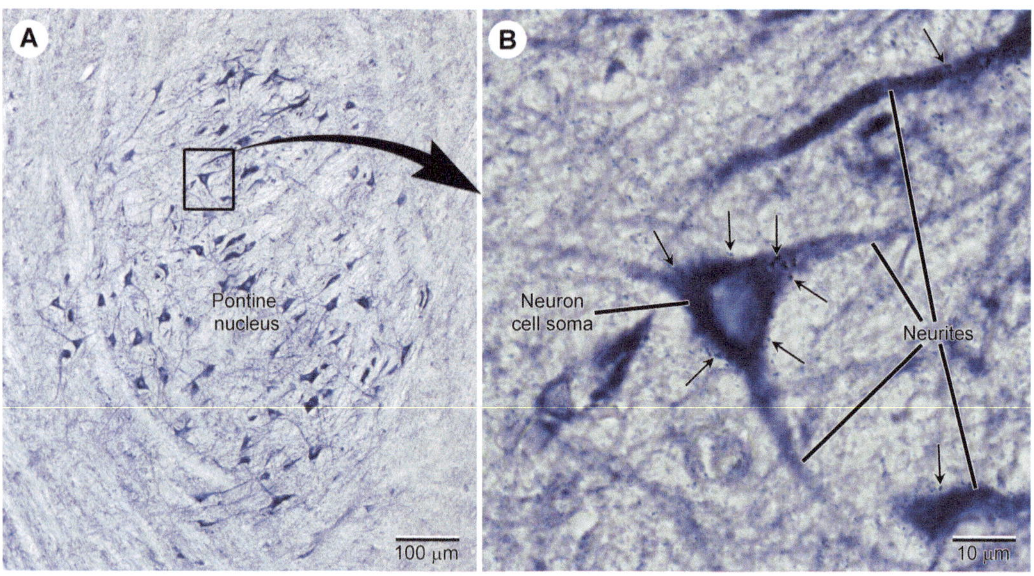

Fig. 2 LM images of NBT-NADH staining of mitochondria in the pontine nucleus of an adult mouse brain. Brain tissue samples were processed as described in Fig. 1. The boxed region in A is shown at high magnification in B. Black arrows point to dark-colored staining puncta that represent mitochondria. Note the high density of staining within cell soma and neurites

(Fig. 3). The OMM and IMM define the mitochondrial compartments, with the IMM separating the matrix from the IMS.

We have studied regional mitochondrial density and size in various brain cell types and have identified subtle but important differences. Using TEM and morphometric analysis, we found that the percent area occupied by mitochondria in the brain ranges from ~2% to 12%. Mitochondria of the cerebellum, somatosensory cortex, and hippocampus occupy 6%–9%, 8%–11%, and 2%–5% of the overall area, respectively. Mitochondria in the cortex and cerebellum appear similar in number and size, while those in the hippocampus are fewer and smaller (Fig. 4). These percentages roughly match skeletal muscle (2%–10%) and are lower than cardiac muscle (~35%) [15, 16].

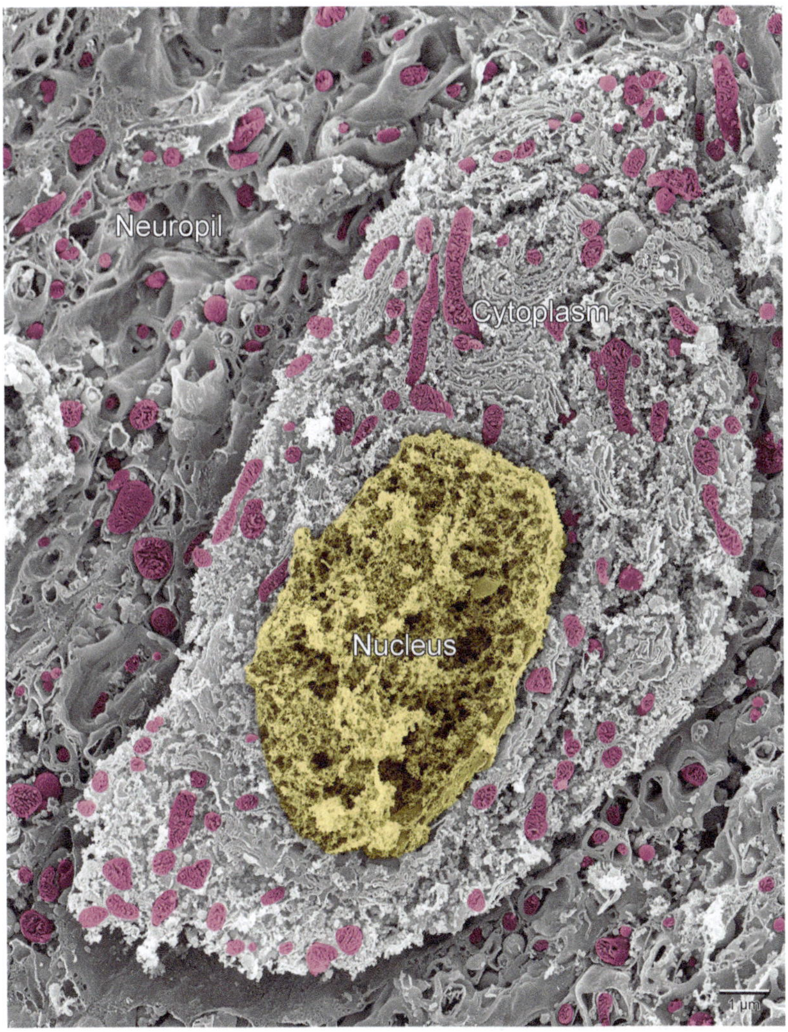

Fig. 3 Mitochondria (pseudo-colored magenta) of the brain viewed by SEM. Using a freeze-fracture and cytosolic extraction method, SEM reveals the internal three-dimensional structure of a large neuron cell soma and surrounding neuropil containing a plethora of mitochondria. Spherical-to-ovoid in shape, mitochondria are scattered throughout the cytoplasm of this neuron and in neighboring axons, dendrites, and glial cells. Fractured open mitochondria reveal their internal worm-like cristae. The chromatin-containing nucleus is pseudo-colored yellow

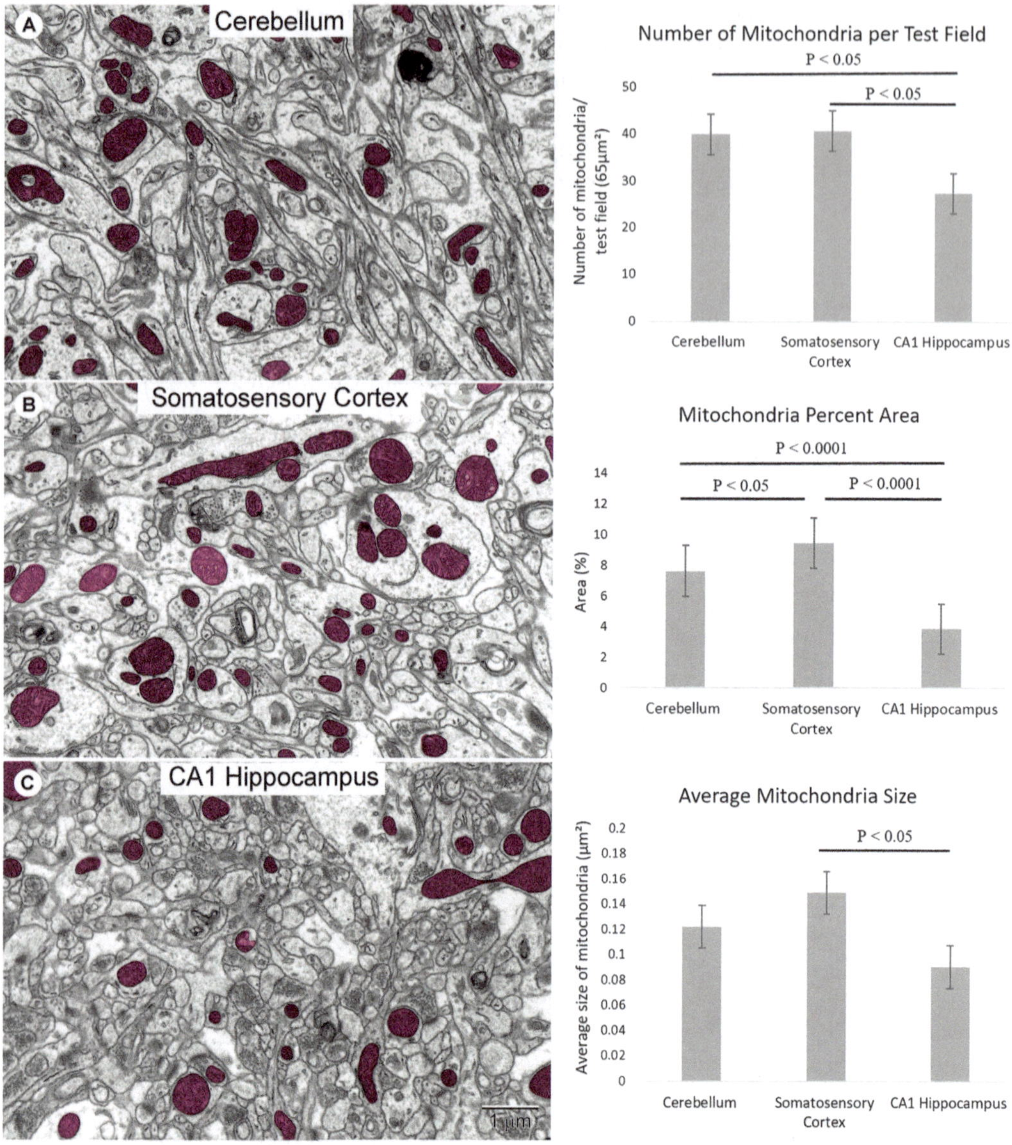

Fig. 4 Morphometric analysis of mitochondria in three regions of the brain. TEM images of the neuropil regions in the cerebellum (**A**), somatosensory cortex (**B**), and the CA1 region of the hippocampus (**C**) of an adult mouse. Fiji was used to trace and quantify the number of mitochondria per test field (65 μm^2; 10 × 6.5 μm, n = 5 test fields), the percent area, and the average size of mitochondria in each brain region (in μm^2, n = 50/region) (graphs shown to the right). The mean of all acquired images was calculated for each type of morphometric analysis per region, and statistical significance was determined with a one-way ANOVA followed by a Post Hoc Tukey Honest Significant Difference (HSD) test. Results indicate that mitochondria are more numerous and occupy a larger area in the neuropil of the cerebellum and somatosensory cortex when compared to the CA1 of the hippocampus. The somatosensory cortex also exhibits significantly larger mitochondria than the CA1 of the hippocampus. p values are indicated on graphs

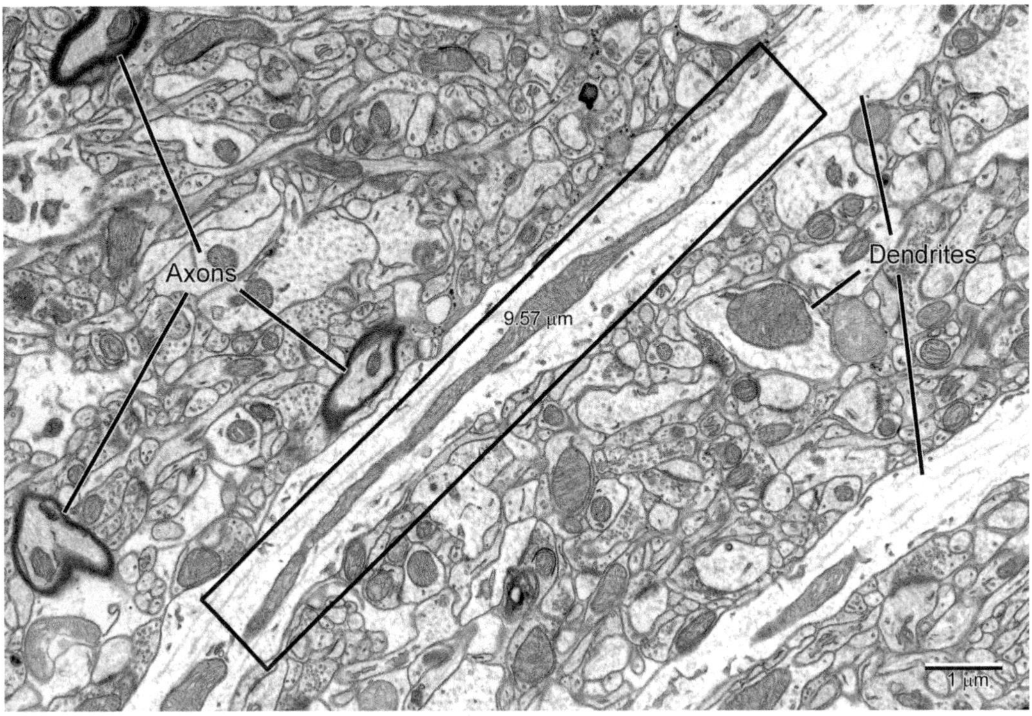

Fig. 5 Elongated mitochondrion in a neuronal dendrite. TEM image of the somatosensory cortex in a 2-month-old mouse showing a mixture of dendrites, myelinated axons, pre- and postsynaptic terminals, and glial cells in the neuropil. Several dendrites are sectioned in multiple planes. The longitudinally sectioned dendrite in the middle of the field contains a mitochondrion (pseudo-colored magenta) measuring an astonishing 9.57 μm in length

Mitochondrial density also varies between neurons and glial cell types and depends on their local environment. Recent cryo-electron tomography has shown that human microglia contain more tightly packed patterns of neighboring mitochondria than astrocytes [17]. Microglial mitochondria also have larger volumes and long-to-short axis ratios than those in astrocytes, reflecting cell-specific energy needs [17].

Location within the cell further correlates with differences in mitochondrial phenotypes. Axonal mitochondria appear short, whereas dendritic mitochondria are long and filamentous (Fig. 5) [18]. Three-dimensional analyses of mitochondria in various brain regions (CA1, nucleus accumbens, somatosensory cortex, and dorsal cochlear nucleus) show that presynaptic mitochondria are small with little size variability, while postsynaptic/dendritic mitochondria are much larger [19]. Larger dendritic mitochondria often span several synapses, possibly helping integrate inputs and buffer calcium ions (Ca^{2+}), while small presynaptic mitochondria support neurotransmission by elevating cytoplasmic Ca^{2+} [19].

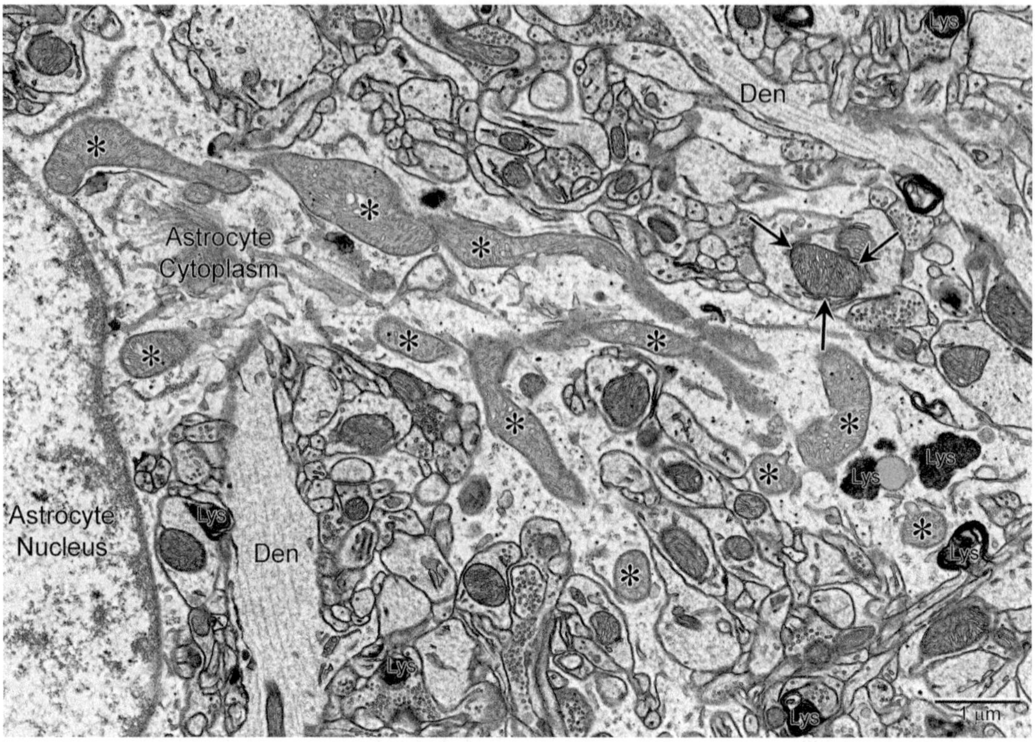

Fig. 6 Mitochondrial phenotypes. TEM image of the somatosensory cortex of an adult mouse brain showing the lighter (less electron dense) astrocytic mitochondria (asterisks) compared to dendritic mitochondria (black arrows). A portion of the astrocyte cell body and nucleus is seen to the left, and its cytoplasmic process emanates to the right. Amorphous darkly stained lysosomes (Lys) are seen throughout the astrocyte cytoplasm. Neuronal dendrites (Den) are visible at the bottom left and upper right of the image and contain straw-like microtubules. Neurotransmitter-filled vesicles are seen in presynaptic terminals

Typically, mitochondria range from 0.25 to 1 µm in diameter, but elongated tubular phenotypes up to ~10 µm in length have been observed in neuronal dendrites (Fig. 5).

Mitochondrial morphology can also be defined by staining differences and cristae phenotypes. In our TEM preparations, mitochondria appear lighter (i.e., less electron dense) in astrocytes than in neurons, with microglial mitochondria intermediate in electron density (Figs. 6 and 7). Neuronal mitochondria often exhibit either dense lamellar cristae or more spacious tubulo-vesicular cristae (Figs. 8 and 9). Tightly packed lamellar cristae possess larger membrane surface area and can accommodate more ETC proteins, suggesting a higher ATP-producing capacity [20, 21]. Experiments show that increased synaptic performance correlates with higher mitochondrial volume, cristae density, lamellarity, and cytochrome c levels [20]. Activity-dependent ultrastructural plasticity of mitochondria is thus implicated. More research is needed to better characterize structure–function relationships among brain cell mitochondria, but the emerging picture is increasingly detailed.

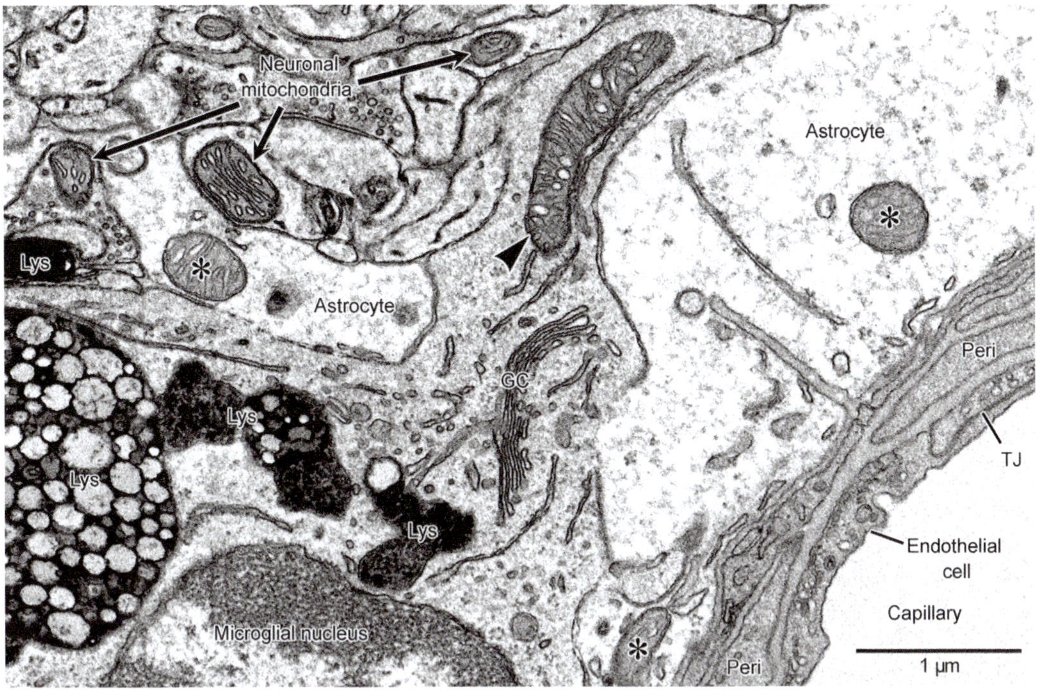

Fig. 7 Mitochondrial phenotypes. Mitochondria of neurons stain darker than astrocytic mitochondria (asterisks), and microglial mitochondria (black arrowhead) stain slightly less than neuronal mitochondria. Lysosomes (Lys) in the microglial cell appear as dark amorphous vacuoles that contain a mixture of electron-dense material. GC, Golgi complex; TJ, tight junction

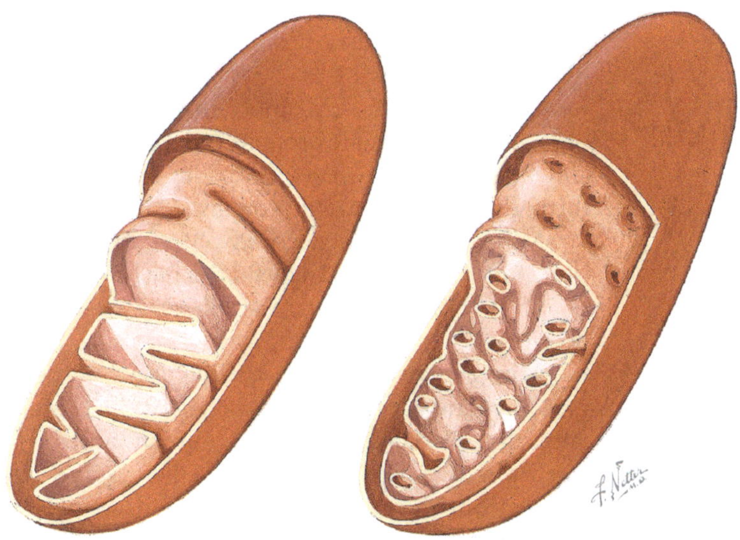

Fig. 8 Basic structure of mitochondria. Mitochondria possess two layers of membranes, an outer smooth membrane and an inner convoluted membrane that is thrown into folds called cristae. Mitochondria can have either shelf-like (left) or tubular (right) inner membranes to adjust their surface area for energy production. An inner and outer space are created by these membranes and allow for electrochemical gradients to be established. Used with permission from Netter's Essential Histology, 3rd ed., Ovalle and Nahirney (2021)

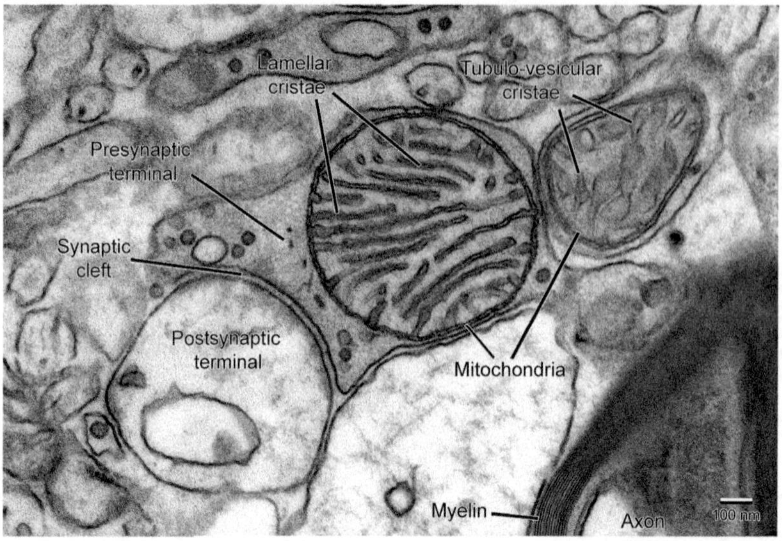

Fig. 9 Mitochondrial cristae differences. Mitochondria can possess either lamellar cristae (seen within the large central mitochondrion in a presynaptic terminal) or tubulo-vesicular cristae (seen within the mitochondrion of an astrocyte at right). Note the presence of presynaptic vesicles and a synaptic cleft between the presynaptic terminal and the dendritic spine. A portion of a myelinated axon is visible at the bottom right

5 Shape-Changing Mitochondria

Mitochondria undergo continuous dynamic processes including fusion, fission, biogenesis, and mitophagy [22, 23]. These events maintain mitochondrial health and function and contribute to morphological diversity.

Fusion unites two or more mitochondria into one elongated organelle, with specific proteins supporting the fusion of the OMM and IMM separately. Fission divides one mitochondrion into two through dynamin-related proteins.

Mitochondrial biogenesis involves growth and division of pre-existing mitochondria and depends on mitochondrial DNA (mtDNA), as well as transcription, translation, and assembly of mitochondrial proteins.

Mitophagy eliminates non-functional mitochondria via autolysosomes and is essential for cell survival [22]. Given their central importance to cellular function, mitochondrial dynamics are noteworthy in the following sections, which explore how mitochondria cooperate with other organelles and how they change following ischemic stroke.

6 Functional Relationships between Mitochondria, Organelles, and Cytoplasmic Elements in the Brain

Mitochondria support numerous cellular roles, including energy production, pH maintenance, cell signaling, apoptosis, and homeostasis of Ca^{2+}, glucose, lipids, and reactive oxygen species (ROS) [24, 25]. To accomplish this, they rely on specialized machinery enabling both independent and coordinated activity with other organelles.

6.1 Mitochondria–Nucleus Relationship: mtDNA Housing and Retrograde Signaling

Mitochondria possess their own DNA (mtDNA), which led Lynn Margulis (Sagan) in 1967 to propose the theory of endosymbiosis, suggesting that mitochondria originated from bacteria phagocytosed by ancestral eukaryotic cells [26, 27]. Supporting this theory, mitochondria contain their own RNA and ribosomes [28]. The coexistence of mitochondria and host cells for millions of years has conferred advantages such as increased ATP synthesis and improved molecular homeostasis.

Have mitochondria evolved since this merger? Yes. mtDNA is maternally inherited because the ovum contains mitochondria to sustain the germ cell before fertilization. Despite conservation across species, mtDNA mutates more frequently than nuclear DNA. Tracking these mutations through the maternal line has provided insights into human population evolution [29]. Whole-genome sequencing of over 53,000 participants has identified recent rare mtDNA segments in the nucleus, suggesting that mtDNA-to-nucleus transfer is ongoing [30].

Mitochondria also influence nuclear activity through retrograde signaling. The nucleus is enclosed by a double-membrane nuclear envelope containing pores that regulate transport (Fig. 10). Mitochondria are often positioned perinuclearly to support energy requirements of mRNA and protein trafficking [31, 32] and to enable rapid modulation of nuclear gene expression in response to environmental changes.

For example, mitochondrial acetyl-CoA is required for histone acetylation, which loosens chromatin structure and enhances transcription [33]. Increased glucose and fatty acid availability likely drives this process. α-ketoglutarate (α-KG), a CAC metabolite, is a cofactor for histone demethylases. Mitochondrial conversion of α-KG to 2-hydroxyglutarate (2-HG) inhibits these enzymes and promotes histone methylation, influencing transcription [24, 32, 34].

Mitochondria also detect genotoxic stressors, such as the anticancer drug doxorubicin, and respond by releasing mtDNA fragments that activate nuclear DNA repair pathways [35]. Additional retrograde signals include ROS, Ca^{2+}, and cholesterol, which promote nuclear stabilization [36].

Many more examples exist, and new pathways continue to be identified.

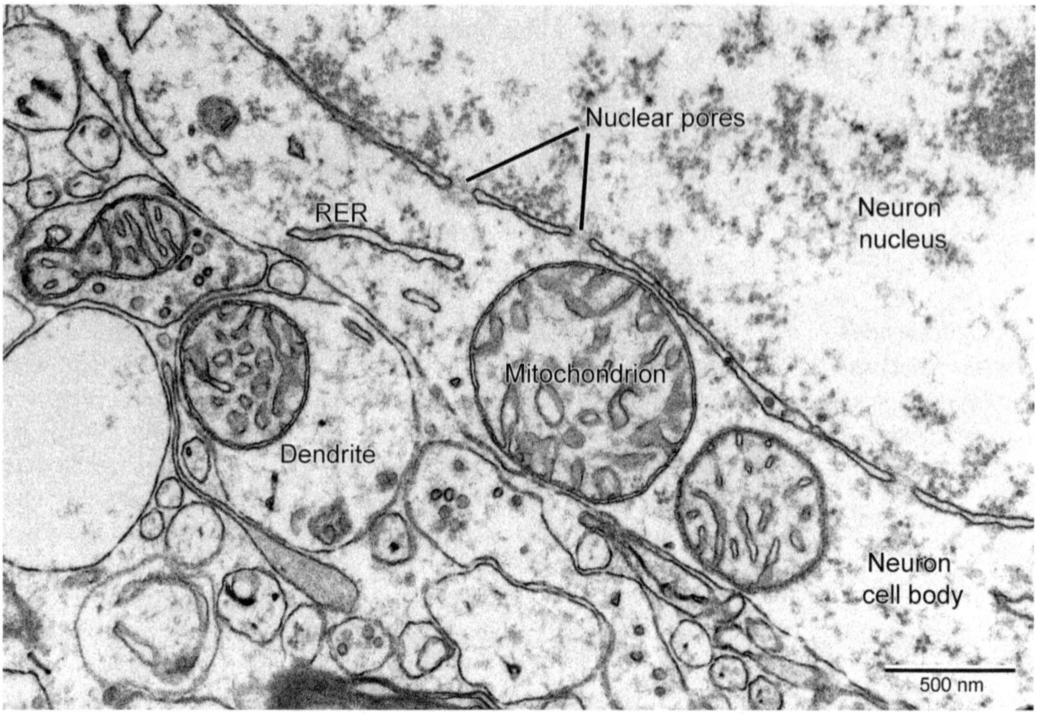

Fig. 10 Perinuclear mitochondria in a neuron of the adult mouse somatosensory cortex. The euchromatic nucleus (top right) is surrounded by the nuclear envelope, which is a double membrane with intermittent nuclear pore openings. Two mitochondria are seen adjacent to the nucleus and the RER. These strategically positioned mitochondria provide ATP for nuclear import and export and facilitate retrograde signaling to the nucleus

6.2 Mitochondria–Cytoskeleton Relationship: Motility and Synaptic Energy Requirements

Mitochondria relocate within cells in response to chemical gradients such as Ca^{2+}, ROS, and nutrient levels. In neurons—which may contain axons up to a meter long—specialized transport mechanisms are essential.

Long-distance mitochondrial motility depends on motor and adaptor proteins that move mitochondria along microtubules. Kinesin-1 drives anterograde (plus-end–directed) transport, while dynein drives retrograde (minus-end–directed) transport [37–40]. Neuronal microtubules are polarized with plus-ends typically oriented away from the soma in axons and mixed orientations in dendrites [41–43]. This polarity explains why mitochondria are frequently observed adjacent to microtubules (Fig. 11).

Adaptor proteins, including the mitochondrial Rho GTPase (Miro) and Milton (Trak1/2 in mammals), mediate motor binding to mitochondria [44]. Ca^{2+} binding to Miro cues the detachment of mitochondria from kinesin, causing mitochondrial arrest—particularly at synaptic terminals where Ca^{2+} is elevated [22]. Syntaphilin also recruits and immobilizes mitochondria in axons in an activity-dependent manner by inhibiting kinesin ATPase activity [44, 45].

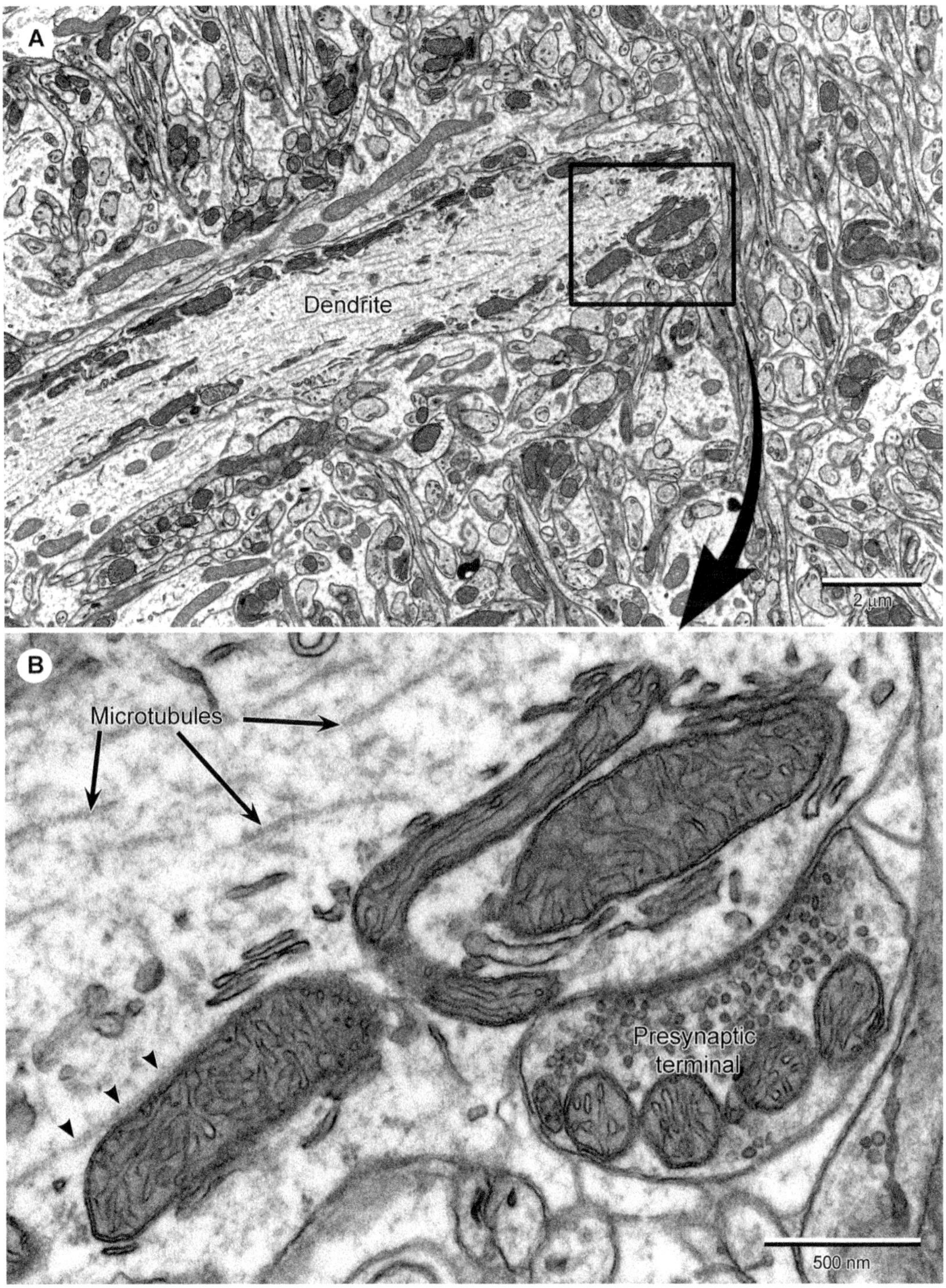

Fig. 11 Mitochondria adjacent to microtubules in an apical dendrite of a cerebellar Purkinje cell. (**A**) Low-magnification view of a large Purkinje cell apical dendrite among parallel fibers in the molecular layer of the cerebellum. (**B**) High-magnification view of the boxed area in A reveals a large GABAergic-type presynaptic terminal with small flattened vesicles synapsing onto the dendrite. The presynaptic terminal contains four mitochondria, while three mitochondria are near the postsynaptic membrane of the dendrite. Note the intimate association of a microtubule (arrowheads) alongside a dendritic mitochondrion in the lower left region of B. Mitochondria use microtubules to move to strategic locations such as synapses where ATP is needed

Along with microtubule transport, mitochondria also associate with neurofilaments and actin filaments (Fig. 12). Actin-based motility governs short-distance movement, especially in axon terminals and dendritic spines, where myosin motors transport mitochondria in response to synaptic activity [46]. An example of high mitochondrial density in synaptic terminals is found in cerebellar mossy fibers (Fig. 13) [47, 48].

Mitochondria also buffer presynaptic Ca^{2+} and influence neurotransmitter release [49]. When action potentials open voltage-gated Ca^{2+} channels, Ca^{2+} influx triggers vesicle fusion. Mitochondria take up a portion of this Ca^{2+} through voltage-dependent anion channels (VDACs) and the mitochondrial Ca^{2+} uniporter, while Ca^{2+} efflux occurs via the $Na^+/Ca^{2+}/Li^+$ exchanger (NCLX) [50, 51]. Equilibrium is restored by Ca^{2+}-ATPases on the presynaptic membrane and mitochondrial exchangers [50, 52]. At rest, mitochondria buffer Ca^{2+} leakage from the ER and spontaneous Ca^{2+} fluctuations.

Once inside the mitochondrion, Ca^{2+} enhances ATP production by stimulating enzymes such as pyruvate dehydrogenase phosphatase (PDP), which increases acetyl-CoA production [53], and S100A1,

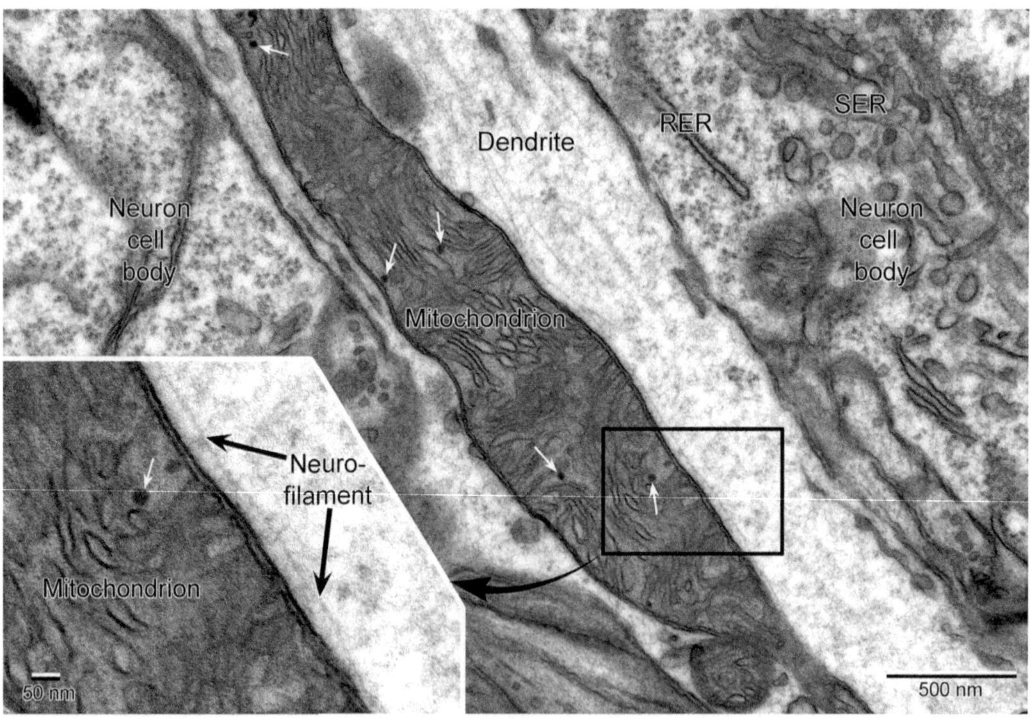

Fig. 12 Longitudinal view of a neuronal dendrite containing an elongated mitochondrion. The large dendrite in the center of the field is located between two neuron cell bodies. Note at high magnification the intimate relationship between the mitochondrion and a neurofilament (inset) that appears to anchor to the mitochondrial surface with small, inconspicuous filamentous bridges. White arrows indicate Ca^{2+} phosphate crystals stored in the mitochondrial matrix. RER, rough endoplasmic reticulum; SER, smooth endoplasmic reticulum

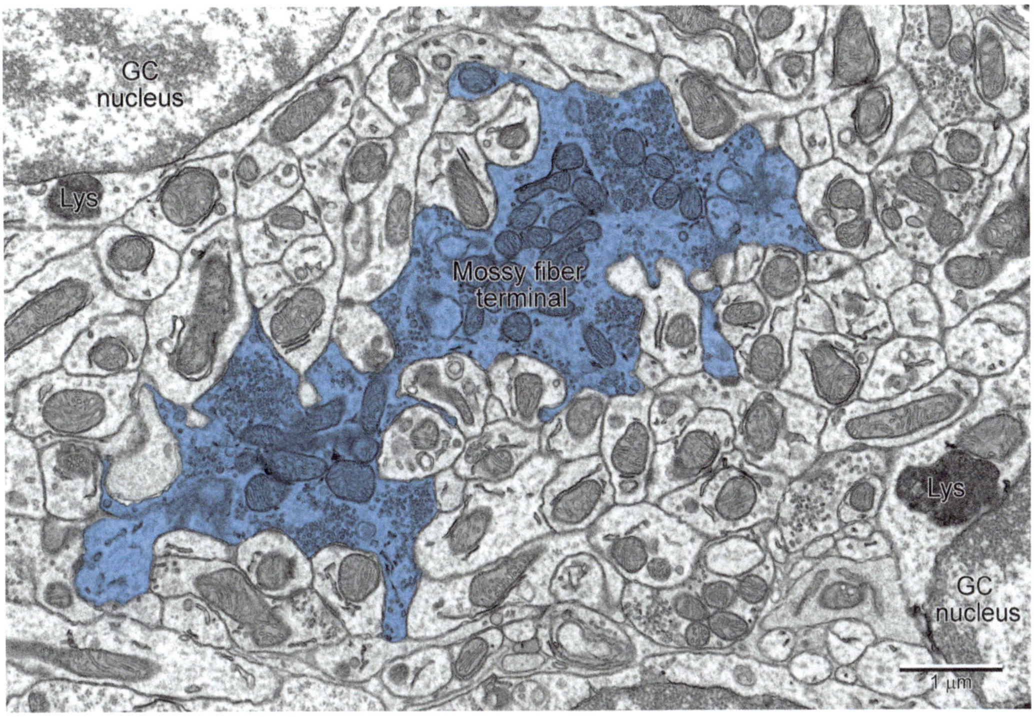

Fig. 13 Mossy fiber terminal between two granule cells in the cerebellum. The mossy fiber terminal (pseudo-colored blue) contains a high density of mitochondria and glutamate-filled vesicles. The neurosecretory vesicles are located peripherally in the terminal, where they can readily fuse with the membrane and be released into the synaptic cleft. Two granule cell (GC) nuclei are visible in the top right and bottom left. Lys, lysosome

which binds ATP synthase to boost ATP synthesis [54]. Ca^{2+} also influences neurotransmission efficiency by regulating vesicle cycling [49] and affects mitochondrial motility through ROS production [55].

6.3 The Mitochondria–Endoplasmic Reticulum Relationship: Maintaining Homeostasis

Mitochondria–ER interactions support Ca^{2+} and lipid homeostasis and apoptosis. These interactions are the most prominent mitochondrial contacts [56], with the organelles frequently positioned adjacent to each other (Figs. 14 and 15). Mitochondria-associated membranes (MAMs) contain specialized proteins that mediate reversible tethering between the organelles [57].

The ER stores large quantities of Ca^{2+}. When overloaded, it releases Ca^{2+} that mitochondria take up through MAM channels. Excess mitochondrial Ca^{2+} triggers pro-apoptotic factors such as cytochrome c [58, 59], a critical consideration in brain pathologies including cerebral ischemia [58].

The ER–mitochondria relationship is also essential in lipid homeostasis. While the ER is the major site of lipid synthesis, MAMs and mitochondrial compartments contain enzymes involved in lipid trafficking and metabolism. For example, diacylglycerol O-acyltransferase

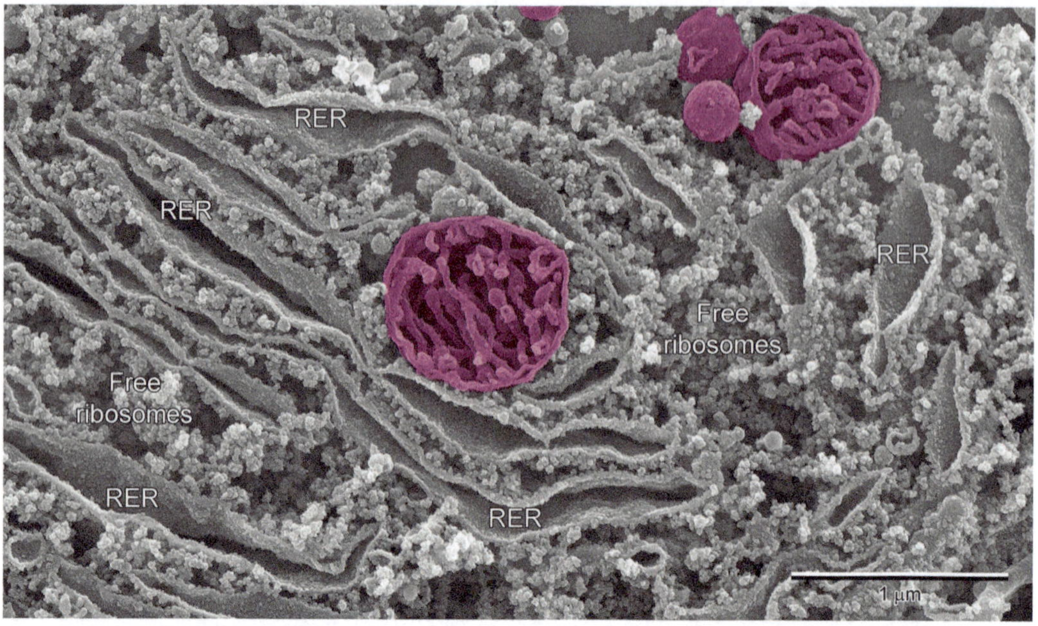

Fig. 14 SEM of spherical mitochondria surrounded by RER and free ribosomes. Mitochondria (pseudo-colored magenta) are nestled between the folds of the RER and have crucial roles in these protein-producing sites. The granular appearance of the ribosomes is evident on the sheet-like membranes of the RER

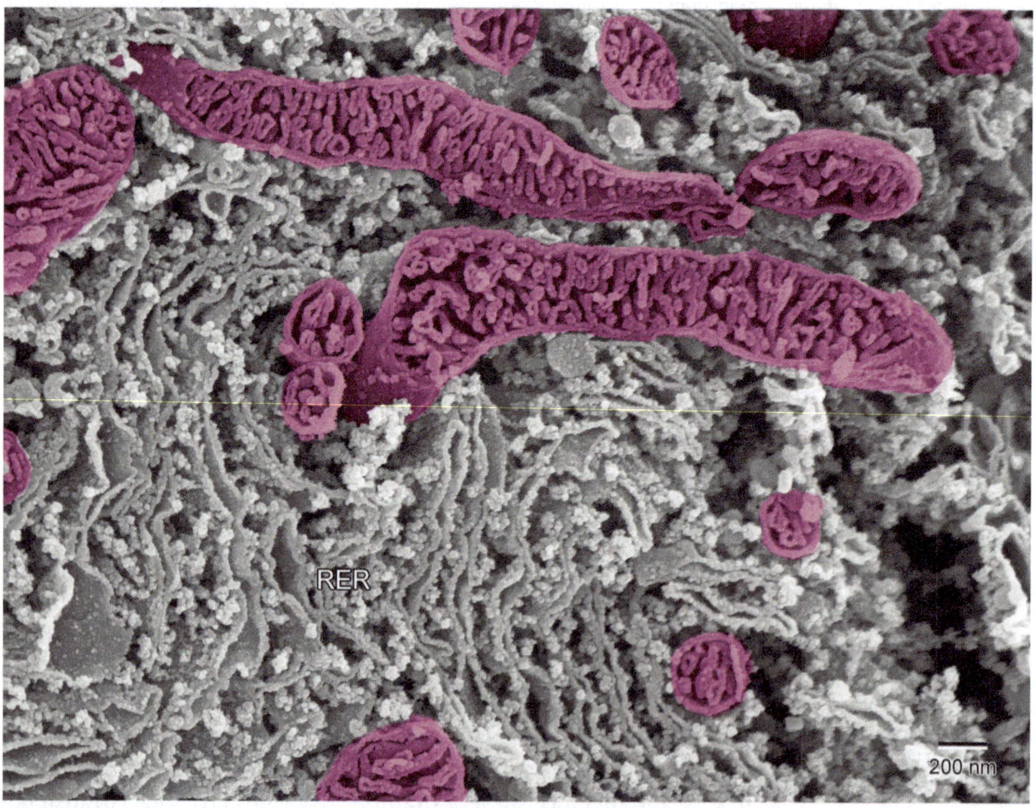

Fig. 15 SEM of elongated tubular-shaped mitochondria surrounded by RER. High-magnification image of the long tubular mitochondria located above the nucleus in Fig. 3. The elongated mitochondria weave among the tightly packed RER lamellae

2 (DGAT2), a major enzyme in triglyceride biosynthesis, is enriched in both the ER and MAMs [60]. ACSL4, another MAM enzyme, activates long-chain fatty acids for the production of complex lipids [61]. Phospholipids synthesized in the ER are transferred to mitochondria, where phosphatidylserine is converted to phosphatidylethanolamine (PE) and subsequently returned to the ER to form phosphatidylcholine (PC) [62]. PE and PC are the most abundant phospholipids in mammalian membranes.

MAMs also contain enzymes involved in cholesterol and ceramide biosynthesis, regulating the balance of membrane-bound and cytoplasmic cholesterol [63, 64]. These lipids are essential for membrane integrity, hormone synthesis, digestion, and brain development.

6.4 The Mitochondria–Golgi Complex Relationship: Regulating Apoptosis

Computational analyses indicate that multiple organelles interact with mitochondria, with the Golgi being the second most frequent partner after the ER [56]. The best-studied aspect of this interaction involves apoptosis.

Apoptosis—programmed cell death—can be triggered by extrinsic (external) or intrinsic (internal) signals. The intrinsic pathway includes both mitochondria-mediated and ER-mediated mechanisms. All apoptotic pathways converge on the caspase cascade, where activated caspases dismantle structural and metabolic proteins before the cell is engulfed by phagocytes [65].

In the mitochondria-mediated pathway, mitochondrial permeability transition pores (mPTP) become irreversibly open in response to pro-apoptotic signals, allowing molecules up to 1.5 kDa to pass freely. This releases cytochrome c, which activates downstream caspases [65].

The Golgi apparatus—composed of stacked membrane-bound cisternae and vesicles (Figs. 16 and 17)—regulates this process. Certain Golgi proteins inhibit apoptosis; for example, BRUCE (baculoviral inhibition of apoptosis protein repeat [BIR] containing ubiquitin-conjugating enzyme), located in the trans-Golgi, binds and inhibits caspases and pro-apoptotic factors [65]. Thus, the Golgi can counteract mitochondrial release of cytochrome c.

The trans-Golgi protein parkin chemically modifies and inhibits Bak and Bax proteins, which promote mitochondrial permeabilization and cytochrome c release [65, 66]. Conversely, cleavage of the Golgi protein P115 enhances Bak and Bax activation, promoting apoptosis [65]. Through these and other pathways, the Golgi tightly regulates mitochondria-mediated cell death.

7 Mitochondrial Response in Pathology: Ischemic Stroke

The human brain consumes approximately 25% and 20% of total glucose and oxygen metabolism, respectively, despite comprising only about 2% of total body weight [67]. Because mitochondrial

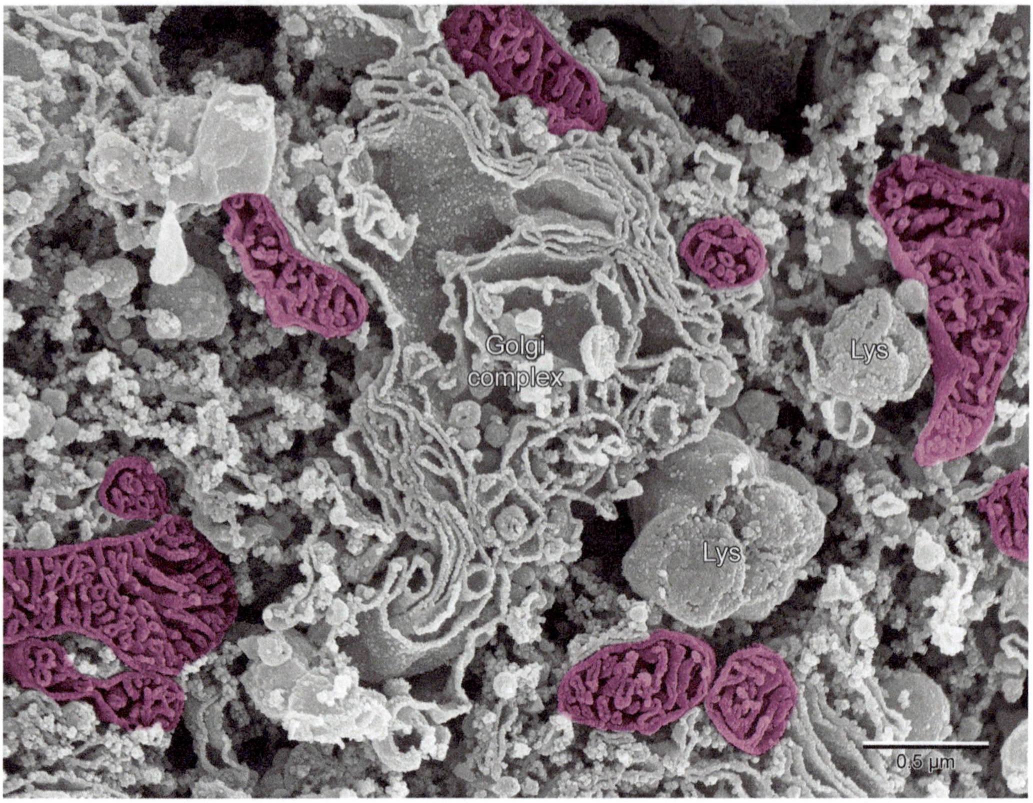

Fig. 16 SEM of the Golgi complex and mitochondria in a neuron cell soma (from the upper portion of Fig. 3). The large Golgi complex is composed of lamellar membrane-enclosed stacks. Vesicles produced by the Golgi complex and several lysosomes (Lys) are seen adjacent to the trans face. Mitochondria are pseudo-colored magenta

respiration is the primary source of this oxygen and glucose usage and is necessary for normal neurological function, mitochondrial dysfunction is a major concern in brain diseases. Stroke is a neurological condition and a leading cause of death and disability worldwide. There are two major categories of stroke: cerebral ischemic stroke and hemorrhagic stroke, both of which involve impaired cerebral blood flow. Of the two, ischemic stroke is more common (~80% of clinical cases) and occurs when blood flow to brain tissue is obstructed, halting oxygen and glucose delivery. Causes of such interruptions include atherosclerosis within vessels or emboli from the peripheral circulation. Although the pathophysiology of ischemic stroke is not fully understood, it is recognized to progress through distinct post-injury phases involving various molecular and cellular cascades. In simplified terms, the cascade of events following ischemia includes focal hypoperfusion, glutamate excitotoxicity, Ca^{2+} overload, oxidative stress, blood–brain barrier dysfunction, vascular and neuronal inflammation, and ultimately cell death [68, 69]. Multiple brain cell

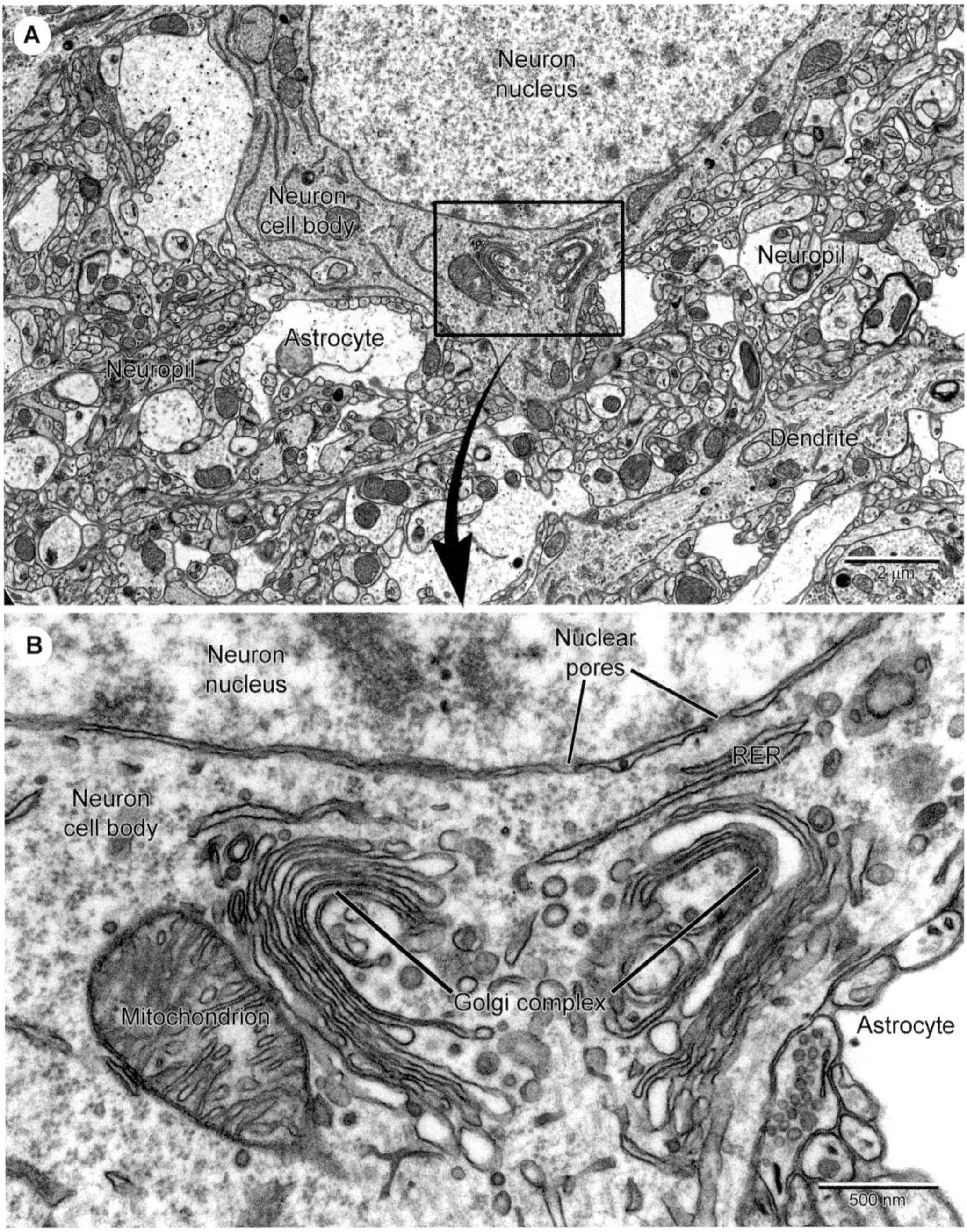

Fig. 17 Golgi complex and mitochondrion in a pyramidal neuron of the mouse somatosensory cortex. (**A**) Low-magnification TEM view of a pyramidal neuron cell soma with surrounding neuropil. The soma contains a large euchromatic nucleus and organelles including ER, Golgi complex, mitochondria, and lysosomes in the cytoplasm. (**B**) High-magnification view of the boxed region in A showing the nucleus, nuclear envelope, nuclear pores, and a large Golgi complex with a strategically located mitochondrion closely associated with its cis face at the left. Numerous vesicles are present at the central trans face of the Golgi. The Golgi complex is curved in an umbrella-like fashion, so two sides of the same organelle are seen in this section

types are affected, including neurons, endothelial cells, pericytes, astrocytes, oligodendrocytes, microglia, and other immune cells [70]. The ultrastructure of these cells, their organelles, and their intracellular processes are also significantly impaired.

EM studies have documented diverse ultrastructural changes in the brain following ischemia, with progressive damage observed from the time of stroke to at least 72 h post-injury and persistent pathology up to 21 days post-injury [71–73]. For example, astrocytes exhibit pathological changes, including abnormally swollen cytoplasmic processes characterized by large clear spaces. At longer time periods (e.g., 72 h), astrocytes display high quantities of glycogen granules in their perivascular end-feet. Figures 18–20 illustrate representative ultrastructural and mitochondrial changes in the peri-infarct zone of adult mice following photothrombotic stroke (an ischemic stroke model; [74]). Mitochondria undergo progressive deterioration, beginning with swelling and advancing to aberrant or absent cristae, hollow occlusions with a "honeycomb" appearance, and ruptured membranes. Such deterioration compromises ATP production and exacerbates stroke pathology. Because most neuronal energy is used to maintain the resting membrane potential via Na+/K+ pumps, insufficient ATP synthesis causes neuronal depolarization, which in turn triggers glutamate release. Sustained glutamate signaling leads to excitotoxicity and, ultimately, cell dysfunction and death [69]. Continued glutamate receptor activation also results in excessive Ca^{2+} influx [75]. Mitochondria normally buffer intracellular Ca^{2+} to maintain homeostasis and prevent excitotoxicity-induced depolarization; however, excessive Ca^{2+} uptake disrupts mitochondrial metabolic pathways and promotes ROS overproduction. Mitochondria located near ER structures—which serve as Ca^{2+} reservoirs under normal conditions but release large quantities of Ca^{2+} post-ischemia—are particularly susceptible to Ca^{2+}-induced damage (Fig. 18). Excess ROS contributes to the oxidative stress characteristic of stroke pathology [69].

Oxidative stress arises when ROS production exceeds the cell's antioxidant capacity. Mitochondria are the primary source of intracellular ROS through aerobic respiration. ROS can damage nucleic acids, proteins, and lipids [76]. As macromolecules are compromised, mitochondrial fusion and fission dynamics can become disrupted, and apoptosis may be initiated through mitochondrial pathways [69]. Ischemic stroke can lead to both programmed cell death and necrosis, because apoptosis requires ATP, ATP depletion favors necrotic cell death. ATP availability likely contributes to the differential tissue damage observed in the ischemic core versus the peri-infarct zone (Figs. 18–20). Proteins in the B-cell lymphoma (BCL-2) family, such as Bak and Bax (discussed in Sect. 6.4), regulate mitochondrial outer membrane permeability [77]. Increases in pro-apoptotic BCL-2 proteins and in mitochondrial membrane permeability after cerebral ischemia underscore the strong involvement of mitochondria in post-stroke apoptosis [69, 77].

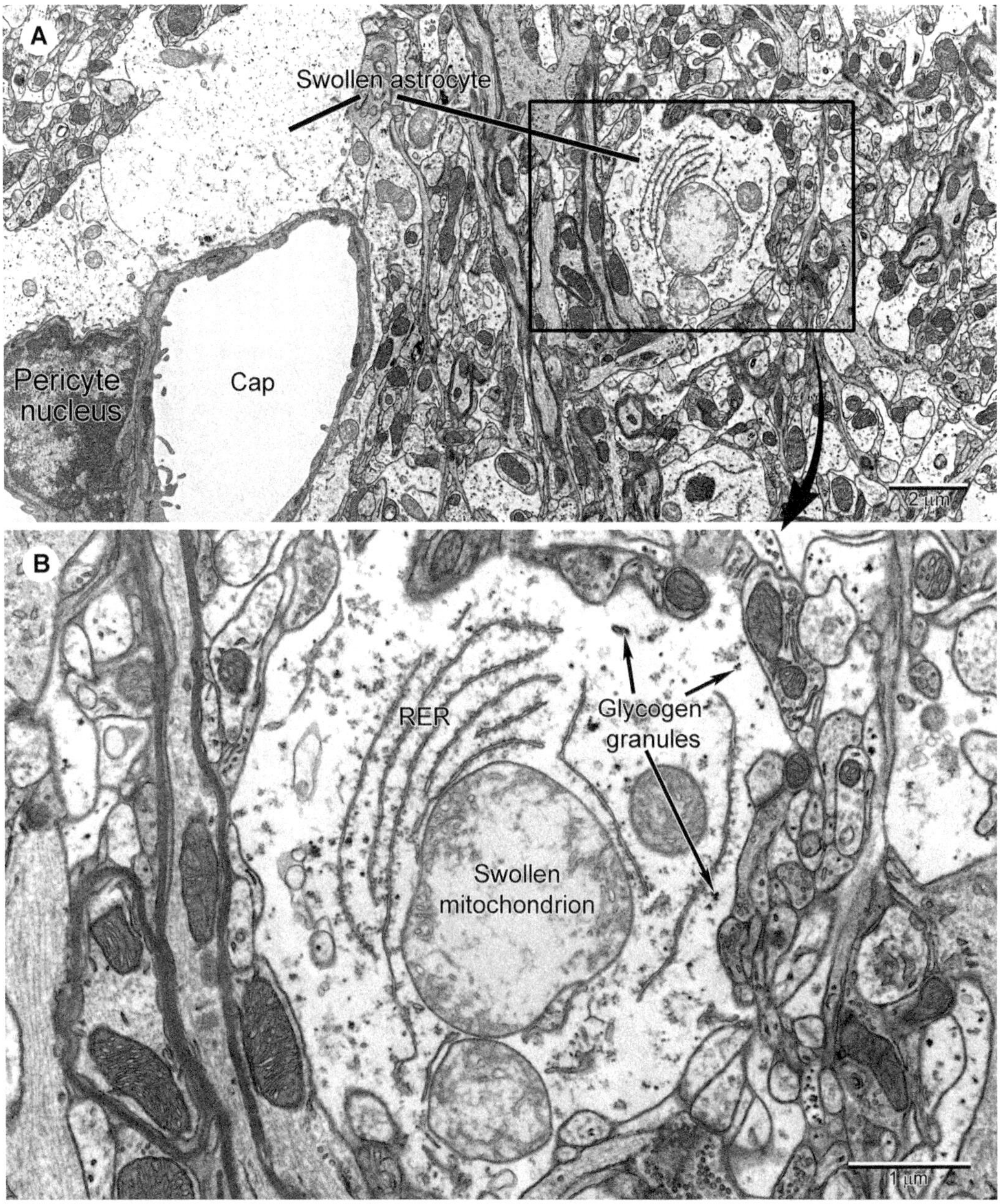

Fig. 18 Peri-infarct zone of the somatosensory cortex 72 h post-ischemia. (**A**) Low-magnification view showing a capillary with its pericyte at the bottom left and surrounding swollen astrocytes evident as large light spaces with sharp perimeters. (**B**) High-magnification view of the boxed area in A. An ischemia-affected astrocyte contains a swollen mitochondrion with its cristae pushed away from its center. These pathological changes are obvious when compared to the dark-staining healthy mitochondria in the axons on the left. Dark glycogen granules are apparent in the astrocyte, which is a response from ischemic stroke

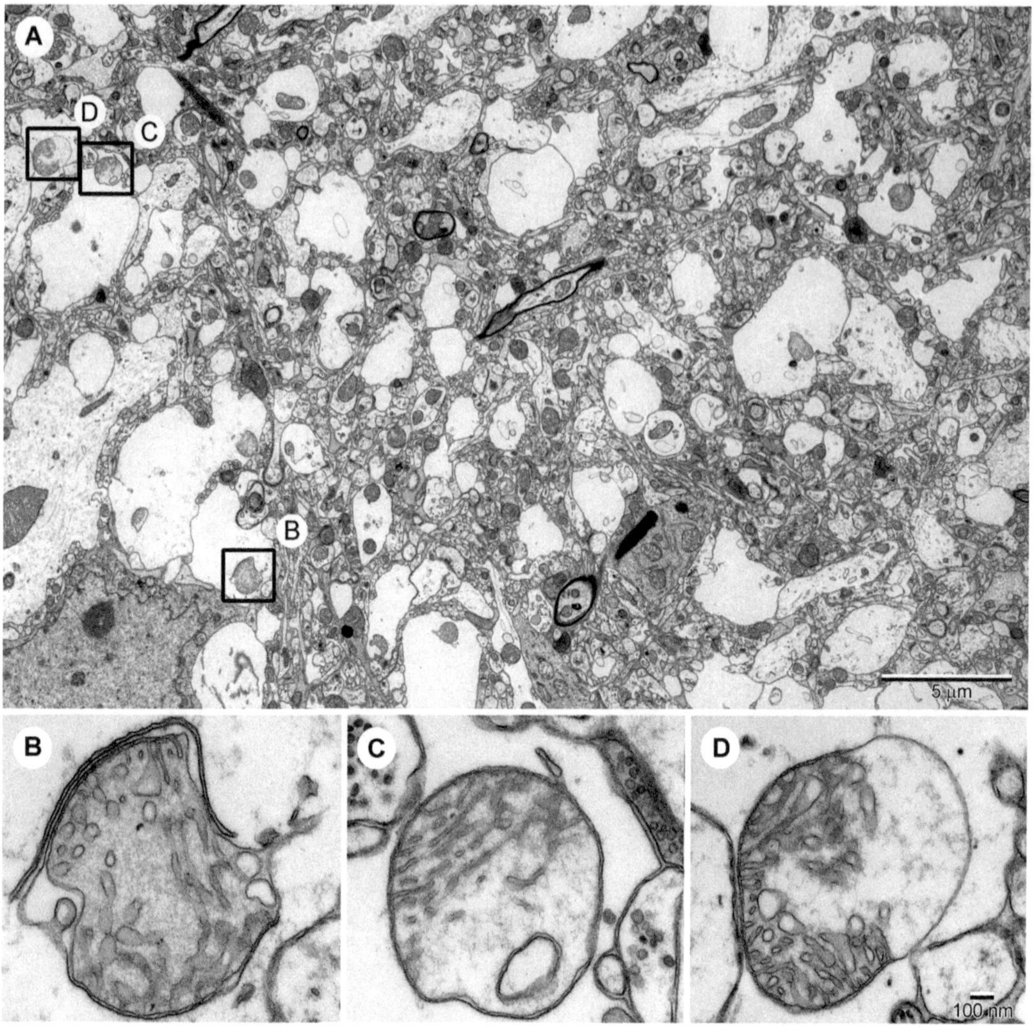

Fig. 19 Mitochondrial ischemic injury pathology. Changes in mitochondria within the peri-infarct zone of a 2-month-old mouse 3 h post-ischemia are shown. (**A**) Low-magnification view of the somatosensory cortex showing enlarged astrocytes containing few organelles, giving the tissue a "spongy" appearance. High-magnification views of the boxed regions with mitochondria marked B-D are seen in the corresponding lower panels. (**B**) The distorted mitochondrion is wrapped on its top half by ER, and the separation between its inner and outer membranes is evident. (**C**) The swollen mitochondrion contains a double-membrane occlusion in its lower half, and its cristae are pushed off to one side. (**D**) The right side of the mitochondrion has been reduced to a single membrane, and cristae are compacted to the left

Studies using the bilateral common carotid artery occlusion (BCCAO) stroke model have shown that dendritic mitochondria undergo swelling and fragmentation, whereas astrocytic mitochondria appear more resilient [78, 79]. Stroke-induced mitochondrial swelling results in measurably larger mitochondrial diameters relative to controls, consistent with observations in Figs. 18 and 20. Neuronal mitochondrial fragmentation has been reported within

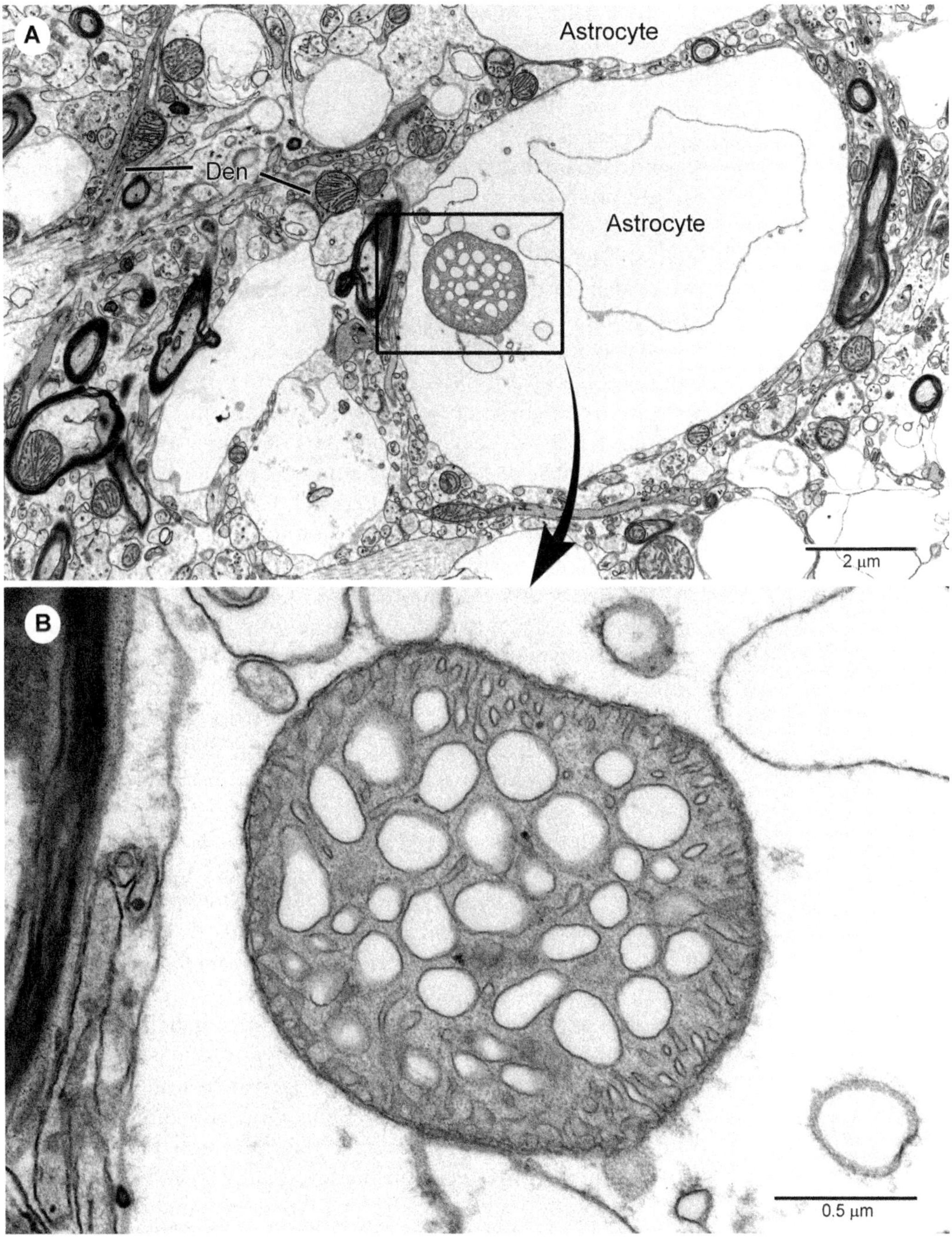

Fig. 20 Swollen central cristae in post-ischemic mitochondria. (**A**) Low-magnification view of the peri-infarct zone of the somatosensory cortex of a three-month-old mouse 3 h post-ischemia. Several enlarged astrocyte processes are visible and contain swollen ER and mitochondria. (**B**) High-magnification view of the boxed region in A reveals swollen cristae centrally and less-affected smaller cristae around the periphery

5 min post-BCCAO [80]. Reperfusion leads to full mitochondrial recovery if blood flow is restored within 5 min or partial recovery if delayed to 8 min [80]. These findings highlight the importance of timely and efficient reperfusion for reversing mitochondrial structural changes [80]. Neurons predominantly produce ATP via mitochondrial oxidative phosphorylation, whereas astrocytes rely largely on glycolysis [81]. This difference likely contributes to the relative resilience of astrocytes under oxygen and glucose deprivation, while neurons remain vulnerable [82, 83]. Nonetheless, astrocytic changes are evident in the peri-infarct zone, and multiple mechanisms may contribute to this degradation.

Astrocytes serve diverse roles in the CNS, including supporting neuronal health. Notably, it has been proposed that astrocytes help injured neurons post-ischemia by donating functional mitochondria and by receiving damaged neuronal mitochondria for degradation [82, 84–86]. One study employing both in vitro and in vivo approaches demonstrated that cortical neurons subjected to oxygen–glucose deprivation, resulting in reduced ATP and cell viability, recovered after treatment with astrocyte-conditioned media [84]. Astrocytes were shown to produce functional extracellular mitochondria that were taken up by neurons, enhancing cell survival and neuroplasticity [84]. Astrocytic mitochondria were detected in both axons and soma, and in some cases fused with endogenous neuronal mitochondria [84]. Additionally, astrocytes supply metabolic precursors—such as glucose and lactate—that fuel the mitochondrial CAC and support neuronal ATP production. Astrocytes contain the majority of brain glycogen and regularly undergo glycophagy and glycogenolysis to convert glycogen into usable glucose [87]. Following ischemic stroke, astrocytic glycogen accumulation increases (Fig. 18). Several hypotheses have been proposed to explain this. One possibility is reduced glycogenolysis due to impaired glucose transport from blood vessels to astrocytes coupled with decreased energy use by neurons and other glia [88]. This aligns with findings that glycogen granules decrease 1 week after ischemia, coinciding with cellular recovery [88]. Another hypothesis is that glycogen accumulation reflects disrupted glycogenolysis following reperfusion—either normal or experimentally induced—due to dysregulation of biochemical pathways required for post-ischemic homeostasis [89]. Although the mechanisms underlying astrocytic metabolic neuroprotection and mitochondria transfer remain incompletely understood, it is clear that astrocytes play a critical supportive role following ischemia.

Mitochondrial pathologies encompass a wide range of structural and functional abnormalities that can profoundly impact cells and organisms. In ischemic stroke, mitochondrial morphology changes drastically, and these structural alterations likely contribute to excitotoxicity, oxidative stress, apoptosis, and necrosis. Ongoing research continues to address knowledge gaps and develop therapeutics.

Future studies that clarify mitochondrial interactions with other organelles, their roles in different brain cell types, and their responses to various physiological and pathological conditions will be essential for advancing treatments for mitochondria-related diseases.

8 Conclusion

Mitochondria are undoubtedly intriguing and essential organelles in eukaryotes and higher-level organisms. Since their discovery, they have captivated the attention of renowned scientists and have become central to advances in genetics, biology, medical sciences, and more. At the ultrastructural level, their distinctive morphology immediately attracts interest, while their diverse phenotypes invite deeper questions about their structural specificity within different cells and locations. Although long recognized as miniature biochemical generators for their hosts, recent research has emphasized the wide diversity of roles they perform. Mitochondria can be likened to a "jack of all trades," collaborating with numerous other organelles—including the nucleus, cytoskeletal elements, endoplasmic reticulum, and Golgi complex—to support a variety of cellular functions.

Yet their capabilities should not be overestimated, as they are not impervious to disease. Mitochondria can undergo profound structural and functional deterioration when under physiological threat, such as during ischemic stroke. Mitochondrial pathologies in the brain—and throughout the body—are serious, but current and emerging research is directed toward developing treatments that may reverse such damage and ultimately save lives. Mitochondria offer a unique and promising path forward.

Conflict of Interest

The authors have no conflicts of interest to declare that are relevant to the content of this chapter.

References

1. Altmann R (1894) Die elementarorganismen und ihre beziehungen zu den zellen. Veit & Comp, Leipzig, p 145
2. Mokranjac D, Perocchi F (eds) (2017) Mitochondria: practical protocols. Methods in molecular biology, vol 1567. Springer, New York, NY
3. Ryan TE (2019) Morphing mitochondria: understanding the development of the mitochondrial reticulum in skeletal muscle. J Physiol 597(10):2619–2620
4. Molnar MJ, Kovacs GG (2017) Chapter 10 - Mitochondrial diseases. In: Kovacs GG, Alafuzoff I (eds) Handbook of clinical neurology, vol 145. Elsevier, pp 147–55
5. Chin HL, Lai PS, Tay SKH (2024) A clinical approach to diagnosis and management of mitochondrial myopathies. Neurotherapeutics 21(1):e00304
6. Das J (2006) The role of mitochondrial respiration in physiological and evolutionary adaptation. BioEssays 28(9):890–901

7. Ernster L, Danielson L, Ljunggren M (1962) Dt diaphorase I. Purification from the soluble fraction of rat-liver cytoplasm, and properties. Biochim Biophys Acta 58:171–188
8. Kiss G, Konrad C, Pour-Ghaz I, Mansour JJ, Németh B, Starkov AA et al (2014) Mitochondrial diaphorases as NAD+ donors to segments of the citric acid cycle that support substrate-level phosphorylation yielding ATP during respiratory inhibition. FASEB J 28(4):1682–1697
9. Vasiliou V, Ross D, Nebert DW (2006) Update of the NAD(P)H:quinone oxidoreductase (NQO) gene family. Hum Genomics 2(5):329–335
10. Judson RN, Gray SR, Walker C, Carroll AM, Itzstein C, Lionikas A et al (2013) Constitutive expression of Yes-associated protein (Yap) in adult skeletal muscle fibres induces muscle atrophy and myopathy. PLoS One 8(3):e59622
11. Klein HH, Puschmann S, Schaper J, Schaper W (1981) The mechanism of the tetrazolium reaction in identifying experimental myocardial infarction. Virchows Arch A Pathol Anat Histopathol 393(3):287–297
12. Balan IS, Fiskum G, Kristian T (2009) Visualization and quantification of NAD(H) in brain sections by a novel histo-enzymatic nitrotetrazolium blue staining technique. Brain Res 1316C:112
13. Palade GE (1953) An electron microscope study of the mitochondrial structure. J Histochem Cytochem 1(4):188–211
14. Nahirney PC, Tremblay ME (2021) Brain ultrastructure: putting the pieces together. Front Cell Dev Biol 9:629503
15. Bleck CKE, Kim Y, Willingham TB, Glancy B (2018) Subcellular connectomic analyses of energy networks in striated muscle. Nat Commun 9(1):5111
16. Page E, McCallister LP (1973) Quantitative electron microscopic description of heart muscle cells: application to normal, hypertrophied and thyroxin-stimulated hearts. Am J Cardiol 31(2):172–181
17. Wang R, Lei H, Wang H, Qi L, Liu Y, Liu Y et al (2024) Dysregulated inter-mitochondrial crosstalk in glioblastoma cells revealed by in situ cryo-electron tomography. Proc Natl Acad Sci U S A 121(9):e2311160121
18. Popov V, Medvedev NI, Davies HA, Stewart MG (2005) Mitochondria form a filamentous reticular network in hippocampal dendrites but are present as discrete bodies in axons: a three-dimensional ultrastructural study. J Comp Neurol 492(1):50–65
19. Delgado T, Petralia RS, Freeman DW, Sedlacek M, Wang YX, Brenowitz SD et al (2019) Comparing 3D ultrastructure of presynaptic and postsynaptic mitochondria. Biol Open 8(8):bio044834
20. Cserép C, Pósfai B, Schwarcz AD, Dénes Á (2018) Mitochondrial ultrastructure is coupled to synaptic performance at axonal release sites. eNeuro 5:1
21. Zick M, Rabl R, Reichert AS (2009) Cristae formation—linking ultrastructure and function of mitochondria. Biochim Biophys Acta 1793(1):5–19
22. Yu SB, Pekkurnaz G (2018) Mechanisms orchestrating mitochondrial dynamics for energy homeostasis. J Mol Biol 430(21):3922–3941
23. Marcuzzo MB, de Andrade Silveira J, Streck EL, Vockley J, Leipnitz G (2025) Disruption of mitochondrial quality control in inherited metabolic disorders. Mol Neurobiol 62(6):6770–6784
24. Casanova A, Wevers A, Navarro-Ledesma S, Pruimboom L (2023) Mitochondria: it is all about energy. Front Physiol 14:1114231
25. Shen X, Sun P, Zhang H, Yang H (2022) Mitochondrial quality control in the brain: the physiological and pathological roles. Front Neurosci 16:1075141
26. Gray MW (2017) Lynn Margulis and the endosymbiont hypothesis: 50 years later. Mol Biol Cell 28(10):1285–1287
27. Sagan L (1967) On the origin of mitosing cells. J Theor Biol 14(3):255–274
28. De Silva D, Tu YT, Amunts A, Fontanesi F, Barrientos A (2015) Mitochondrial ribosome assembly in health and disease. Cell Cycle 14(14):2226–2250
29. Posth C, Wißing C, Kitagawa K, Pagani L, van Holstein L, Racimo F et al (2017) Deeply divergent archaic mitochondrial genome provides lower time boundary for African gene flow into Neanderthals. Nat Commun 8(1):16046
30. Wei W, Schon KR, Elgar G, Orioli A, Tanguy M, Giess A et al (2022) Nuclear-embedded mitochondrial DNA sequences in 66,083 human genomes. Nature 611(7934):105–114
31. Prachar J (2003) Intimate contacts of mitochondria with nuclear envelope as a potential energy gateway for nucleo-cytoplasmic mRNA transport. Gen Physiol Biophys 22(4):525–534
32. Walker BR, Moraes CT (2022) Nuclear-mitochondrial interactions. Biomolecules 12(3):427
33. Martínez-Reyes I, Chandel NS (2020) Mitochondrial TCA cycle metabolites control physiology and disease. Nat Commun 11(1):102
34. Picard M, Shirihai OS (2022) Mitochondrial signal transduction. Cell Metab 34(11):1620–1653
35. Wu Z, Oeck S, West AP, Mangalhara KC, Sainz AG, Newman LE et al (2019) Mitochondrial DNA stress signalling protects the nuclear genome. Nat Metab 1(12):1209–1218
36. Desai R, East DA, Hardy L, Faccenda D, Rigon M, Crosby J et al (2020) Mitochondria form

contact sites with the nucleus to couple prosurvival retrograde response. Sci Adv 6(51):eabc9955
37. Pilling AD, Horiuchi D, Lively CM, Saxton WM (2006) Kinesin-1 and dynein are the primary motors for fast transport of mitochondria in Drosophila motor axons. Mol Biol Cell 17(4):2057–2068
38. Drerup CM, Herbert AL, Monk KR, Nechiporuk AV (2017) Regulation of mitochondria-dynactin interaction and mitochondrial retrograde transport in axons. eLife 6:e22234
39. Vale RD, Reese TS, Sheetz MP (1985) Identification of a novel force-generating protein, kinesin, involved in microtubule-based motility. Cell 42(1):39–50
40. Davis K, Basu H, Izquierdo-Villalba I, Shurberg E, Schwarz TL (2022) Miro GTPase domains regulate the assembly of the mitochondrial motor–adaptor complex. Life Sci Alliance 6(1):e202201406
41. van Haren J, Wittmann T (2019) Microtubule plus end dynamics – do we know how microtubules grow?. BioEssays 41(3):e1800194
42. Tas RP, Chazeau A, Cloin BMC, Lambers MLA, Hoogenraad CC, Kapitein LC (2017) Differentiation between oppositely oriented microtubules controls polarized neuronal transport. Neuron 96(6):1264–1271.e5
43. Barnhart EL (2016) Mechanics of mitochondrial motility in neurons. Curr Opin Cell Biol 38:90–99
44. Chen Y, Sheng ZH (2013) Kinesin-1–syntaphilin coupling mediates activity-dependent regulation of axonal mitochondrial transport. J Cell Biol 202(2):351–364
45. Kang JS, Tian JH, Pan PY, Zald P, Li C, Deng C, Sheng Z (2008) Docking of axonal mitochondria by syntaphilin controls their mobility and affects short-term facilitation. Cell 132(1):137–148
46. Boldogh IR, Pon LA (2006) Interactions of mitochondria with the actin cytoskeleton. Biochim Biophys Acta 1763(5-6):450–462
47. Palay SL, Chan-Palay V (1974) Cerebellar cortex. Springer, Berlin, Heidelberg
48. Xu-Friedman MA, Regehr WG (2003) Ultrastructural contributions to desensitization at cerebellar mossy fiber to granule cell synapses. J Neurosci 23(6):2182–2192
49. Billups B, Forsythe ID (2002) Presynaptic mitochondrial calcium sequestration influences transmission at mammalian central synapses. J Neurosci 22(14):5840–5847
50. Carafoli E, Tiozzo R, Lugli G, Crovetti F, Kratzing C (1974) The release of calcium from heart mitochondria by sodium. J Mol Cell Cardiol 6(4):361–371
51. Datta S, Jaiswal M (2021) Mitochondrial calcium at the synapse. Mitochondrion 59:135–153
52. Brini M, Carafoli E (2011) The plasma mitochondrial calcium at the synapse. mitochondrion.membrane Ca^{2+} ATPase and the plasma membrane sodium calcium exchanger cooperate in the regulation of cell calcium. Cold Spring Harb Perspect Biol 3(2):a004168
53. Tarasov AI, Griffiths EJ, Rutter GA (2012) Regulation of ATP production by mitochondrial Ca^{2+}. Cell Calcium 52(1):28–35
54. Boerries M, Most P, Gledhill JR, Walker JE, Katus HA, Koch WJ et al (2007) Ca^{2+}-dependent interaction of S100A1 with F1-ATPase leads to an increased ATP content in cardiomyocytes. Mol Cell Biol 27(12):4365–4373
55. Debattisti V, Gerencser AA, Saotome M, Das S, Hajnóczky G (2017) ROS control mitochondrial motility through p38 and the motor adaptor Miro/Trak. Cell Rep 21(6):1667–1680
56. Valm AM, Cohen S, Legant WR, Melunis J, Hershberg U, Wait E et al (2017) Applying systems-level spectral imaging and analysis to reveal the organelle interactome. Nature 546(7656):162–167
57. Vance JE (2014) MAM (mitochondria-associated membranes) in mammalian cells: lipids and beyond. Biochim Biophys Acta 1841(4):595–609
58. Andreyev A, Tamrakar P, Rosenthal RE, Fiskum G (2018) Calcium uptake and cytochrome c release from normal and ischemic brain mitochondria. Neurochem Int 117:15–22
59. Smaili SS, Hsu YT, Youle RJ, Russell JT (2000) Mitochondria in Ca^{2+} signaling and apoptosis. J Bioenerg Biomembr 32(1):35–46
60. Stone SJ, Levin MC, Zhou P, Han J, Walther TC, Farese RV (2009) The endoplasmic reticulum enzyme DGAT2 is found in mitochondria-associated membranes and has a mitochondrial targeting signal that promotes its association with mitochondria. J Biol Chem 284(8):5352–5361
61. Chen F, Kang R, Liu J, Tang D (2023) The ACSL4 network regulates cell death and autophagy in diseases. Biology 12(6):864
62. van der Veen JN, Kennelly JP, Wan S, Vance JE, Vance DE, Jacobs RL (2017) The critical role of phosphatidylcholine and phosphatidylethanolamine metabolism in health and disease. Biochim Biophysa Acta Biomembr 1859(9, Part B):1558–1572
63. Annunziata I, Sano R, d'Azzo A (2018) Mitochondria-associated ER membranes (MAMs) and lysosomal storage diseases. Cell Death Dis 9(3):328
64. Rusiñol AE, Cui Z, Chen MH, Vance JE (1994) A unique mitochondria-associated membrane fraction from rat liver has a high capacity for lipid synthesis and contains pre-Golgi secretory proteins including nascent lipoproteins. J Biol Chem 269(44):27494–27502

65. He Q, Liu H, Deng S, Chen X, Li D, Jiang X et al (2020) The Golgi apparatus may be a potential therapeutic target for apoptosis-related neurological diseases. Front Cell Dev Biol 8:830
66. Bernardini JP, Brouwer JM, Tan IK, Sandow JJ, Huang S, Stafford CA et al (2019) Parkin inhibits BAK and BAX apoptotic function by distinct mechanisms during mitophagy. EMBO J 38(2):e99916
67. Attwell D, Laughlin SB (2001) An energy budget for signaling in the grey matter of the brain. J Cereb Blood Flow Metab 21(10):1133–1145
68. Turner RC, Dodson SC, Rosen CL, Huber JD (2013) The science of cerebral ischemia and the quest for neuroprotection: navigating past failure to future success: a review. J Neurosurg 118(5):1072–1085
69. Zhou X, Chen H, Wang L, Lenahan C, Lian L, Ou Y et al (2021) Mitochondrial dynamics: a potential therapeutic target for ischemic stroke. Front Aging Neurosci 13:721428
70. Gao Y, Fang C, Wang J, Ye Y, Li Y, Xu Q et al (2023) Neuroinflammatory biomarkers in the brain, cerebrospinal fluid, and blood after ischemic stroke. Mol Neurobiol 60(9):5117–5136
71. Nahirney PC, Reeson P, Brown CE (2016) Ultrastructural analysis of blood–brain barrier breakdown in the peri-infarct zone in young adult and aged mice. J Cereb Blood Flow Metab 36(2):413–425
72. Solenski NJ, diPierro CG, Trimmer PA, Kwan AL, Helms GA (2002) Ultrastructural changes of neuronal mitochondria after transient and permanent cerebral ischemia. Stroke 33(3):816–824
73. Han CW, Lee KH, Noh MG, Kim JM, Kim HS, Kim HS et al (2017) An experimental infarct targeting the internal capsule: histopathological and ultrastructural changes. J Pathol Transl Med 51(3):292–305
74. Watson BD, Dietrich WD, Busto R, Wachtel MS, Ginsberg MD (1985) Induction of reproducible brain infarction by photochemically initiated thrombosis. Ann Neurol 17(5):497–504
75. Baranov SV, Jauhari A, Carlisle DL, Friedlander RM (2021) Two hit mitochondrial-driven model of synapse loss in neurodegeneration. Neurobiol Dis 158:105451
76. Meo SD, Reed TT, Venditti P, Victor VM (2016) Role of ROS and RNS sources in physiological and pathological conditions. Oxid Med Cell Longev 2016:1245049
77. Yang JL, Mukda S, Chen SD (2018) Diverse roles of mitochondria in ischemic stroke. Redox Biol 16:263–275
78. Fomitcheva IV, Sword J, Shi Y, Kirov SA (2022) Plasticity of perisynaptic astroglia during ischemia-induced spreading depolarization. Cereb Cortex 33(9):5469–5483
79. Kirov SA, Fomitcheva IV, Sword J (2020) Rapid neuronal ultrastructure disruption and recovery during spreading depolarization-induced cytotoxic edema. Cereb Cortex 30(10):5517–5531
80. Kislin M, Sword J, Fomitcheva IV, Croom D, Pryazhnikov E, Lihavainen E et al (2017) Reversible disruption of neuronal mitochondria by ischemic and traumatic injury revealed by quantitative two-photon imaging in the neocortex of anesthetized mice. J Neurosci 37(2):333–348
81. Beard E, Lengacher S, Dias S, Magistretti PJ, Finsterwald C (2022) Astrocytes as key regulators of brain energy metabolism: new therapeutic perspectives. Front Physiol 12:825816
82. Collyer E, Blanco-Suarez E (2023) Astrocytes in stroke-induced neurodegeneration: a timeline. Front Mol Med 3:1240862
83. Gürer G, Gursoy-Ozdemir Y, Erdemli E, Can A, Dalkara T (2008) Astrocytes are more resistant to focal cerebral ischemia than neurons and die by a delayed necrosis. Brain Pathol 19(4):630–641
84. Hayakawa K, Esposito E, Wang X, Terasaki Y, Liu Y, Xing C et al (2016) Transfer of mitochondria from astrocytes to neurons after stroke. Nature 535(7613):551–555
85. Davis ChaO, Kim KY, Bushong EA, Mills EA, Boassa D, Shih T et al (2014) Transcellular degradation of axonal mitochondria. Proc Natl Acad Sci U S A 111(26):9633–9638
86. Lee EH, Kim M, Ko SH, Kim CH, Lee M, Park CH (2023) Primary astrocytic mitochondrial transplantation ameliorates ischemic stroke. BMB Rep 56(2):90–95
87. Guo H, Li Y, Wang S, Yang Y, Xu T, Zhao J et al (2024) Dysfunction of astrocytic glycophagy exacerbates reperfusion injury in ischemic stroke. Redox Biol 74:103234
88. Kajihara H, Tsutsumi E, Kinoshita A, Nakano J, Takagi K, Takeo S (2001) Activated astrocytes with glycogen accumulation in ischemic penumbra during the early stage of brain infarction: immunohistochemical and electron microscopic studies. Brain Res 909(1):92–101
89. Cai Y, Guo H, Fan Z, Zhang X, Wu D, Tang W et al (2020) Glycogenolysis is crucial for astrocytic glycogen accumulation and brain damage after reperfusion in ischemic stroke. iScience 23(5):101136

Check for updates

Chapter 6

Ultrastructure of Glial Cells in iPSC-Derived Unguided Neural Organoids from *APOE4/4* Patients with Alzheimer's Disease

Eva Šimončičová, Marloes Verkerke, Fernando González Ibáñez, Raphaël Bourgeois, Katherine Picard, Mohammadparsa Khakpour, Julia TCW, Elly M. Hol and Marie-Ève Tremblay

Abstract

Progress in the treatment of neurodegenerative disorders, such as Alzheimer's disease (AD), has become largely obstructed by a lack of successful transition between drug discovery at the preclinical level to clinical application. Traditional animal models are valuable but often replicate only part of the pathological spectrum of human neurodegenerative cellular processes. There is a pressing need for additional translational models to reveal human-specific disease signatures while accounting for patient variability. Three-dimensional (3D) in vitro stem cell-derived neural cell models, termed neural organoids, mimic the complex architecture of brain tissue, including dynamic and heterogeneous cell-to-cell interactions and signaling within an in vivo-like brain microenvironment. The integration of this strategy with patient-derived induced pluripotent stem cell (iPSC) approach provides a renewable in vitro model that enables patient-specific disease modeling and patient-specific drug testing. While this 3D methodology is continuously advancing to address various limitations, the research application of these models is gaining significant traction. So far, the characterization of brain organoids has heavily relied on molecular techniques, leaving important morphological and structural features underexplored. In the current chapter, we address the ultrastructural features of 120-day-old unguided neural organoids (UNOs) with the use of a scanning electron microscopy (SEM). These organoids are derived from iPSCs of human patients with AD carrying an *APOE4/4* risk gene variant or its isogenic (*APOE3/3*) control. We describe the ultrastructural profile of neural rosettes and their organoid-grown radial glia cell (*oRGC*) component. Using classically utilized cell markers, GFAP and IBA1, we further focus on the ultrastructure of organoid-grown astrocytes (*oASTRO*) and microglia (*oMG*), respectively. Throughout the last section, we discuss limitations and considerations for experimental design, cell sampling and analysis of UNOs using electron microscopy.

Keywords Neural organoid, Ultrastructure, Scanning electron microscopy, Microglia, Astrocyte, Organoid rosette

---

The authors "Eva Šimončičová and Marloes Verkerke" contributed equally.

Marie-Ève Tremblay (ed.), *Ultrastructural Neuroscience Methods*, Neuromethods, vol. 238,
https://doi.org/10.1007/978-1-0716-5456-9_6,

1 Introduction

Alzheimer's disease (AD) is the most common form of dementia, characterized by extracellular deposits of amyloid beta (Aβ) into parenchymal plaques, formation of intracellular neurofibrillary tangles consisting of hyperphosphorylated and misfolded protein tau [1], a microtubule-associated protein, and reactive gliosis [2, 3]. These neuropathological changes are most notably associated with synaptic dysfunction, neuronal loss, angiopathy, and overall brain atrophy resulting in clinical symptoms of memory loss, problems with learning, dysfunctional executive functioning, changes in mood and personality. Recent research has shown that glial cells play an important role in the pathophysiology of the disease [1, 2, 4, 5].

The largest genetic risk factor for late-onset AD (LOAD) is a polymorphism in the *apolipoprotein E* (*APOE*) gene, which is predominantly expressed in the central nervous system (CNS) by astrocytes and to a lesser extent by microglia [6–8]. The *APOE* gene contains two single-nucleotide polymorphisms that give rise to three main alleles: *APOE2*, *APOE3*, and *APOE4*. The *APOE3* allele has the highest frequency among the global population [9]. The *APOE4* allele increases the risk of developing AD, whereas *APOE2* decreases this risk [10–13]. APOE's main function is to maintain lipid homeostasis in the brain through facilitation of intercellular cholesterol and phospholipid transport [14]. The hypotheses on how *APOE4* increases the risk of developing AD are plentiful, and it is most likely a combination of several *APOE4*-induced disturbances and other genetic or environmental factors. First of all, there are structural and biochemical differences between the APOE isoforms [15, 16]. The amount of secreted APOE is genotype-dependent, with lower levels of secreted APOE4 compared to APOE3 [17–19]. Compared to APOE3, the APOE4 isoform has been shown to enhance Aβ accumulation [17, 20, 21], tau pathology [22, 23], and neuroinflammation [24–27]. Furthermore, APOE4 disrupts synaptic functioning [28, 29], mitochondrial functioning [8, 30], lipid homeostasis [17, 19], and blood–brain barrier integrity [31–33].

To study the complex role of human astrocytes and microglia in AD, post-mortem brain tissue allows for the examination of these cells in their natural environment. Post-mortem tissue, however, often displays the end-stage of a disease, thereby increasing the difficulty of investigating the sequence of events in the pathogenesis. Variable post-mortem interval and character of sample preservation further introduce variability into experimental conditions. Many studies were conducted to study astrocyte and microglia physiology and pathology in rodent models of AD pathology. However, along the way, important differences were identified between rodent and human glia, in terms of transcriptional, morphological, and functional properties. For example, a human astrocyte can cover more synapses

(270,000–2 million) than a rodent astrocyte (20,000–120,000), display significantly faster calcium signaling compared to rodent astrocytes, and respond differently to stressors [34–39]. Additionally, a comparison between astrocytes and microglia from an AD pathology mouse model (5×FAD) and those from human patients with AD revealed significant transcriptomic differences [40].

The development of human induced pluripotent stem cells (iPSC) technology led to an entirely new set of human cell models [41, 42]. These iPSC-derived models are currently an invaluable tool for studying the underlying pathophysiological mechanisms of human diseases and the identification and characterization of patient-specific phenotypes. One of these iPSC-derived three-dimensional (3D) models is the unguided neural organoid (UNO) model [43]. UNOs are generated through a relatively unguided differentiation process, whereby iPSCs are cultured in a favorable environment for neural differentiation without additional patterning factors [44, 45]. This approach results in the formation of a complex 3D neural structure that comprises multiple cell types, including neurons, astrocytes, and microglia [45–47]. As all cell types in this model have evolved in the same environment, it enables the investigation of their dynamic interactions in a 3D context. In comparison to other organoid models using patterning factors, a unique feature of UNO is the presence of microglia, which are of a mesodermal origin. Therefore, this model offers a novel option for a study of glial-mediated disease processes, such as the role of glial APOE in the pathophysiology of AD.

Considering the novelty of 3D organoid models, UNOs included, previous studies have predominantly used molecular and fluorescent microscopy techniques to describe cell-specific molecular signatures, cytoarchitecture and morphology, signaling, and functional capacities [44, 45]. So far, electron microscopy (EM) studies, which provide valuable information on the ultrastructure of cells and subcellular organelles, remain rare. EM is an advanced imaging technique allowing for the study of complex ultrastructural features within the sampled tissue at a high resolution [48–50]. Two principal EM methodologies include transmission electron microscopy (TEM) and scanning electron microscopy (SEM). These methods use a focused electron beam to image a sample by detecting either transmitted electrons passing through the sample (TEM) or backscattered and secondary electrons generated at the sample surface (SEM). Osmium is a heavy metal utilized in the preparation of samples for EM, acting as a conductive that strongly binds to lipids and thus efficiently enhances the electron density of lipid-rich structures such as cellular membranes. With the addition of thiocarbohydrazide (THC) that increases the opportunity for additional osmium binding, the osmium-THC-osmium (OTO) method is particularly beneficial for improved membrane contrast and delineation of cellular structures [51]. In practice, this multistep technique provides crucial information on both the surface topography and composition of the imaged sample. High resolution of the imaging, on a nanometric scale, allows for a study of fine changes to intracellular components that are instrumental for the assessment of cellular functions and stress.

In this chapter, we present a guide for immunohistochemistry-EM processing adapted for the specific features of human iPSC-derived UNO samples using SEM. We describe the ultrastructural profile of neural rosettes, a prominent organoid histoarchitectural feature, and delve into its glial components. We utilize classical expression markers glial fibrillary acidic protein (GFAP), an intermediate filament protein, and allograft inflammatory factor 1 (AIF1)/ionized calcium-binding adapter molecule 1 (IBA1), to visualize astrocytes and microglia, respectively. We further discuss experimental and analytical considerations for ultrastructural analysis of organoid cells, such as those related to cell line- and organoid-specific variability. This work lays the foundation for future in-depth exploration of organoid-grown glia ultrastructure, which is instrumental for further progress in the development of brain organoid models and their utilization for diagnostic and treatment purposes.

2 Methods

2.1 Preparation of Unguided Neural Organoid (UNO)

Induced Pluripotent Stem Cell Lines

Human isogenic APOE iPSC lines were generated and provided by Dr. J. TCW as described by [19]. The parental iPSC lines originate from two AD patients with an *APOE4/4* genotype. To achieve isogenic iPSC pairs, the *APOE4/4* genotype was corrected to an *APOE3/3* genotype with a CRISPR/Cas9 approach.

For the purposes of this chapter, we used iPSCs generated from patient 1 (female, 72), specifically clones TCW1E33-1C6 (referred to as **MV 8.1**) and TCW1E44-2C2 (referred to as **MV 8.4**), and from patient 2 (male, 80), specifically clones TCW2E33-2E3 (referred to as **MV 10.1**) and TCW2E44-4B1 (referred to as **MV 10.4**).

2.2 Materials

1. *Culture dishes* (35 and 60 mm).
2. *Aggrewell 800 microwell plate* (StemCell Technologies, 27865).
3. *Ultra-low attachment 96-well plate* (Corning, 3474).
4. *Belly dancer* (IBI Scientific BDRLS0001).
5. *StemFlex medium* (Thermo Fisher, A3349401).

6. *Ethylenediaminetetraacetic acid* (*EDTA*; Thermo Fisher, 15575020).
7. *Accutase* (Thermo Fisher, A1110501).
8. *Geltrex* (Thermo Fisher, A1413202).
9. *Matrigel* (Corning, 356234).
10. *ROCK inhibitor Y-27632* (Axon Medchem, 1683).
11. *Basic Fibroblast Growth Factor* (*bFGF*; Thermo Fisher, AA10155).
12. ***HES0 medium***: DMEM/F-12 (Thermo Fisher, 11320074), 20% KnockOut Serum Replacement (KOSR; Thermo Fisher, 10828028), 3% Fetal Bovine Serum (FBS; Sigma-Aldrich, F7524), 1% L-glutamine (Thermo Fisher, 25030024), 1% Non-Essential AminoAcids (NEAA; Thermo Fisher, 11140035), 0.0007% β-mercaptoethanol (Merck, 805740).
13. ***Neural Induction Medium***: DMEM/F-12, 1% N2 (Thermo Fisher, 17502048), 1% L-glutamine, 1% NEAA, 0.5 μg/mL heparin (Sigma-Aldrich, H3149-10KU).
14. ***Differentiation medium without vitamin A***: 1:1 DMEM/F-12:Neurobasal (Thermo Fisher, 21103049), 1% L-glutamine, 1% penicillin/streptomycin (P/S; Thermo Fisher, 15140122), 1% B27 without vitamin A (Thermo Fisher, 12587010), 0.5% NEAA, 0.5% N2, 0.025% insulin (Sigma-Aldrich, I9278-5ML), 0.0003% β-mercaptoethanol.
15. ***Differentiation medium with vitamin A***: 1:1 DMEM/F-12:Neurobasal (Thermo Fisher, 21103049), 1% L-glutamine, 1% penicillin/streptomycin (P/S; Thermo Fisher, 15140122), 1% B27 with vitamin A (Thermo Fisher, 17504001), 0.5% NEAA, 0.5% N2, 0.025% insulin (Sigma-Aldrich, I9278-5ML), 0.0003% β-mercaptoethanol.

2.3 Protocol

Unguided neural organoid (UNO) differentiation (as described in [44, 45] with minor adaptations):

1. Culture the iPSCs on Geltrex-coated dishes in StemFlex medium at 37°C with 5% CO_2.
2. When 80%–90% confluent, detach the colonies with 0.5 mM EDTA and seed on a new Geltrex-coated dish. Supplement the medium with 5 μM Y-27632 for 24 hours (h) after passaging to reduce cell death.
3. To start differentiation of unguided neural organoids (UNOs), detach the iPSCs with 0.5 mM EDTA followed by Accutase (cell detachment solution with proteolytic and collagenolytic enzymes).

4. Resuspend detached iPSCs in HES0 medium supplemented with 50 μM Y-27632 and 4 ng/mL basic fibroblast growth factor (bFGF).
5. Count the cells and seed 3.5 million iPSCs in one well of an Aggrewell 800 plate. This is considered day 0.
6. On day 2, manually select embryoid bodies (EBs) based on size and cellular density. Transfer selected EBs to an ultra-low attachment 96-well plate in HES0 medium with Y-27632 and bFGF.
7. On day 4, change medium to HES0 without Y-27632 and bFGF to facilitate differentiation.
8. From day 6 to day 12, the formation of neuroectoderm is facilitated by changing the medium to Neural Induction Medium.
9. On day 13, manually select UNOs based on the formation of neuroectoderm and embed each organoid in 30 μl Matrigel. Culture 16 organoids in a 60 mm dish in differentiation medium without vitamin A until day 17.
10. From day 17 to day 120, culture UNOs in differentiation medium with vitamin A on a belly dancer (speed 3). As UNOs grow, decrease the number of UNOs to 8 per dish.

3 Immunohistochemistry and OTO Protocol for Scanning Electron Microscopy

3.1 Materials

Organoid Fixation and Sectioning

1. Glass vial, 50 mL.
2. Forceps, sterile.
3. Spatulas, sterile.
4. Weighing boats.
5. Cylinder, 1 L.
6. Glass bottles, 1 L.
7. Stir/Heat plate.
8. Filter paper.
9. Delicate-task wipers.
10. Microwave.
11. Six-well plate, sterile (or a small Petri dish).
12. Vibratome.
13. Ice packs.
14. Razor blades, sterile.

15. Fast Glue.
16. ***Agarose**, 4% (Molecular biology grade, low melting temperature)*: Add 4 g of agarose powder (FroggaBio, A87-500G) in 100 mL of phosphate saline buffer (PBS). Heat up using a microwave until the solution becomes clear. Let the mixture cool down to skin temperature before use.
17. ***Phosphate buffer** (PB), 0.1 M, pH 7.4*: for 2 L, dissolve 23.48 g sodium phosphate dibasic anhydrous (Na_2HPO_4; Fisher Scientific, S374500) and 4.8 g of sodium phosphate monobasic monohydrate ($NaH_2PO_4*H_2O$; Fisher Scientific, S369500) in 1.9 L of double-distilled (dd)H_2O. Adjust the pH with HCl/NaOH and top up to 2 L with ddH_2O.
18. ***Phosphate buffer saline** (PBS), 0.05 M, pH 7.4*: for 1 L, dissolve 5.87 g of Na_2HPO_4, 1.2 g NaH_2PO_4, and 9 g of sodium chloride (NaCl; Fisher Scientific, S2713) in 800 mL of ddH_2O. Adjust the pH with HCl/NaOH and top up to 1 L with ddH_2O.
19. ***Paraformaldehyde** (PFA), 2%*: for 1 L, dissolve 20 g of PFA (EMS, 19210) in 1 L of PB heated to a temperature of 55°C–60°C (maximum) in a fume hood. Filter the room-temperature solution using filter paper (*see* **Note 1**).
20. **Fixative solution**—***PFA**, 2%/**Glutaraldehyde** (GLUT, EM grade), 2%*: for 1 L, gradually add 40 mL of 50% GLUT (EMS, 16320) to the 2% PFA solution and gently swirl the glass bottle to facilitate dissolving. Handle in a fume hood (*see* **Note 2**).
21. ***Cryoprotectant solution***: for 1 L, mix 400 mL of PBS, 300 mL of glycerol (Fisher Scientific, BP229-1), and 300 mL of ethylene glycol (Avantor Sciences, 0852-NBB-20L).

Organoid Post-Fixation and Immunohistochemical Staining for IBA1 or GFAP

1. 24-well plate, sterile.
2. Plastic pipette, sterile.
3. Fine-tipped plastic pipette, sterile.
4. Orbital shaker.
5. **PB**, 0.1 M, pH 7.4.
6. **PBS**, 0.05 M, pH 7.4.
7. ***PBS with Triton x100** (PBS-T), 0.01%*: add Triton ×100 (Sigma-Aldrich, T8787) to PBS.
8. ***Tris Base** (TB), 0.05 M, pH 8.0*: for 1 L, add 60.55 g of Tris Base Buffer (Trizma® Base C4H11NO3; Fisher Scientific, BP152-1) to 900 mL of ddH_2O. Adjust the pH with HCl/NaOH and top up to 1 L with ddH_2O.
9. ***Tris-buffered saline** (TBS), 0.05 M, pH 7.4*: for 1 L, add 100 mL of TB and 8.7 g NaCl to 900 mL of ddH_2O.

10. ***PFA**, 4%:* for 1 L, dissolve 40 g of PFA (EMS, 19210) in 1 L of PB heated to a temperature of 55 °C–60 °C (maximum) in a fume hood. Filter the room-temperature solution using filter paper.
11. **Fixative solution—*PFA***, *4%/**GLUT**, 0.6% (EM grade)*: for 1 L, gradually add 12 mL of 50% GLUT (EMS, 16320) to the 4% PFA solution and gently swirl the glass bottle to facilitate dissolving. Handle in a fume hood.
12. ***Hydrogen peroxide** (H_2O_2), 0.3%*: dilute H_2O_2 (Fischer Scientific, H325) in PBS. Prepare fresh.
13. ***Sodium borohydride** ($NaBH_4$), 0.1%*: dissolve $NaBH_4$ (Biochemika, 71997) in PBS. Prepare fresh in a fume hood (*see* **Note 3**).
14. ***Blocking buffers***:

IBA1 protocol	*10% Fetal bovine serum (FBS)*	*3% BSA*	*0.01% Triton ×100*	*Dilute in TBS*
GFAP protocol	*5% Goat serum (GS)*	*3% BSA*	*0.03% Triton ×100*	*Dilute in PBS*

15. **Antibody (Ab) solutions**:

Type	Species	Company	Concentration	Diluent
IBA1 primary Ab	Rabbit (polyclonal)	FUJIFILM WAKO Chemical (019-19741)	1:1000	Blocking buffer
GFAP primary Ab	Mouse (monoclonal)	Cell Signaling (3670X)	1:400 OR 1:1000 (*see* **Note 4**)	Blocking buffer
IBA1 secondary Ab	Biotinylated goat anti-rabbit	Jackson ImmunoResearch (111-066-046)	1:300	TBS
GFAP secondary Ab	Biotinylated goat anti-mouse	Jackson ImmunoResearch (115-065-003)	1:300	PBS-T

16. ***ABC Complex***: dilute component A (1:100) and B (1:100) (Vectastain, VECTPK6100) in either TBS (**IBA1** protocol) or PBS-T (**GFAP** protocol).
17. ***3,3′-Diaminobenzidine** (DAB), 0.05%*: dissolve 1 DAB pellet (Sigma-Aldrich, D5905) in 20 mL of TB. Immediately before use, add 1 μL of H_2O_2 per 2 mL of solution to activate the DAB. Handle in a fume hood (*see* **Note 5**).

OTO Protocol and Resin Embedding

1. 24-well plate, sterile.
2. Plastic pipette, sterile.
3. Fine-tipped plastic pipette, sterile.

4. Glass pipette, sterile.
5. Syringe, 20 mL.
6. Syringe filter, 0.22/0.45 μm pore diameter (VWR, 76479-034/76479-032).
7. Glass vials, 50 mL.
8. Falcons, 20 mL.
9. Glass beaker, 50 mL.
10. Fine-tipped brush.
11. Weighing boats.
12. Aluminum boats.
13. Plastic cup, 50 mL.
14. Oven.
15. Fluoropolymer sheets (ACLAR Film; EMS, 50425-25).
16. Slide warmer.
17. ddH_2O.
18. **PB**, 0.1 M, pH 7.4.
19. **Ethanol**, 35%, 50%, 70%, 80%, 90%, 100%.
20. ***Propylene oxide**, 100%* (Sigma-Aldrich, 110205-18L-C).
21. ***Potassium ferrocyanide** solution, 3%*: dissolve 0.03 g of potassium ferrocyanide (Sigma-Aldrich, P9387) in 1 mL of PB. Handle in the fume hood.
22. ***Potassium ferrocyanide**, 3%/**Osmium tetroxide**, 2%*: mix equal volumes of 3% potassium ferrocyanide solution in PB and 4% osmium tetroxide (EMS, 19190). Handle in the fume hood (*see* **Note 6**).
23. ***Thiocarbohydrazide** (TCH) solution*: dissolve 0.1 g of TCH powder (Sigma-Aldrich, 223220) in 10 mL of ddH_2O and gently shake. Place the solution in the oven at 55°C for 1 h. Agitate by gentle swirling of the solution container every 10 min to facilitate dissolving. After resting at room temperature (RT) for 10–15 min, filter the solution using a syringe and a Millipore filter (0.22/0.45 μm diameter) to remove any precipitates. Handle in the fume hood (*see* **Note 7**).
24. ***Osmium tetroxide**, 2%*: dilute an equal volume of 4% osmium tetroxide (stock) and ddH_2O. Handle in the fume hood.
25. ***Durcupan resin***: in a plastic cup, mix resin components (Sigma-Aldrich, 44610) in the following order and volume: 20 g A + 20 g B + 0.6 g C + 0.4 g D. Stir and let rest for at least 30 min before use. Handle in the fume hood (*see* **Note 8**).

Block Mounting, Ultramicrotomy, and Scanning Electron Microscopy

1. Resin blocks.
2. Razor blades (EM grade), sterile.
3. Super glue.
4. Eyelash brush.
5. Duster.
6. ddH_2O.
7. Syringe, 20 mL.
8. Syringe filter, 0.22/0.45 µm pore diameter (VWR, 76479-034/76479-032).
9. Bibulous paper.
10. Ultra-fine tweezers.
11. Stereoscope.
12. Ultramicrotome (Leica ARTOS 3D).
13. Diamond knife.
14. Xylene, 100%.
15. Silicon wafers (Ted Pella, 16008).
16. Conductive carbon adhesive disk tape (EMS, 77825-06).
17. Specimen mount (Ted Pella, 16111/16119).
18. Scanning electron microscope (Zeiss Crossbeam 350, software Zeiss Atlas 5 (Fibics)).

3.2 Protocol

Following the protocol in [50] with further modifications:

Organoid Fixation and Sectioning

1. Using the forceps, place the organoid in a glass vial with the fixative solution (*PFA, 2%/GLUT, 2% in PB*). Incubation time depends on the size of the organoid: calculate an approximate ratio of 1 h per 1 mm of tissue at 4°C.
2. Rinse the sample three times for 10 min with PBS.
3. Dry the excess liquid off the organoid using delicate-task wipers.
4. Using the 6-well plate as a mold, pour a layer of liquid agarose.
5. Before the agarose mixture solidifies, place the organoid in the center and completely cover the tissue with another layer of agarose (*see* **Note 9**).
6. Let cool down at 4°C until the agarose block solidifies.
7. Using a spatula, remove the agarose block with a sample from the well-plate and carefully trim excessive agarose with a razor blade.
8. Mount the agarose block on the vibratome and proceed with sectioning at 60 µm (at a moderate speed) in ice-cold PBS.

9. Collect sections into a clean glass vial (or a well-plate) filled with cryoprotectant solution and store at −20°C.

Organoid Post-Fixation and Immunohistochemical Staining

1. Rinse the sample three times for 10 min with PB on an orbital shaker at RT.
2. Perform a secondary fixation with fixative solution (*PFA, 4%/GLUT, 0.6% in PB*) for 1 h.
3. Rinse the sample three times for 10 min with PBS.
4. Quench with H_2O_2 in PBS, 0.3%, for 5 min.
5. Rinse the sample three times for 10 min with PBS.
6. Incubate with $NaBH_4$ in PBS, 0.1%, for 30 min.
7. Rinse the sample three times for 10 min with PBS.
8. Incubate with the IBA1- or GFAP-specific blocking buffer for 1 h.
9. Incubate with the IBA1- or GFAP-specific primary antibody diluted in its respective blocking buffer on a shaker at 4°C overnight.
10. On the next day, let the sample rest at RT for 15 min.
11. Rinse the sample three times for 10 min with TBS (**IBA1** protocol) or PBS-T (**GFAP** protocol).
12. Incubate with antibody-specific secondary antibody on a shaker at RT for either 1.5 h (**IBA** protocol) or 2 h (**GFAP** protocol).
13. Rinse the sample three times for 10 min with TBS (**IBA1** protocol) or PBS-T (**GFAP** protocol).
14. Incubate with ABC complex diluted in TBS (**IBA1** protocol) or PBS-T (**GFAP** protocol).
15. Rinse the sample three times for 10 min with TBS (**IBA1** protocol) or PBS-T (**GFAP** protocol).
16. Incubate with DAB solution in TB, 0.05%, until brown staining is visible (*see* **Note 10**).
17. Immediately after the incubation, wash the sample rapidly two times in PB to deactivate DAB.
18. Confirm the staining using a light microscope.
19. Rinse the sample three times for 10 min with PB.
20. Store the sample in PB on a shaker and at 4°C overnight or continue with the next steps.

OTO Protocol and Resin Embedding

1. Let the sample rest at RT for 15 min.
2. Incubate with freshly prepared potassium ferrocyanide in PB, 3%/osmium tetroxide, 2%, solution for 1 h. *All the following*

procedures need to be performed in a fume hood. For Steps 2–9, protect the samples from prolonged light exposure.

3. During the incubation in Step 2, prepare the TCH solution in ddH_2O.
4. Rinse the sample in the following order: one time for 10 min with PB, one time for 10 min with PB:ddH_2O (ratio 1:1), and one time for 10 min with ddH_2O. This course of action protects the sample from an osmotic shock.
5. Incubate with filtered TCH for 20 min.
6. Rinse the sample three times for 5 min with ddH_2O.
7. Incubate with osmium solution in ddH_2O, 2%, for 30 min.
8. During the incubation in Step 7, prepare the Durcupan resin mixture for embedding.
9. Rinse the sample three times for 5 min with ddH_2O.
10. Using a fine-tipped brush, gently transfer the sample from the plastic well plate to a glass vial with ddH_2O (*see* **Note 11**).
11. Dehydrate the sample in ethanol solutions of increasing concentrations for 5 min each in the following order: 35%, 35%, 50%, 70%, 80%, 90%, 100%, 100%, 100%. *After dehydration, samples will become very fragile for handling.*
12. Immerse the sample three times for 5 min in propylene oxide (*see* **Note 12**).
13. In aluminum boats, carefully mix propylene oxide with resin at increasing ratios and incubate in the following order: 70:30, 50:50, 30:70, and 10:90. Incubate the sample for ~30 min in each mixture (*see* **Note 13**).
14. Using a brush, carefully transfer the sample to resin (with no propylene oxide) and let it rest overnight.
15. On the next day, prepare a pair of the same-sized ACLAR sheets that will be used for sandwich-like embedding of the sample. Using a permanent marker, add sample information on the outer sides of each sheet.
16. On a slide warmer set at ~55°C, spread a thin layer of resin evenly on the inner side of the bottom sheet.
17. Using a brush, carefully transfer the tissue to the sheet. Avoid creating bubbles with the brush and leave space between sections.
18. Carefully place the top sheet on the bottom sheet, starting from the midline and towards the sides. Apply pressure evenly and slowly to avoid the formation of bubbles (*see* **Note 14).**
19. Place the ACLAR sheet on a flat surface, cover it with evenly distributed small weights and transfer the sheet to the oven for 5 days at 55°C.

Block Mounting, Ultramicrotomy, and Scanning Electron Microscopy

1. Remove the ACLAR sheet from the oven and carefully open it.
2. Using a binocular microscope and a razor, cut out the region of interest (ROI).
3. Glue the ROI on top of the resin block. The sample needs to lie flat.
4. Transfer the blocks to the oven (55°C) to allow the glue to solidify for a minimum of 2 h. Avoid leaving the blocks in the oven for more than 24 h.
5. Mount the block with the sample on the ultramicrotome.
6. Using a razor, trim the resin on top of the block in a pyramid-like fashion with the ROI in the center. Then, using a new razor, trim the tissue to achieve a trapezoid shape (*see* **Note 15**).
7. Fill the boat of a diamond knife with ddH_2O and move the knife into the proximity of the vertically mounted sample.
8. Carefully adjust the angle of the knife and of the block to achieve a parallel positioning.
9. Proceed with sectioning at the semi-thin side of the diamond knife (thickness: 300 nm, speed 1 μm/s).
10. Upon reaching the tissue, carefully clean the boat and the knife, and switch to the ultra-thin side of the diamond knife. Proceed with sectioning (thickness: 75 – 90 nm, speed 0.8 μm/s).
11. Add a small volume of xylene on a strip of bibulous paper and briefly come into the proximity of the boat surface (*see* **Note 16**).
12. With ultra-fine tweezers, submerge the silicon wafer into the boat and, using an eyelash brush, gently guide the sections on its reflective side.
13. Let the wafers dry in an enclosed space to avoid the accumulation of dust particles.
14. On a metallic specimen mount, place a conductive carbon adhesive disk tape and then mount the dry wafer.
15. Insert the specimen mount into a scanning electron microscope and proceed with the imaging setup.

4 Results and Discussion

This study provides a protocol for immunohistochemistry-EM processing of UNOs, which can be used to obtain insights into the ultrastructural profile of UNOs and characterization of their glial cell populations. The UNOs, cultured for a period of 120 days, were derived from iPSCs of two AD patients with an *APOE4/4* genotype (MV 8.4; MV 10.4, respectively) and their isogenic *APOE3/3* control lines (MV 8.1; MV 10.1, respectively) generated using the CRISPR/Cas9 methodology [19]. In view of their in vitro origin and the inherent sensitivity of glial cells, particularly microglia, to cues from their microenvironment [52], we employ the terminology of *organoid-derived* radial glia cells (*oRGC*), -astrocytes (*oASTRO*), and -microglia (*oMG*). On this basis, we want to recognize that while organoid-derived cells bear a striking similarity in several features, ranging from morphology to transcriptome to tissue glial cells observed in vivo and/or in situ [44, 45, 53–55], they may also attain features of purely organoid-specific origin per our ultrastructural observations. These differences can be driven by the source cell line; type, level, and length of their access to nutrients in the culture media; the size and distribution of the cellular population within the organoid; cell-to-cell interactions; as well as variable metabolic and functional demands within the microenvironment.

UNOs investigated in this study are cultured based on a protocol provided by Ormel and colleagues [45], which utilizes the Lancaster method [44, 45, 53]. Through the lack of patterning factors, this approach permits concurrent development of neuroectodermal lineage cells and mesenchymal cells like microglia (Fig. 1). Absence of

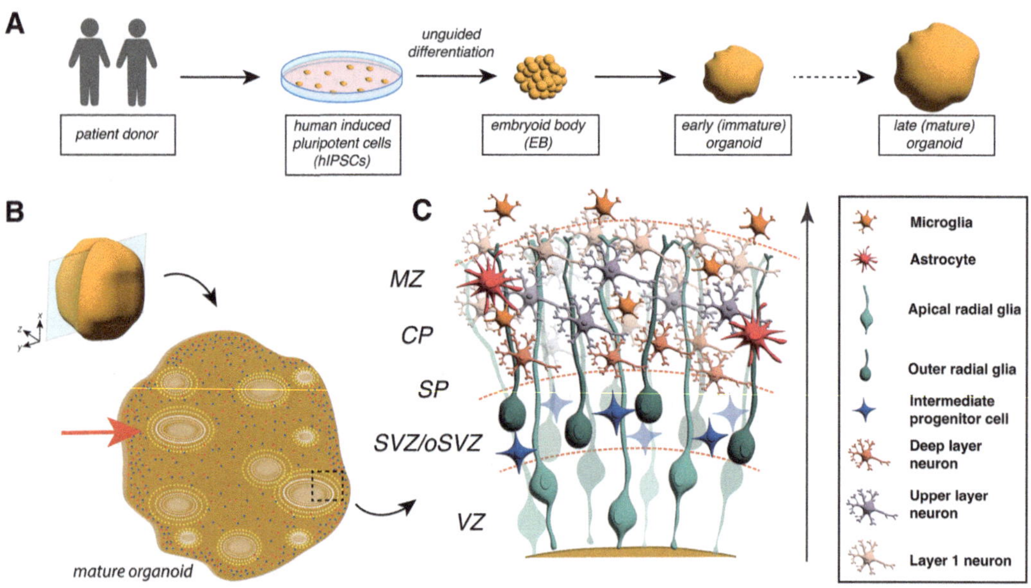

Fig. 1 Generation of an unguided neural organoid (UNO). (**A**) Schematic illustration of individual steps in preparation of a mature UNO. (**B**) A simplified illustrative representation of a mature neural rosette of a neural organoid with layer-specific cell populations of neuroectoderm lineage (as reviewed before by [59]). The figure has been modified to include putative position of organoid-derived microglia (*oMG*) based on previous immunohistological findings [45] and current ultrastructural observations. *VZ ventricular zone*, *SVZ* subventricular zone, *oSVZ* outer subventricular zone, *SP* subplate, *CP* cortical plate, *MZ* marginal zone

immune cells of mesenchymal origin within 3D brain organoids has been a limitation in organoid models, given the crucial role of these cells in CNS development, maintenance and disease [56]. Therefore, several protocols exist to add microglia to organoids. Subsequent integration of human iPSC-derived microglia into midbrain organoids was shown to enhance neuronal maturation and excitability and impact synaptic remodeling [57]. Other method for microglial inclusions included a coculture of neural organoids with human iPSC-derived primitive macrophages, partially mimicking the first wave of microglial seeding of embryonic brain [58]. This study further highlights the importance of macrophages/microglia for organoid neural progenitor cell maturation through cholesterol transfer, underscoring the need to investigate the underlying metabolic processes that regulate cellular growth, interactions and region differentiation in UNOs. These strategies offer distinct advantages, yet are also not without limitations from a translational standpoint—which, however, can also be addressed through a deeper understanding of their cellular and subcellular properties.

While the field has advanced in defining the molecular signature of UNOs, the qualitative foundation for the ultrastructure of either neuronal or non-neuronal cells or the spatial ultrastructural profile of the organoid brain-mimicking regions remains lacking. This prevents further advances in any quantitative measurements. Previous studies provided early ultrastructural descriptions of a variety of 3D neural organoid and spheroid models, and in some cases of organoid-derived glial cells [60–63]. The cellular identification was, however, performed without the use of immunohistochemical markers, leaning on ultrastructural features standardized in rodent and human brain tissue in situ (for a review see [64]). While immunofluorescent marker-based spatial profiling offers deep insight, the resourcefulness of these data for the recognition of cell types in organoids at the level of ultrastructure is limited. An exception would be the utilization of combined techniques such as correlative light/fluorescence and electron microscopy (CLEM/CFEM), which would allow explorations of specific cells. This consideration accounts for various limitations at the level of ultrastructure. In comparison to the highly organized and compartmentalized "typical" human or animal brain cytoarchitecture, the spatial composition of UNOs—given their unguided self-organization—lacks cohesive character and exhibits an area-specific type and degree of organization. In a single plane, whole UNO ultrastructure scanning shows (a) multiple systematically arranged neural tube-like areas (neural rosettes) (Figs. 1 and 2), (b) compact cell body-dense areas with high levels of cell-to-cell contacts; and (c) sparsely cell-populated areas displaying minimum to no cell-to-cell contacts (containing particularly glial cell bodies). These areas can be differently sized and shaped, with margins not clearly delineated and populated by varying cell type clusters.

In part, this disorderly organization reflects UNO's recapitulation of developmental periods during which brain circuits and regions undergo significant remodeling. In parallel, the presence of cell populations at a varying level of maturation/differentiation across different UNO regions likely influences this aspect. Secondly, with UNO cells having the potential to develop specific organoid-induced features, standardized ultrastructural cell features predominantly founded on preclinical animal data may not be sufficient for cell characterization in these human-derived UNOs.

5 Neural Rosette and *oRGC* Ultrastructure in Human AD Patient iPSC-Derived UNOs

On a 25 nm scale, neural rosettes occurring in a variety of shapes and sizes are undeniably prominent cytoarchitectural hallmarks of UNO ultrastructure. Characterized by a distinctive radial organization of cells around a centralized lumen, these neuroepithelial structures were proposed to closely resemble the neural tube to ventricle-like structure formation occurring in the early brain development and were described in a range of neural organoids [54, 59, 61–63, 65, 66]. These pseudostratified zones are composed of a diverse cellular population resembling neuroepithelial/neural stem cells (NSCs) and—at later stages of rosette maturation—radial glia cells (RGCs) of the developing cortex. Presumably, rosettes act as organizational centers and multipotent progenitor pools for neuroectodermal lineage cell differentiation and maturation of the organoid. Quantitative live imaging showed rosette maturation stage and size may be some of the contributing factors to their capability to adopt a radial organization and maintain NSC pool, with progressive loss of rosette integrity and an elevated number of differentiated neurons at later stages of in vitro cultivation [67]. As observed with immunohistochemical methods, depending on their stage of maturation, neural rosettes in neural organoids can generate distinct cell layers that mimic the cytoarchitecture of specific brain regions (for a review see [59]) (Fig. 1B), such as cortex, hippocampus and/or choroid plexus [54, 65, 67]. As compared to spatial immunofluorescent profiling, this layered cytoarchitecture can only be visualized partially at the level of ultrastructure and without the use of specific immunohistochemical markers. Based on our qualitative observations, GFAP-immunopositive (+) and IBA1+ glial cells tend to accumulate outside of rosettes and towards the outer edges of the UNO (*data not shown*). These glial cells are interspersed with neuronal cells at various stages of maturation, as well as cells showing ultrastructural features of cellular stress or apoptosis. Compared to fixed human or rodent post-mortem tissue, this cellular heterogeneity prevents accurate ultrastructural identification of glial cells solely on spatial location and without immunohistochemical labeling. Another crucial limitation for the ultrastructural analysis of UNOs using EM is the loss of tissue during processing. The combination of frequent mechanical handling and multiple dehydration steps can be detrimental to fragile organoid sections, causing dissolution of the edge areas (abundant in mature glia cells) and cell shrinkage. While PFA-based fixation is generally suitable for Matrigel-embedded cells for microscopy, the penetration of this gel-like extracellular matrix may depend on its density and composition [68, 69], producing variable results in organoid preparations. Furthermore, the absence of vasculature causes a growing organoid to become necrotic at the core after a prolonged time in culture, which alters the molecular landscape and further affects the integrity of the organoid. Performing a second round of aldehyde fixation, minimizing the tissue handling (e.g., by performing the staining on glass slides instead of well plates), and preserving the agarose enclosure can significantly improve the quality and quantity of the organoid sample during EM processing.

In early human development, a single neural tube forms from a neural plate and later differentiates into brain and spinal cord. In comparison, 3D neural organoids possess multiple of these organizational center-like regions capable of producing new neuronal and glial cells [46, 62, 65]. Lacking a centralized regulatory entity, these rosettes likely operate independently. Their localization within the organoid impacts nutrient access, which can affect proliferation, differentiation, maturation, and functional specialization of associated cells. It is still unclear whether these structures affect each other, especially, whether they interfere with the development of nearby regions [54]. Interestingly, a study performed a 3D reconstruction of an entire forebrain organoid, showing its neural rosettes were not necessarily single entities—as a single-plane view would suggest—but formed a complex tubular structure of lumen-interconnected compartments [46]. In the context of UNOs with unregulated growth and maturation, developing regions may also be variably represented along their dorso-ventral axis [46]. When designing the individual cell sampling approach for a quantitative analysis of EM images, these possibilities need to be carefully considered. To bypass the morphological variability associated with the presence of multiple rosettes at a different maturation stage, size and shape, different strategies have been utilized, including the development of a 3D organoid from a 2D monolayer of pluripotent stem cells, which produced a self-organizing cortical organoid with a single rosette [62].

In addition to their variability in shape and size, the ultrastructure of individual neural rosettes also exhibits substantial variation across individual UNOs and imaged 2D planes, as reflected in the following examples (Fig. 2). The ventricular zone (VZ) surrounding the lumen of an example rosette in the UNO of patient 1 isogenic control (MV 8.1) shows high cell heterogeneity, with electron density of cytoplasm and nucleoplasm ranging from low (light) to high (dark gray) between

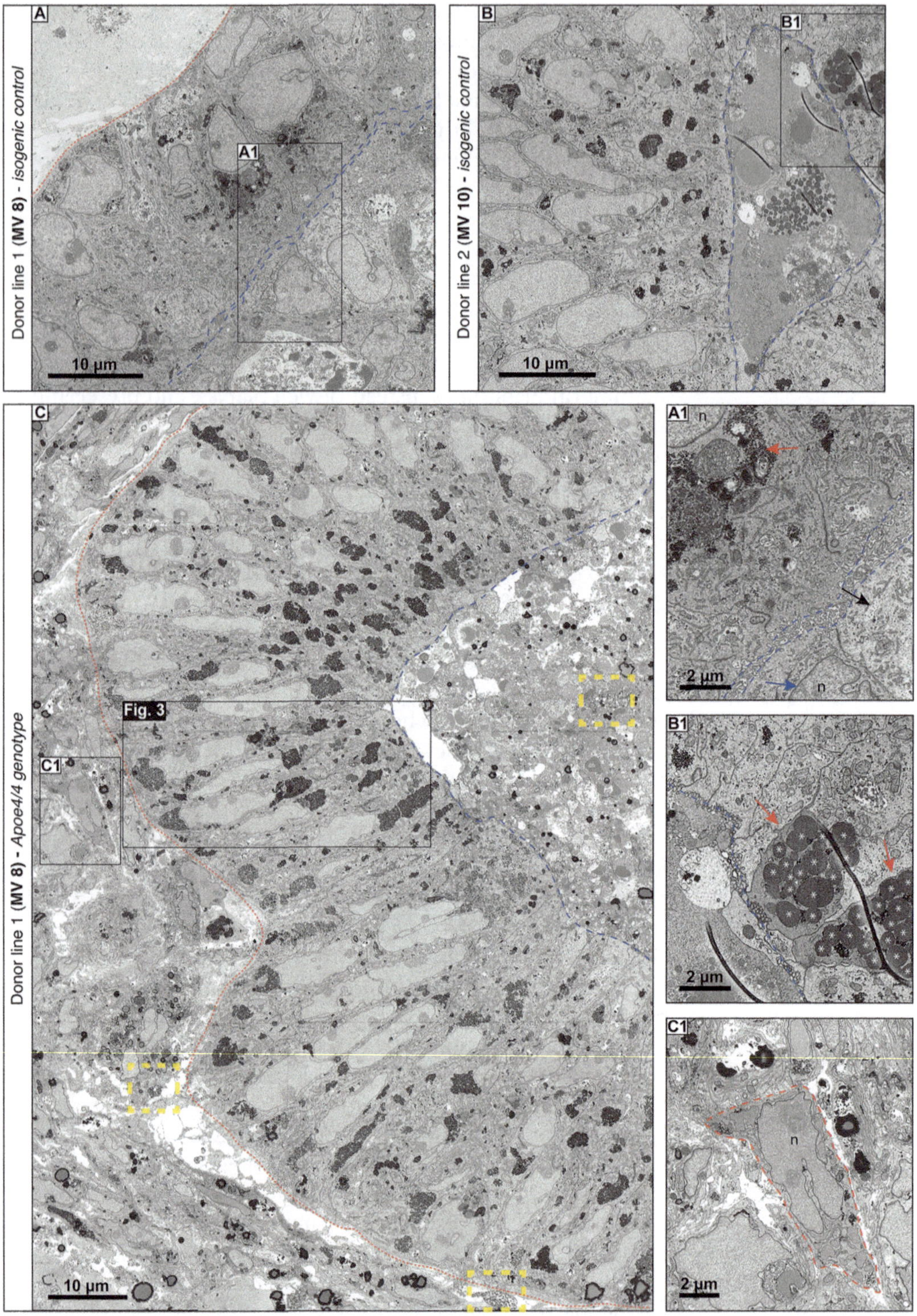

Fig. 2 Ultrastructural profile of a neural rosette in human patient iPSC-derived unguided neural organoid (UNO) using scanning electron microscopy. The cellular composition of a neural rosette consisting of neural stem-like cells (neuroepithelial cells and radial glia cells) in (**A**) *APOE3/3* isogenic control derived from patient

cells (Fig. 2A, A1). The example rosette VZ in the UNO of patient 2 isogenic control (MV 10.1) is more homogenous, with only a sparse presence of such electron-dense cells. In this case, however, darker cells were notably rich in round, electron-dense subcellular structures reminiscent of phagosomes/lysosomes (Fig. 2B, B1), and their density was higher outside of the immediate VZ. In terms of cell electron density, the VZ profile of an example bigger-sized rosette in UNO of patient 1 with *APOE4/4* genotype is more homogenous, lacking darker cells (Fig. 2C). Whether these observations denote a representation of various *oRGC* populations in UNOs; rosette maturation/size stage; varying functional and/or metabolic output within a same cell population; an EM artifact; or are associated with a cell line of origin, remains to be investigated. In the future, 3D EM reconstruction of a bigger sample of similarly sized rosettes per organoid would be necessary to determine any genotype impacts. Previously, there were reports of "dark" and "light" cells (based on their electron density) within the lumen-surrounding VZ in rosettes of choroid plexus-like neural organoids [60, 70]. These cells were found to differentially express genes associated with mitochondrial activity, ATP synthesis and cilium assembly, and with their ultrastructure lacking ciliary features. Higher electron density of the cellular cytoplasm/nucleoplasm was also associated with cellular stress and altered functional capability of affected glial cells (e.g., on the level of immune response, cell–cell interaction, or phagocytosis) in conditions of cellular stress, aging, and pathology [50, 71]. Additionally, a recent publication investigating human cerebral organoids using TEM identified mature neuron cells localized in the vicinity of the VZ, which also displayed a high electron opacity [63]. The determining criteria for cellular classification included a large nucleus with indentations, electron-dense heterochromatin lining the inner nuclear membrane, an electron-dense nucleolus, numerous Golgi apparatus (GA) and mitochondria (as described previously in [64]). While we also observe dark cells closely to VZ (Fig. 2C1), both with nuclear indentations and similar heterochromatin pattern, their neuronal character is not confirmed. Interestingly, these cells are prone to accumulate large quantities of yet unidentified electron-dense single-membrane organelles, partially reminiscent of altered/swollen endoplasmic reticulum (ER) compartments. Beyond the ventricle-like niche, we also see

Fig. 2 Continued donor cell line (MV 8), (**B**) *APOE3/3* isogenic control derived from patient donor cell line (MV 10), and (**C**) a patient donor cell line with *APOE4/4* genotype (MV 8). Dashed blue lines mark the apical surface of the lumen and dashed red lines delineate the basal membrane of this ventricle-like structure. Dashed yellow squares denote the presence of the IBA1 immunostaining in the vicinity of the neural rosette. (**A1**) Inset points out the heterogeneity of cells with respect to the electron density of their cytoplasm—dark (red arrow), intermediate (blue arrow), or light (**black** arrow)—in the ventricular zone of the neural rosette from **A**. (**B1**) Inset shows the sporadic presence of electron-dense (dark), phagosome/lysosome-rich cells in the ventricular zone of the neural rosette from **B**. Lysosome/phagosome structures are highlighted with a white asterisk (*). (**C1**) Inset shows an electron-dense (dark) cell (dashed red line delineates the cytoplasm) with indented nuclei and located near basal end of ventricular zone of the neural rosette from **C**. *n* nucleus

a range of electron density differences of cytoplasm/nucleoplasm among the glial cells. Whether this is a histological artifact or implies an existing state/subtype heterogeneity within their population remains an open question.

In both *APOE4/4* and isogenic control *APOE3/3* UNOs, *oRGCs* display an elongated shape of cytoplasm and nucleus, with the latter displaying several invaginations of the nuclear membrane and none to multiple circular nucleoli (Figs. 2A–C and 3.A1). Lumen-associated *oRGCs* feature larger quantities of free and aggregated electron-dense glycogen granules (Fig. 3.A2, A3). These glycogen aggregates are without a visible membranous enclosure and frequently contact lipid droplets with a less electron-dense core. Membranous round structures laden with dense gray material and a range of vesicles were identified as endo-phagolysosomal compartments with content in various stages of degradation (Fig. 3.A4, A5). A preliminary qualitative observation of the *APOE4/4* UNOs and its isogenic control (patient 1, MV 8) indicated a potential for increased glycogen load associated with the AD risk genotype. As glycogen synthase kinase 3 (GSK3) was previously found to regulate Aβ-mediated tau phosphorylation in a 3D human neural cell culture [72], it could suggest a possible role for glycogen pathway alteration in AD UNOs. However, in contrast, we observed high glycogen deposits also within the rosette of the other isogenic control (patient 2, MV 10). Rather than a genotype impact, this indicates baseline donor line-specific differences in glycogen concentration. Additionally, the rosette maturation stage may also impact this aspect, highlighting the importance of selective rosette sampling within an organoid. Although not systematically investigated, UNOs derived from patient 2 tend to develop fluid-filled cysts that conceptually resemble choroid plexus. Glycogen-rich columnar epithelial cells were described as one of the stages of the lateral ventricular choroid plexus development originating from modified neuroepithelium in human embryos; with glycogen declining towards the later stages when cells acquire a cuboidal to squamous shape and nuclei become centralized [60, 73]. In this manner, glycogen, as a primary source of energy, could serve as a tentative histological marker of an organoid-derived rosette maturation and metabolic status; however, this hypothesis will require further experiments that combine imaging modalities, -omics, and molecular approaches across 3D organoid models. Following this line of thought, *oRGCs* in UNOs (regardless of the genotype) possess numerous mitochondrial organelles (Fig. 3.A5) with variably visible cristae and of different shapes—typical bean-shaped, elongated, round, and donut-shaped—potentially indicating varying demands for ATP production and supply in these cells. These cells also display one to two densely layered narrow stretches of GA, and multiple short, narrow, or swollen vesicles of ER cisternae (Fig. 3.A6). Atypically shaped mitochondria (e.g., donut-shaped or elongated) and swollen ER or GA are generally considered markers of cellular stress [74].

Prominent structural and morphological features of *oRGCs* include their apico-basal polarity and ciliated apical/luminal surface (Figs. 2 and 3). In a human embryo, the macrostructure of the NSC-comprising VZ is characterized as pseudostratified, meaning that while individual nuclei seem to be organized in a multiple layer-like fashion, the cytoplasm of each cell extends to both the apical and basal ends of the VZ, forming a single layer [67]. The biologically functional relevance of this cellular organization lies in providing a progenitor mitotic pool and an anatomical scaffolding for an apical-basal migration of nuclei in the process of further differentiation and maturation [67]. As most mitochondria and glycogen deposits localize at the luminal end of *oRGC*, this may indicate a locally increased metabolic demand in the present example of a neural rosette (Figs. 2 and 3). The luminal surface of *oRGCs* is abundantly populated with villi and primary cilia boasting a microtubule-based cytoskeleton (Fig. 3A, B). We also identify centrioles, organizing centers for microtubule bundles of primary cilia (Fig. 3.A7). The presence of primary cilia was found to influence cortical development, allowing apical radial glia to maintain their cell shape and sense embryonic cerebrospinal fluid cues [75]. In later stages of human development, motile cilia of ventricular ependymal cells participate in the modulation of cerebrospinal fluid flow [76]. Ciliary signaling axis (Hedgehog signaling) was reported to be involved in the determination of the dorsal/ventral regional fate in a human iPSC-derived neural organoid [77]. While the ciliated character of *oRGCs* was reported in choroid plexus-forming organoid [60] or colorectal cancer organoid [78], the cells forming rosettes in human-derived neural organoids (as shown by [63]) do not seem to develop this attribute. This difference can stem from organoid culture length, rosette maturation stage [63], culture methodology, cell line source, and EM processing. Consistent with previous research, we observe clathrin-coated pits at the luminal surface (Fig. 3C), which are small, round, membranous cytoplasmic membrane invaginations involved in endocytosis [63, 79]. We note a junction-based network connecting neighbor *oRGCs* within the apical zone comprising tight junctions (TJ), and more distally positioned adherens junctions (AJ) and desmosomes (Fig. 3C, A8), partially in line with previous observations [63]. Additionally, *oRGCs* formed frequent physical cell-to-cell contacts through finger-like shaped cytoskeletal protrusions into the membrane of the neighbor cell present close to the basal (Fig. 3.A9) or apical end of the cell (Fig. 3.A10). Calcium-activated production of filopodia that penetrate neighboring cells was described previously in epithelial cells as a physical force underlying the catalysis of AJ formation and the formation of a continuous cell layer [80]. Additionally, their potential contribution to

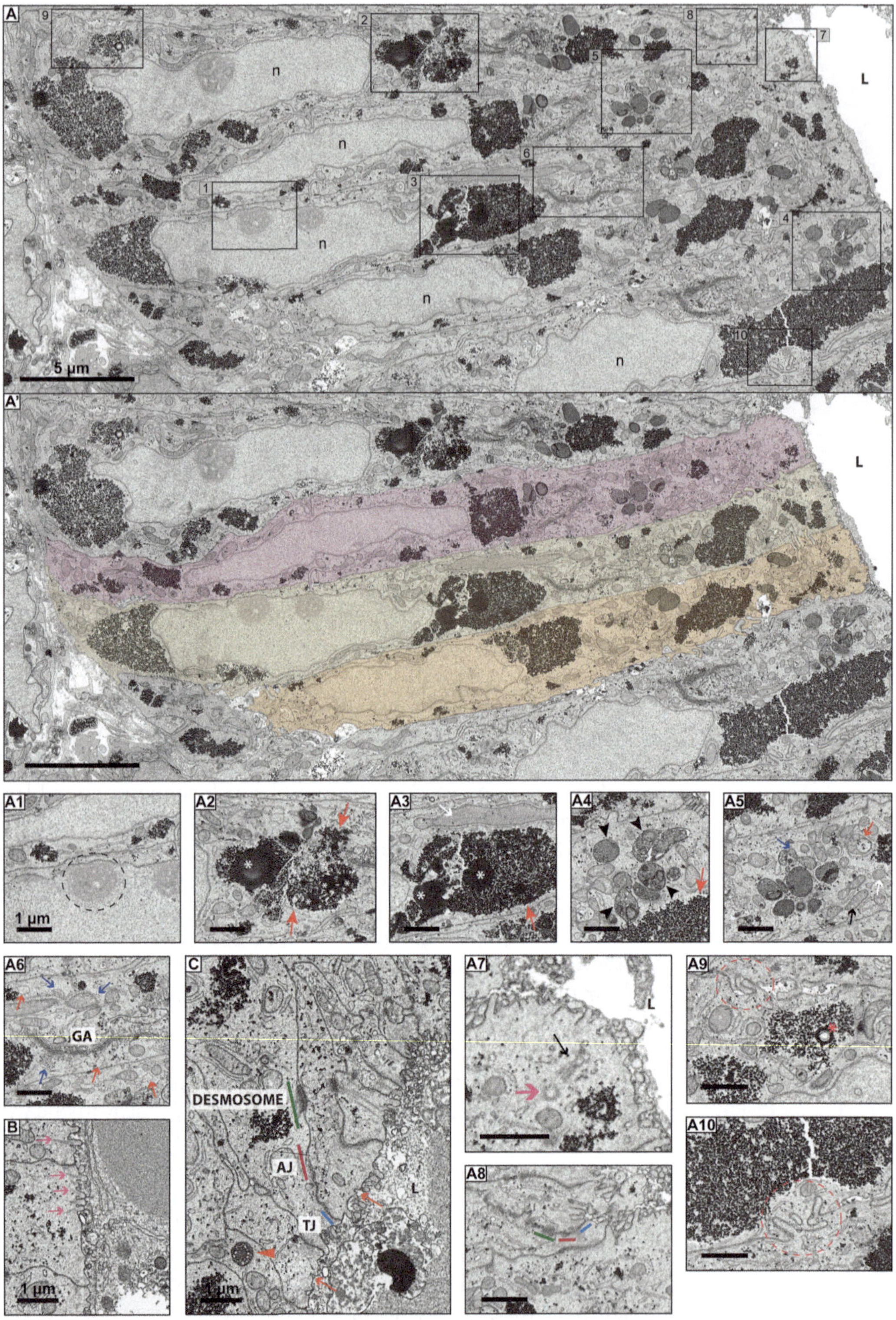

Fig. 3 Ultrastructure of the organoid-derived radial glia cells (*oRGC*) in the neural rosette of a human iPSC-derived unguided neural organoid (UNO). **(A)** Inset from (Fig. 2) illustrating radially organized *oRGCs* facing rosette lumen (L) on their apical end in UNO with *APOE4/4* genotype (patient donor line MV 8).

mitotic spindle assembly and growth of primary cilia was shown in the embryo of a sea squirt [81].

6 Ultrastructural Features of GFAP+ *oASTRO* in Human Patient iPSC-Derived UNO Isogenic Control (*APOE3/3* Genotype)

In ~45-day-old human neural organoids [63] or 5-month-old embryonic stem cell-derived human cortical spheroids [61], *oASTROs* displayed ultrastructure featuring electron-lucent and irregular nuclei, intermediate filament bundles and glycogen granules, frequently marked by contact with surrounding neurons and their processes. In the current model, GFAP+ *oASTROs* with *APOE3/3* genotype are large cells with narrow and angular extensions emanating into their surroundings, and a thin rim of nuclear heterochromatin (Fig. 4A–E) [64]. In a single plane, *oASTROs* are frequently surrounded by other GFAP+ *oASTRO* cell extensions (Fig. 4A–E). This may be related to a heterogeneous distribution of *oASTROs* and their tendency to form clusters within UNOs. Alternatively, ITGA6+ astrocytes in ~130-day-old UNOs derived from healthy donor iPSCs, cultured using the same methodology as in this protocol, displayed a bushy morphology [53]. Henceforth, this could be representative of their conceptual stellated morphology with complex extensions in a 2D plane. With a transcriptomic profile similar to mature human astrocytes, these ITGA6+ astrocytes were assumed to partake in the maintenance of organoid homeostasis [53]. Molecular maturation of rodent astrocytes and adoption of an arborized morphology with finer processes ensheathing synapses are regulated by glutamatergic synaptic activity in vivo [82]. While we observe only infrequent direct contacts of *oASTRO* cells and synapses, multiple functional synaptic connections are visible in their proximity (Fig. 4A–C, A1). With a distinctive double-membrane whose inner part forms characteristic

Fig. 3 Continued (**A'**) Pseudo-colored version of A), with pink, yellow, and orange highlighting individual neighbor cells. (**B**) An example of the microtubular infrastructure of the cilia (pink arrow) that populate the apical surface of *oRGCs* along with multiple microvilli in isogenic UNO with *APOE3/3* gene (patient donor line MV 10). (**C**) Junctional network at the apical end of *oRGCs* in isogenic UNO with *APOE3/3* gene (patient donor line MV 10) displaying tight junctions (TJ, blue line) closer to lumen, more distally located adherens junctions (AJ, red line) and desmosomes (green line). Red arrows point to putative clathrin-coated pits and caveoles within the luminal membrane. Red arrowhead shows a multivesicular body. Maximized insets at 1 μm scale: (**A1**) nucleolus (dashed **black** line); (**A2–A3**) lipid droplet with a pale hydrophobic core (white asterisk, *) surrounded by extensive deposits of round glycogen granules (red arrow) and elongated mitochondria (white arrow); (**A4**) lysosome/phagosome-like structures (**black** arrowheads) near glycogen aggregates (red arrow); (**A5**) mitochondria with elongated (blue arrow), round (white arrow), donut shape (red arrow), and dystrophic mitochondria with cristae pockets (**black** arrow); (**A6**) Golgi apparatus (GA) and typical (blue arrow) or altered/swollen endoplasmic reticulum (red arrow); (**A7**) Centriole (pink arrow) and microtubule bundle of primary cilia (**black** arrow). (**A8**) Junctional connection within apical end of *oRGCs*. (**A9–A10**) Finger-like cytoplasmic invaginations (dashed red line) in basal and apical end of *oRGCs*, respectively. Lipid droplet with a clear (empty) core is marked with red asterisk, *. *n* nucleus, *GA* Golgi apparatus

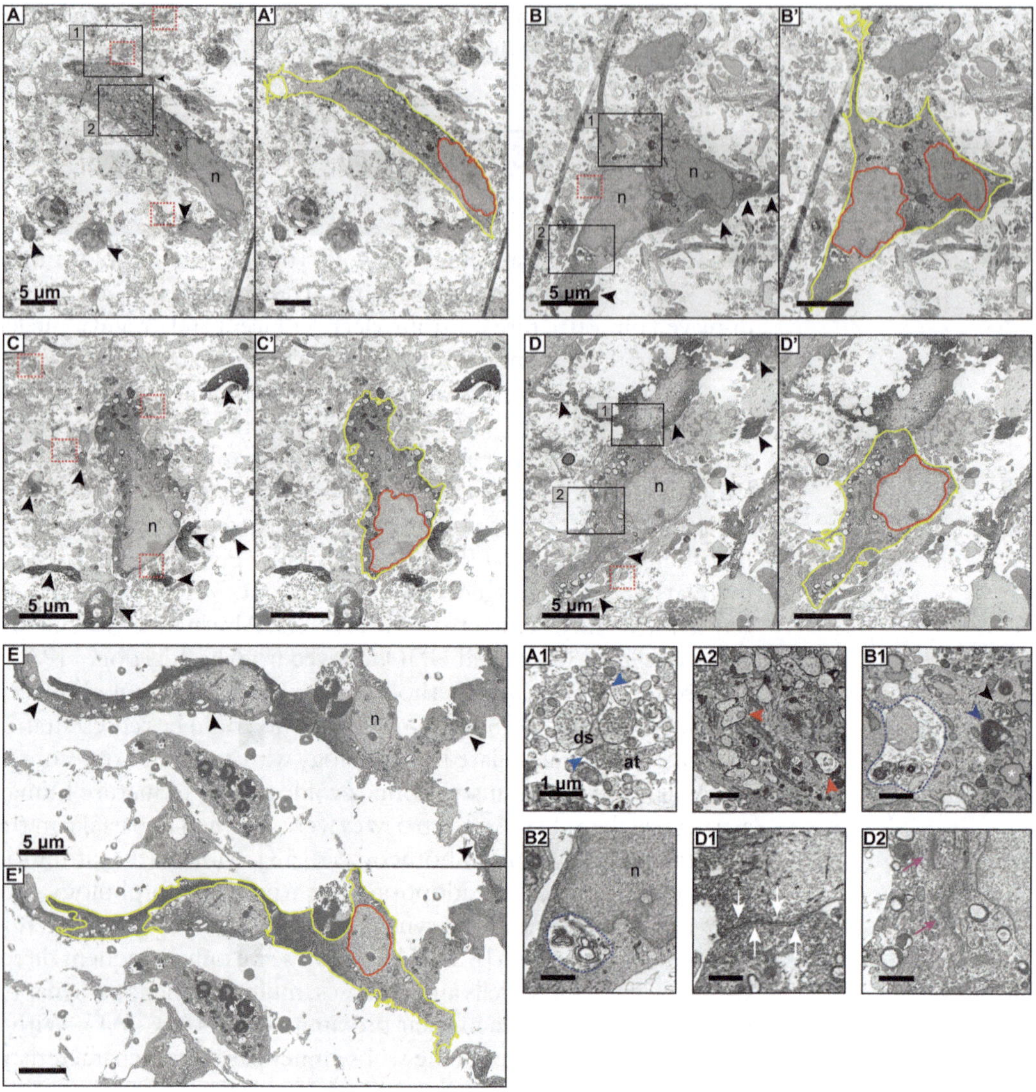

Fig. 4 Ultrastructure of GFAP-immunopositive organoid-derived astrocytes (*oASTRO*) in human iPSC-derived unguided neural organoids (UNO) with *APOE3/3* genotype using scanning electron microscopy. Example images of *oASTRO* ultrastructure in (**A–B**) isogenic UNO (patient donor line MV 8, GFAP antibody concentration at 1:400); (**C–D**) isogenic UNO (patient donor line MV 8, GFAP antibody concentration at 1:1000); and (**E**) isogenic UNO (patient donor line MV 10, GFAP antibody concentration at 1:1000). (**A'–E'**) Pseudo-coloured version of **A–E**), with yellow line delineating cell membrane and red line delineating nuclear membrane. **Black** arrowheads point to GFAP-positive staining of individual *oASTRO* cell processes. Dashed red square signalizes a synaptic connection. Maximized insets at 1 µm scale: (**A1**) Synaptic connections: synaptic cleft (blue arrowhead), axon terminal with presynaptic vesicles (at) and dendritic spine (ds); (**A2**) Examples of vesicles with unevenly thick double-membrane and heterogenous content (red arrowhead), Endoplasmic reticulum, Golgi apparatus (GA), and mitochondria; (**B1–B2**) Large autophagosome-like organelles with an undigested vesicular and organellar content (dashed blue line), phagosome with fibrillar content (black arrowhead), lysosome (blue arrowhead), and lipid droplet (white asterisk, *); (**D1**) Contact interface between two neighbour astrocytic cells (white arrowhead). (**D2**) GA and empty vesicles with unevenly thick double-membrane. *n* nucleolus

cristae, *oASTRO* mitochondria exhibit diverse morphologies, with prevalent shapes including small and round or narrow and cylindrical. Qualitatively, mitochondria in *oASTROs* are less abundant than in *oRGCs* (Fig. 4A2). ER cisternae are short and narrow (Fig. 4A2). In comparison to elongated GA structures in *oRGCs*, the GA tubules in *oASTROs* are shorter and more tightly packaged (Fig. 4A2, D2).

oASTROs contain multiple smaller, rounded, or bean-shaped vesicles with an unevenly thick double-membrane, occasionally perturbed by circular structures, and an electron-lucent core containing material remnants (Fig. 4A2, D2). There is a variety of single-membrane phagosomal vesicles that are empty or contain fibrous material, electron-dense lysosomes, and large autophagosome-like structures with heterogeneous undigested content and electron-lucent core (Fig. 4B1, B2). In comparison to *oMGs*, lipid droplets (with a more electron-dense hydrophobic core) are present but at a lesser scale (Figs. 4 and 5). The nucleus exhibits indentations of the nuclear membrane (Fig. 4B2). Besides contacting other cells of the neuronal lineage (at various stages of maturation) or glia [63, 64], *oASTROs* also form prominent physical contacts with other *oASTROs* (Fig. 4A–D, D1). In typical tissue, such a contact interface exhibited an abundance of gap junctions, which contribute to syncytial coupling in astrocyte networks [83–85].

Prior to osmium embedding, we observed numerous cells with positive GFAP immunoperoxidase staining on a brightfield microscope at 20× magnification in both *APOE4/4* and isogenic *APOE3/3* UNO samples (*data not shown*). As mentioned above, the distribution of the *oASTROs* in UNOs can be rather heterogeneous, forming areas with high or low cell densities in an organoid-specific manner. The lack of observed GFAP+ *oASTROs* in the ultrastructure of UNOs with *APOE4/4* genotype in the current protocol is likely due to random region sampling and a small sample size. Furthermore, the peripheral layers of UNOs, which are enriched in mature astrocytes, appear to be particularly susceptible to damage during the EM preparation. The range of genotype impact, while not excluded, is yet unclear and could also manifest in differences in density and distribution of *oASTRO* within the organoid.

7 Ultrastructural Features of IBA1+ *oMG* in Human Patient iPSC-Derived UNO

Immunostaining-based investigation of *oMGs* ultrastructure in UNOs with either *APOE4/4* or *APOE3/3* genotype revealed a set of distinctive features that facilitate a more precise discrimination from other present cell types. Regardless of genotype, IBA1+ *oMGs* adopt a rounded or elongated shape (Fig. 5A, B, D, E), characterized by thinner cell extensions and processes of a shorter length. Prior description of mature *oMGs* in UNOs derived from iPSCs of human

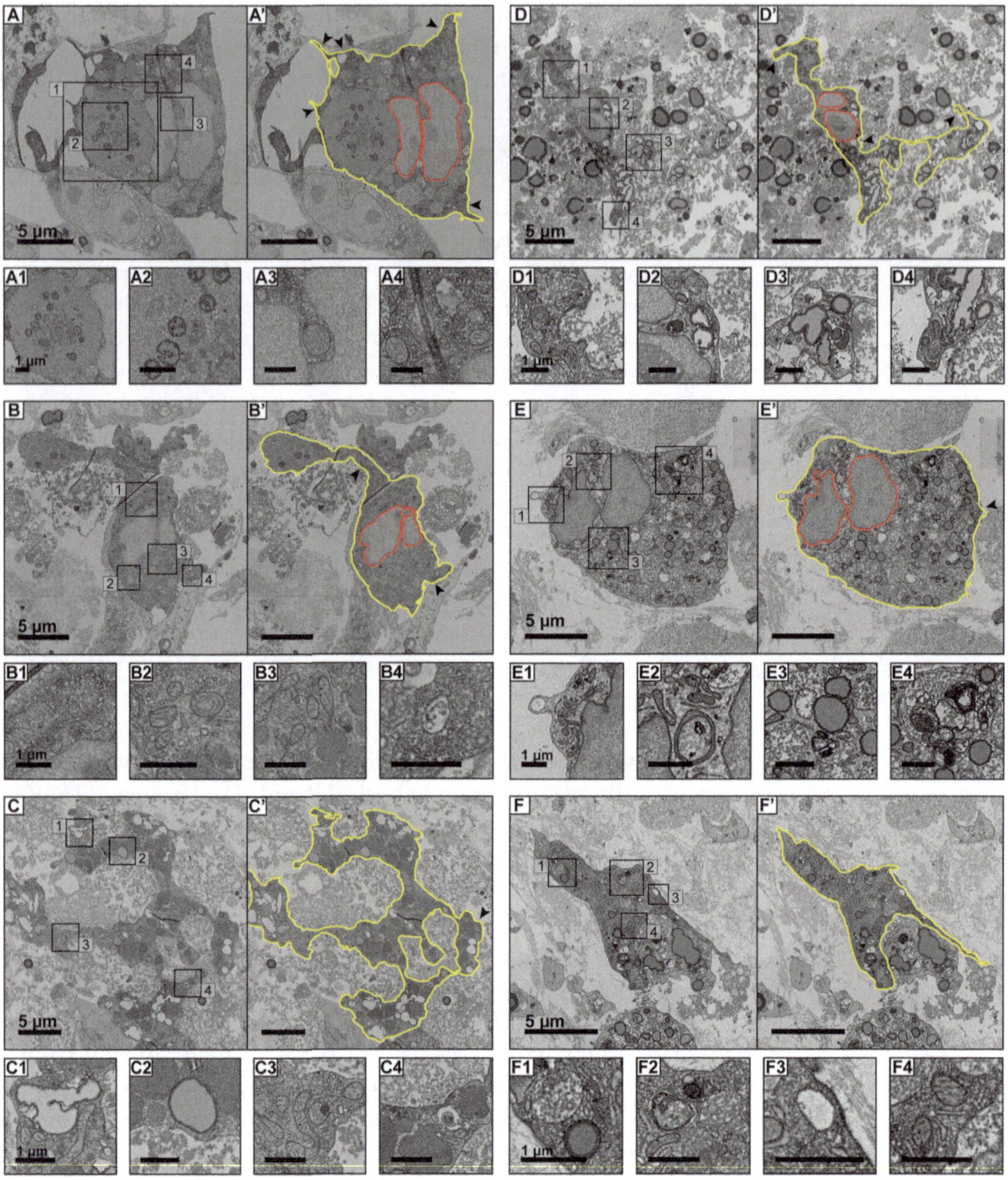

Fig. 5 Ultrastructure of IBA1-immunopositive organoid-derived microglia (*oMG*) in human iPSC-derived unguided neural organoid (UNO) using scanning electron microscopy. Example images of *oMG* ultrastructure in (**A–C**) isogenic UNO with *APOE3/3* genotype (patient donor line MV 10, IBA1 antibody concentration at 1:1000); (**D–F**) UNO with *APOE4/4* genotype (patient donor line MV 10, IBA1 antibody concentration at 1:1000). (**A–B, D–E**) Examples of *oMG* cell bodies. (**C, F**) Examples of IBA1 positive *oMG* processes. (**A'–F'**) Pseudo-colored version of **A–F**), with yellow line delineating cell membrane and red line delineating nuclear membrane. Maximized insets at 1 µm scale: (**A1–A2**) Lipid droplets; (**A3–A4**) Mitochondria, endoplasmic reticulum (ER) and Golgi apparatus (GA). Notice the ER in close contact with the mitochondria; (**B1–B3)** Examples of mitochondria, ER and GA; (**B4**) Example of phagocytic inclusion containing material; (**C1**) Lipid droplet in proximity to mitochondria; (**C2**) Lipid droplet; (**C3**) Mitochondria in proximity to ER; Dystrophic mitochondrion with inclusion; (**C4**) Phagocytic inclusion containing material; (**D1**) Mitochondria in close contact to ER;

donors outlined their ramified morphology and close recapitulation of the immune responsiveness and transcriptomic signatures of adult microglia acutely isolated from post-mortem human brain samples [45]. Furthermore, *oMGs* were capable of C3 receptor-mediated phagocytosis and were positive for PSD-95, a postsynaptic density marker, after 2 months in culture [45]. At the level of ultrastructure, we observe *oMGs* cytoplasm to be notably packaged with large and round single-membrane subcellular compartments containing material and resembling organelles of the phagolysosomal pathway. This contrasts with GFAP+ *oASTRO* cytoplasm that is populated by a variety of rather smaller-sized organelles (Fig. 4). Notably these phagosomes contain varying densities of fibrous material and electron-dense globular and membrane-enclosed structural remnants (Fig. 5B, C, F). Occasionally, these structures are devoid of any material, displaying an electrolucent core (Fig. 5E, F). Through their cytoplasmic extensions and extensive cell body-mediated contacts, *oMGs* can interact with neighboring cells and phagocytose material from their surroundings, consistent with previous findings. During development, human and rodent microglia interact with NPCs via phagocytosis, regulating their population [86]. Human iPSC-derived primitive macrophage progenitors, which developed a microglia-like signature in a coculture with primitive NPCs, showed immunopositivity for CD68 and caspase-3, indicating the ability of organoid microglia to remove apoptotic cells [87]. This relative abundance of phagosomal organelles in *oMGs* could represent a phagocytic hallmark of the developmental stage recapitulated by this organoid model, a response to cellular stress, or even reflect an incomplete degradation of internalized material leading to the accumulation of these vesicles. Unlike prior studies that described the ultrastructure of *oMGs* as smaller-sized cells with larger nucleus, prominent heterochromatin pattern and a thinner rim of cytoplasm containing few organelles [61, 63], this larger organelle-dense phenotype may also be unique to UNOs. This effect may also be related to the increased and unguided cellular heterogeneity of UNOs resulting from the absence of dual patterning factors. Additional factors may include cell line source, varying lengths of the organoid culture, or the absence of immunomarker-based identification in EM in previous studies, all of which could underlie the observed differences.

While maintaining the heterochromatin pattern characteristic of human or rodent tissue microglia [64], the nucleus of *oMGs* appears angular and folded (Fig. 5A–D). Although the potential for the

Fig. 5 Continued (**D2**) Examples of mitochondria, ER, lipid droplet and lysosome; (**D1–D4**) Examples of lipid droplets and mitochondria. Examples of mitochondria containing concentric structures in D1 and D4; (**E1–E2**) Examples of mitochondria, ER and empty inclusion. Empty phagosome can be seen in E1. Notice mitochondrion in donut shape in E2; (**E3–E4**) Examples of lipid droplets and phagosomes containing material at various stages of digestion; (**F1–F4**) Examples of mitochondria, lipid droplets, ER, filled and empty phagosomes

presence of multiple nuclei in *oMGs* cannot be excluded; the exact configuration is challenging to resolve in a single 2D plane. In some *oMGs*, a round and electron-dense nucleolus is observed, which can serve as an additional indicator of cell's protein synthesis capacity and metabolic status (Fig. 5A, D). Mitochondria occur in a variety of shapes and sizes, including small, round, and elongated (Fig. 5B, C), and altered forms as donut-shaped mitochondria or with inclusions (Fig. 5C, E). A percentage of mitochondria shows abnormal layering of inner membrane, whereas the cristae organize in a concentric fashion (Fig. 5D). Resulting from genetic disruptions of mitochondrial proteins and/or related to oxidative stress, such alterations may indicate a disruption of cristae organizing systems, respiratory chain function and ATP synthesis [88]. *oMGs* also contain short and long narrow ER cisternae and shorter stretches of tubular GA (Fig. 5B).

In addition, UNO *oMGs* frequently exhibit lipid bodies/droplets, with substantial morphological variability in phospholipid monolayer thickness and circularity, as well as in the electron density of the hydrophobic core (Fig. 5D). These reflect variability in lipid composition, packing, presence of associated proteins in the core and their availability as signaling mediators on its surface. Beyond their capacity for lipid storage, lipid bodies were identified as dynamic intracellular organelles that actively participate in the regulation of homeostasis, autophagy or cell immune responses [89]. In a coculture with neural organoid, microglial cells, differentiated from iPSC-derived primitive human macrophage precursors, were loaded with lipid droplets positive for perilipin 2 (PLIN2) [58]. This macrophage/microglia integration into organoid was accompanied by atrophy of neural rosettes, reduced NPC proliferation, increased axonogenesis and synaptogenesis, further expanding the range of known microglial impact on brain organoid maturation. Functionally, this lipid-loaded microglial phenotype—concurrently observed in mouse and human embryonic brain samples—was associated with the capacity to store and export cholesterol and its esters, subsequently utilized by NPCs and neuronal cells. The authors proposed this function to putatively serve as a compensatory mechanism in the absence of astrocytes, mediated via APOE, highly expressed by PLIN2+ microglia [58]. A mature astrocyte signature was reported for ITGA6+ astrocytes in healthy donor iPSC-derived UNOs, prepared using the same methodology, after 130 days of culture. In the currently investigated 120-day-old UNOs, the microglia-mediated mechanism for cholesterol transport could be occurring as a compensation for yet functionally not fully mature *oASTROs* [53]. Alternatively, *oASTROs* heterogeneous density and distribution throughout the UNO may also be contributing to this effect. Analysis of lipid droplet morphology, composition and contacts with other intracellular compartments in *oMGs* is needed to determine whether similar mechanisms are occurring. Subsequently, we note lipid droplets are qualitatively more present alongside a lower frequency of phagosomes in *oMGs* of UNOs with *APOE4/4* genotype compared to their isogenic control (*data not shown*). Although lipid inclusions in brain glial cells of human patients with AD were already described by Alzheimer [90], lipids have only recently been recognized as important drivers of AD pathology. Human APOE is involved in lipid transport, and its AD risk-related isoform (*APOE4* allele) was associated with impaired lipid metabolism [91, 92] and accumulation of cholesterol and triacylglyceride-abundant lipid droplets in human iPSC-derived microglia and astrocytes [17, 93, 94]. In vitro microglial cells harboring the *APOE4* gene variant display less efficient uptake of extracellular Aβ when cocultured with 2-month-old neural organoids generated from iPSCs with familial AD mutation (Amyloid precursor protein (APP) duplication) [17], indicative of an altered phagocytic behavior. Our preliminary experiments, however, did not show positive immunofluorescent staining for 6E10 or methoxy-X04 (Aβ and APP markers) at this maturation stage of the *APOE4/4* UNO (*data not shown*). We report no visible ultrastructural signs of extracellular Aβ deposits near IBA1+ *oMGs*. Henceforth, the ability of human AD patient-derived *oMGs* to phagocytose Aβ remains to be determined in this organoid model.

One caveat is that our preliminary investigation employed a region-nonspecific randomized sampling approach within a limited sample size, which may have obscured any intrapopulation differences. Similar to human and animal brain tissue [95], the complex architecture of UNO may give rise to heterogeneity of glial states with distinct features driven by their region-specific microenvironmental cues and various stages of its growth [53, 56, 96].

8 Conclusions

The limited availability of accessible and translatable human disease models, combined with limited access to human brain tissue, has spurred the development of novel tools, with 3D neural organoids being at the forefront of human brain modeling endeavors. This in vitro technology offers a great potential to study individual cell mechanisms and improves research on glia-neuron interactions in a more dynamic and biologically relevant environment. As these mechanisms lie at the core of the onset and development of neuropathological processes, 3D neural organoids are expected to become an invaluable tool for personalized medicine. While these models are yet to achieve their peak potential by overcoming various technical limitations, including a high individual variability and the absence of various immune cell populations, it is equally important for them to be thoroughly defined at each stage of their development. To our knowledge, this study is the first to define the ultrastructure of human-derived UNOs and their glial cells using immunohistochemical markers and SEM. While the unguided strategy allows UNOs to develop microglial cells, it also leads to random growth and expansion of brain-like structures, lacking well-defined spatial boundaries and exhibiting areas of ultrastructural disorganization. This level of unstructured organization and presence of heterogeneous cell population consisting of various cell lineages at various stages of differentiation and/or maturation, in most cases, prevents an efficient identification of cells in UNO without immunohistochemical markers at the level of ultrastructure. Next, combining imaging modalities such as correlative light or fluorescence microscopy with electron microscopy—CLEM or CFEM—and 3D EM reconstruction will be key towards defining the organoid cytoarchitecture and precise spatial localization of glia within. Furthermore, as glial cell populations are innately heterogeneous, with brain regional differences among them as one driver of this heterogeneity, it will be most intriguing to investigate how this aspect replicates within the rather non-compartmentalized ultrastructural UNO architecture. Employing state-of-the-art techniques, such as cryo-EM, will improve preservation of fragile, easily soluble tissue structures, including those of organoids.

9 Notes

1. Avoid preparation of the solution at temperatures >60°C to prevent the formation of unwanted intermediates. The solution should be stored at 4°C and used for up to a maximum of 14 days after preparation. The exposure to PFA, in its powder or liquid form, presents a serious health hazard, particularly through skin contact and inhalation. All handling should be performed in the fume hood while wearing appropriate personal protective equipment (PPE; protective garment, gloves, safety googles) and following the respective Safety Data Sheets (SDS). Any waste should be disposed of according to the facility regulations.
2. Aldehydes provide a stable fixation of proteins by cross-linking via methylene bridges. High degree of fixation, particularly with GLUT or acrolein, is advantageous for preservation of tissue quality for EM. However, it increases the autofluorescence, which may cause the sample to be less suitable for other, e.g., immunofluorescent-based, techniques. Minimum GLUT concentration for electron microscopy would be at 0.4%–0.6%. The solution should be stored at 4°C and used for up to a maximum of 14 days after preparation. The exposure to aldehydes poses a serious health hazard, particularly through skin contact and inhalation. All handling should be performed in the fume hood while wearing appropriate PPE

and following the respective SDS. Any waste should be disposed of according to the facility regulations. This solution can be also commercially purchased (EMS, 16536-06).

3. Sodium borohydride reacts with water, forming a flammable and explosive hydrogen gas. While the reaction is slow at lower concentrations (<1%–2%), the solution should be prepared and handled in the fume hood and following the respective SDS. After preparation, the container should remain partially open to avoid buildup of gas pressure and explosion. Any waste should be disposed of according to the facility regulations.
4. Stock of non-activated DAB (in TB) can be aliquoted and stored at −20°C for later use. If sample incubation with H_2O_2-activated DAB exceeds 10 min, the DAB solution must be reactivated again. Avoid reactivating, the same solution more than three times to not affect the staining quality. DAB is a known carcinogen, and all handling should be performed in the fume hood while wearing appropriate PPE and following the respective SDS. Any waste should be disposed of according to the facility regulations.
5. During staining testing, the GFAP antibody at all listed concentrations produced visible staining.
6. Osmium tetroxide is light-sensitive, with prolonged exposure to light impacting its use in EM staining and causing the formation of hazardous byproducts. After opening an unsealed vial, for a short-term storage, the osmium stock can be kept at 4°C in a tight double-sealed dark container to prevent leakage of the osmium fumes (a potent oxidizing agent) and contamination. Osmium tetroxide is acutely toxic and presents a serious health risk. All handling should be performed in the fume hood while wearing appropriate PPE and following the respective SDS. Any waste should be disposed of according to the facility regulations.
7. As THC powder dissolves in ddH_2O, the solution gains a light purple color. THC is a toxic compound upon skin contact or if swallowed. All handling should be performed in the fume hood while wearing appropriate PPE and following the respective SDS. Any waste should be disposed of according to the facility regulations.
8. Before polymerization, individual components of epoxy resin are harmful upon skin exposure or inhalation (particularly Durcupan components C and D, which present with a stronger odor). All handling should be performed in the fume hood while wearing appropriate PPE and following the respective SDS. Any waste should be disposed of according to the facility regulations.

9. By embedding in agarose, the small-sized organoids can be efficiently sectioned using vibratome into sections of equal thickness and protected from mechanical impacts during later processing. Vibratome sectioning at too slow/high speed or suboptimal frequency may cause the organoids to either get dislodged from the agarose or produce unequally thick sections. As immunohistochemical labeling protocols for EM have limited penetration into tissue, it is advised not to produce sections above 60 μm in thickness.
10. Even with the same protocol, each experimental sample may need a distinct incubation period for optimal staining. It is recommended to perform a pilot test at the start of each project, starting with 20–30 s of incubation followed by a staining check using a light microscope. Prolonging the staining by 15–20 s periods until a complete cell-specific brown staining occurs will allow for finding the optimal incubation time in a controlled manner. Avoid over-incubating, the tissue with DAB to prevent non-specific deposits.
11. After incubation with ethanol in Step 11, the already fragile sample becomes even more prone to damage by mechanical manipulation, therefore, any sample transfers should be done prior this step. As propylene oxide (Step 12) corrodes plastic, plastic well plates are not suitable at this stage of the protocol; however, this step can be skipped if a glass well plate or glass slide is used.
12. Propylene oxide corrodes plastics, all handling should be performed with glass instruments. As an acutely toxic and carcinogenic compound, all handling should be performed in the fume hood while wearing appropriate PPE and following the respective SDS. Any waste should be disposed of according to the facility regulations.
13. Because propylene oxide dissolves resin and evaporates rapidly, it is important to both cover the aluminum boat and add sufficient volume to ensure the sample is sufficiently covered throughout the sample incubation.
14. After completing the sheet sandwich, the blunt end of the brush can be used to apply pressure and guide bubbles to the sides of the sheet. Avoid applying pressure directly on the tissue. Any leftover resin can be used to prepare resin blocks for ultramicrotome cutting.
15. Use the whole length of the razor; avoid cutting the tissue with a razor edge that has already been used. The trapezoid

shape of the tissue will allow for section organization in the form of a long ribbon, which will ease its collection onto the silicon wafer.

16. Brief exposure to xylene promotes tissue flattening. This step can be specifically beneficial for organoid tissues that have tendency to form folds and wrinkle, which obscure the ultrastructure in EM. Strictly avoid any direct contact of xylene and the boat contents to prevent contamination and creation of false immunostaining signals. If contamination occurs, discard the sections and immerse the boat in soapy water overnight prior to cleaning the knife and resuming cutting. Xylene poses an acute adverse health risk and requires careful handling while wearing appropriate PPE in a well-ventilated room. Avoid prolonged exposure to the contents of the stock vial, and in the event of spill, vacate the room. Follow the proper SDS and facility guidelines.

Acknowledgments

We acknowledge the contributions of Dr. J. TCW and Dr. A.M. Goate (Icahn School of Medicine at Mount Sinai, New York, NY, USA), who provided human APOE iPSC lines. E.S., M.V., E.M.H., and M.E.T. wrote the manuscript. E.S. and F.G.I. prepared the figures. M.V. and E.M.H. generated the organoid samples. E.S., F.G.I., R.B., P.K., and K.P. participated in the electron microscopy experiments. All the authors contributed to the article. All the authors read and approved the final version of the manuscript.

This work was supported by an ELA International grant (2023-017C3), co-jointly awarded to the Hol and Tremblay labs. E.S. is a recipient of the Branch Out Foundation Graduate Grant and Faculty of Graduate Studies (University of Victoria) Graduate Scholarships. M.V. was supported by the Dutch Research Council (NWO) Graduate Program (022.006.001). F.G.I. holds a Michael Smith Health Research British Columbia (BC) postdoctoral fellowship. The Hol lab is supported by the BRAINS Initiative of Alzheimer Nederland, EpilepsieNL, Hersenstichting and Stichting MS Research (GLIODEM), Alzheimer Nederland (ASTRODEM-WE.03-2022-04), and ZonMW (MODEM-10510032120006). M.E.T. holds a Canada Research Chair (Tier 1) in *Neurobiology of Healthy Cognitive Aging*. The Tremblay Lab's Zeiss Crossbeam 350 scanning electron microscope was acquired with funding from a Canada Foundation for Innovation John R. Evans Leaders Fund grant (39965 Laboratory of ultrastructural insights into the neurobiology of aging and cognition).

Conflict of Interest The authors have no conflicts of interests to declare that are relevant to the content of this chapter.

References

1. Serrano-Pozo A, Frosch MP, Masliah E, Hyman BT (2011) Neuropathological alterations in Alzheimer disease. Cold Spring Harb Perspect Med 1:a006189
2. Osborn LM, Kamphuis W, Wadman WJ, Hol EM (2016) Astrogliosis: an integral player in the pathogenesis of Alzheimer's disease. Prog Neurobiol 144:121–141
3. Huffels CFM, Middeldorp J, Hol EM (2023) Aß pathology and neuron-glia interactions: a synaptocentric view. Neurochem Res 48:1026–1046
4. Arranz AM, Strooper BD (2019) The role of astroglia in Alzheimer's disease: pathophysiology and clinical implications. Lancet Neurol 18:406–414
5. Heneka MT et al. (2015) Neuroinflammation in Alzheimer's disease. Lancet Neurol 14:388–405
6. Olah M et al. (2020) Single cell RNA sequencing of human microglia uncovers a subset associated with Alzheimer's disease. Nat Commun 11: 6129
7. Zhang X et al. (2016) Quercetin stabilizes apolipoprotein E and reduces brain Aβ levels in amyloid model mice. Neuropharmacology 108:179–192
8. Xu P-T et al. (2006) Differences in apolipoprotein E3/3 and E4/4 allele-specific gene expression in hippocampus in Alzheimer disease. Neurobiol Dis 21:256–275
9. Singh PP, Singh M, Mastana SS (2006) APOE distribution in world populations with new data from India and the UK. Ann Hum Biol 33:279–308
10. Corder EH et al. (1993) Gene dose of apolipoprotein E type 4 allele and the risk of Alzheimer's disease in late onset families. Science 261:921–923
11. Corder EH et al. (1994) Protective effect of apolipoprotein E type 2 allele for late onset Alzheimer disease. Nat Genet 7:180–184
12. Reiman EM et al. (2020) Exceptionally low likelihood of Alzheimer's dementia in APOE2 homozygotes from a 5,000-person neuropathological study. Nat Commun 11: 667
13. Sando SB et al. (2008) APOE epsilon 4 lowers age at onset and is a high risk factor for Alzheimer's disease; a case control study from central Norway. BMC Neurol 8: 9
14. Shi Y, Holtzman DM (2018) Interplay between innate immunity and Alzheimer disease: APOE and TREM2 in the spotlight. Nat Rev Immunol 18:759–772
15. Hatters DM, Peters-Libeu CA, Weisgraber KH (2006) Apolipoprotein E structure: insights into function. Trends Biochem Sci 31:445–454
16. Nelson TJ, Sen A (2018) Apolipoprotein E particle size is increased in Alzheimer's disease. Alzheimers Dement (Amst) 11:10–18
17. Lin Y-T et al. (2018) APOE4 causes widespread molecular and cellular alterations associated with Alzheimer's disease phenotypes in human iPSC-derived brain cell types. Neuron 98:1141–1154.e7
18. Giannisis A et al. (2022) Plasma apolipoprotein E levels in longitudinally followed patients with mild cognitive impairment and Alzheimer's disease. Alzheimers Res Ther 14:115
19. Tcw J et al. (2022) Cholesterol and matrisome pathways dysregulated in astrocytes and microglia. Cell 185:2213–2233.e25
20. Castellano JM et al. (2011) Human apoE isoforms differentially regulate brain amyloid-β peptide clearance. Sci Transl Med 3:89ra57
21. Liu C-C et al. (2017) ApoE4 accelerates early seeding of amyloid Pathology. Neuron 96:1024–1032.e3
22. Shi Y et al. (2017) ApoE4 markedly exacerbates tau-mediated neurodegeneration in a mouse model of tauopathy. Nature 549:523–527
23. Shi Y et al. (2019) Microglia drive APOE-dependent neurodegeneration in a tauopathy mouse model. J Exp Med 216:2546–2561
24. Lynch JR et al. (2003) APOE genotype and an ApoE-mimetic peptide modify the systemic and central nervous system inflammatory response. J Biol Chem 278:48529–48533
25. Vitek MP, Brown CM, Colton CA (2009) APOE genotype-specific differences in the innate immune response. Neurobiol Aging 30:1350–1360
26. Arnaud L et al. (2022) APOE4 drives inflammation in human astrocytes via TAGLN3 repression and NF-κB activation. Cell Rep 40:111200
27. de Leeuw SM et al. (2022) APOE2, E3, and E4 differentially modulate cellular homeostasis, cholesterol metabolism, and inflammatory response in isogenic iPSC-derived astrocytes. Stem Cell Reports 17:110–126
28. Chung W-S et al. (2016) Novel allele-dependent role for APOE in controlling the rate of synapse pruning by astrocytes. Proc Natl Acad Sci U S A 113:10186–10191

29. Tensaouti Y, Stephanz EP, Yu T-S Kernie SG (2018) ApoE regulates the development of adult newborn hippocampal neurons. eNeuro 5:ENEURO.0155-18.2018
30. Schmukler E et al. (2020) Altered mitochondrial dynamics and function in APOE4-expressing astrocytes. Cell Death Dis 11: 578
31. Suri S et al. (2015) Reduced cerebrovascular reactivity in young adults carrying the APOE ε 4 allele. Alzheimers Dement 11:648–657.e1
32. Halliday MR et al. (2016) Accelerated pericyte degeneration and blood-brain barrier breakdown in apolipoprotein E4 carriers with Alzheimer's disease. J Cereb Blood Flow Metab 36:216–227
33. Montagne A et al. (2020) APOE4 leads to blood–brain barrier dysfunction predicting cognitive decline. Nature 581:71–76
34. Oberheim NA et al. (2009) Uniquely hominid features of adult human astrocytes. J Neurosci 29:3276–3287
35. Bayraktar OA et al. (2020) Astrocyte layers in the mammalian cerebral cortex revealed by a single-cell in situ transcriptomic map. Nat Neurosci 23:500–509
36. Ben Haim L, Rowitch DH (2017) Functional diversity of astrocytes in neural circuit regulation. Nat Rev Neurosci 18:31–41
37. Zhang Y et al. (2016) Purification and characterization of progenitor and mature human astrocytes reveals transcriptional and functional differences with mouse. Neuron 89:37–53
38. Holst CB, Brøchner CB, Vitting-Seerup K, Møllgård K (2019) Astrogliogenesis in human fetal brain: complex spatiotemporal immunoreactivity patterns of GFAP, S100, AQP4 and YKL-40. J Anat 235:590–615
39. Li J et al. (2021) Conservation and divergence of vulnerability and responses to stressors between human and mouse astrocytes. Nat Commun 12: 3958
40. Zhou Y et al. (2020) Human and mouse single-nucleus transcriptomics reveal TREM2-dependent and TREM2-independent cellular responses in Alzheimer's disease. Nat Med 26:131–142
41. Gurdon JB (1962) Adult frogs derived from the nuclei of single somatic cells. Dev Biol 4:256–273
42. Takahashi K, Yamanaka S (2006) Induction of pluripotent stem cells from mouse embryonic and adult fibroblast cultures by defined factors. Cell 126:663–676
43. Paşca SP et al. (2022) A nomenclature consensus for nervous system organoids and assembloids. Nature 609:907–910
44. Lancaster MA, Knoblich JA (2014) Generation of cerebral organoids from human pluripotent stem cells. Nat Protoc 9:2329–2340
45. Ormel PR et al. (2018) Microglia innately develop within cerebral organoids. Nat Commun 9: 4167
46. Renner M et al. (2017) Self-organized developmental patterning and differentiation in cerebral organoids. EMBO J 36:1316–1329
47. Matsui TK et al. (2018) Six-month cultured cerebral organoids from human ES cells contain matured neural cells. Neurosci Lett 670:75–82
48. Savage JC, Picard K, González-Ibáñez F, Tremblay M-È (2018) A brief history of microglial ultrastructure: distinctive features, phenotypes, and functions discovered over the past 60 years by electron microscopy. Front Immunol 9:803
49. Carrier M, Robert M-È, González Ibáñez F, Desjardins M, Tremblay M-È (2020) Imaging the neuroimmune dynamics across space and time. Front Neurosci 14:903
50. St-Pierre M-K, Šimončičová E, Carrier M, Tremblay M-È (2023) Microglia in human postmortem brain samples: quantitative ultrastructural analysis of scanning electron microscopy images. In: Chun J (ed) Alzheimer's disease: methods and protocols. Springer, US, New York, NY, pp 63–85. 10.1007/978-1-0716-2655-9_4
51. Willingham MC, Rutherford AV (1984) The use of osmium-thiocarbohydrazide-osmium (OTO) and ferrocyanide-reduced osmium methods to enhance membrane contrast and preservation in cultured cells. J Histochem Cytochem 32:455–460
52. Nimmerjahn A, Kirchhoff F, Helmchen F (2005) Resting microglial cells are highly dynamic surveillants of brain parenchyma in vivo. Science 308:1314–1318
53. Verkerke M et al. (2024) Transcriptomic and morphological maturation of human astrocytes in cerebral organoids. GLIA 72:362–374
54. Lee C-T, Bendriem RM, Wu WW, Shen R-F (2017) 3D brain organoids derived from pluripotent stem cells: promising experimental models for brain development and neurodegenerative disorders. J Biomed Sci 24:59
55. Camp JG et al. (2015) Human cerebral organoids recapitulate gene expression programs of fetal neocortex development. Proc Natl Acad Sci U S A 112:15672–15677
56. Sabogal-Guaqueta AM et al. (2024) Brain organoid models for studying the function of iPSC-derived microglia in neurodegeneration and brain tumours. Neurobiol Dis 203:106742

57. Sabate-Soler S, Bernini M, Schwamborn JC (2022) Immunocompetent brain organoids—microglia enter the stage. Prog Biomed Eng 4:042002
58. Park DS et al. (2023) iPS-cell-derived microglia promote brain organoid maturation via cholesterol transfer. Nature 623:397–405
59. Di Lullo E, Kriegstein AR (2017) The use of brain organoids to investigate neural development and disease. Nat Rev Neurosci 18:573–584
60. Pellegrini L et al. (2020) Human CNS barrier-forming organoids with cerebrospinal fluid production. Science 369:eaaz5626
61. De Kleijn KMA et al. (2023) Human cortical spheroids with a high diversity of innately developing brain cell types. Stem Cell Res Ther 14:50
62. Tidball AM et al. (2023) Deriving early single-rosette brain organoids from human pluripotent stem cells. Stem Cell Reports 18:2498–2514
63. Mateos-Martínez P et al. (2024) Human cerebral organoids: cellular composition and subcellular morphological features. Front Cell Neurosci 18:1406839
64. Nahirney PC, Tremblay M-E (2021) Brain ultrastructure: putting the pieces together. Front Cell Dev Biol 9:629503
65. Lee J-H, Sun W (2022) Neural organoids, a versatile model for neuroscience. Mol Cells 45:53–64
66. Malatesta P, Appolloni I, Calzolari F (2008) Radial glia and neural stem cells. Cell Tissue Res 331:165–178
67. Ziv O et al. (2015) Quantitative live imaging of human embryonic stem cell derived neural rosettes reveals structure-function dynamics coupled to cortical development. PLoS Comput Biol 11:e1004453
68. Borries M et al (2020) Quantification of visco-elastic properties of a matrigel for organoid development as a function of polymer concentration. Front Phys 8:579168
69. Zaman MH et al. (2006) Migration of tumor cells in 3D matrices is governed by matrix stiffness along with cell-matrix adhesion and proteolysis. Proc Natl Acad Sci U S A 103:10889–10894
70. Dohrmann GJ, Bucy PC (1970) Human choroid plexus: a light and electron microscopic study. J Neurosurg 33:506–516
71. Bisht K et al. (2016) Dark microglia: a new phenotype predominantly associated with pathological states. GLIA 64:826–839
72. Choi SH et al. (2014) A three-dimensional human neural cell culture model of Alzheimer's disease. Nature 515:274–278
73. Saunders NR, Dziegielewska KM, Fame RM, Lehtinen MK, Liddelow SA (2023) The choroid plexus: a missing link in our understanding of brain development and function. Physiol Rev 103:919–956
74. González Ibáñez F, VanderZwaag J, Deslauriers J, Tremblay M-È (2024) Ultrastructural features of psychological stress resilience in the brain: a microglial perspective. Open Biol 14:240079
75. Zaidi D, Chinnappa K, Francis F (2022) Primary Cilia Influence progenitor function during cortical development. Cells 11:2895
76. Kumar V et al. (2021) The regulatory roles of motile cilia in CSF circulation and hydrocephalus. Fluids Barriers CNS 18:31
77. Schembs L et al. (2022) The ciliary gene INPP5E confers dorsal telencephalic identity to human cortical organoids by negatively regulating Sonic hedgehog signaling. Cell Rep 39:110811
78. D'Imprima E et al. (2023) Light and electron microscopy continuum-resolution imaging of 3D cell cultures. Dev Cell 58:616–632.e6
79. Hansen CG, Nichols BJ (2010) Exploring the caves: cavins, caveolins and caveolae. Trends Cell Biol 20:177–186
80. Vasioukhin V, Bauer C, Yin M, Fuchs E (2000) Directed actin polymerization is the driving force for epithelial cell-cell adhesion. Cell 100:209–219
81. Negishi T, Miyazaki N, Murata K, Yasuo H, Ueno N (2016) Physical association between a novel plasma-membrane structure and centrosome orients cell division. eLife 5:e16550
82. Morel L, Higashimori H, Tolman M, Yang Y (2014) VGluT1+ neuronal glutamatergic signaling regulates postnatal developmental maturation of cortical protoplasmic astroglia. J Neurosci 34:10950
83. Kiyoshi CM, Zhou M (2019) Astrocyte syncytium: a functional reticular system in the brain. Neural Regen Res 14:595
84. Bennett MVL, Contreras JE, Bukauskas FF, Sáez JC (2003) New roles for astrocytes: Gap junction hemichannels have something to communicate. Trends Neurosci 26:610–617
85. Dermietzel R, Spray DC (1993) Gap junctions in the brain: where, what type, how many and why?. Trends Neurosci 16:186–192
86. Wolf SA, Boddeke HWGM, Kettenmann H (2017) Microglia in physiology and disease. Annu Rev Physiol 79:619–643
87. Xu R et al. (2021) Developing human pluripotent stem cell-based cerebral organoids with a controllable microglia ratio for modeling brain development and pathology. Stem Cell Reports 16:1923–1937

88. Vincent A et al. (2016) The spectrum of mitochondrial ultrastructural defects in mitochondrial myopathy. Sci Rep 6: 30610
89. Onal G, Kutlu O, Gozuacik D, Dokmeci Emre S (2017) Lipid droplets in health and disease. Lipids Health Dis 16:128
90. Foley P (2010) Lipids in Alzheimer's disease: a century-old story. Biochim Biophys Acta 1801:750–753
91. Yang LG, March ZM, Stephenson RA, Narayan PS (2023) Apolipoprotein E in lipid metabolism and neurodegenerative disease. Trends Endocrinol Metab 34:430–445
92. Lazar A-N et al. (2022) Lipid dys-homeostasis contributes to APOE4-Associated AD pathology. Cells 11:3616
93. Sienski G et al. (2021) APOE4 disrupts intracellular lipid homeostasis in human iPSC-derived glia. Sci Transl Med 13:eaaz4564
94. Victor MB et al. (2022) Lipid accumulation induced by APOE4 impairs microglial surveillance of neuronal-network activity. Cell Stem Cell 29:1197–1212.e8
95. Stratoulias V, Venero JL, Tremblay M, Joseph B (2019) Microglial subtypes: diversity within the microglial community. EMBO J 38:e101997
96. Song L et al. (2019) Functionalization of brain region-specific spheroids with isogenic microglia-like cells. Sci Rep 9: 11055

Check for updates

Chapter 7

A Guide for Microglial Ultrastructural Analysis

Marianela Evelyn Traetta, Bianca Caroline Bobotis and Marie-Ève Tremblay

Abstract

Microglia, the resident immune cells of the central nervous system, exhibit diverse and dynamic states that are critical for maintaining neural homeostasis, regulating synaptic remodeling, supporting axonal myelination, and responding to injury or disease. Electron microscopy provides a nanoscale-resolution visualization of microglial cell bodies, ramified processes, organelles such as phagocytic inclusions, and contacts with other parenchymal elements such as synapses, axonal myelin, and oligodendrocytes in situ. In this chapter, we describe how to identify microglial cells based on their ultrastructure and their interactions with different brain cells and compartments. We describe 2D analysis using an open-source imaging processing software and the quantification of key parameters. Overall, this chapter provides a framework to analyze microglial ultrastructure and their cell–cell interactions.

Keywords Microglia, Electron microscopy, Ultrastructural analysis, QuPath

1 Introduction: Microglia

1.1 Role in Homeostasis

Microglia, the resident immune cells of the central nervous system (CNS), perform essential homeostatic functions across the lifespan. They actively survey the CNS milieu via the continuous extension and retraction of their ramified processes, thereby maintaining neuronal circuit integrity [1]. During both development and adulthood, they participate in synaptic remodeling and pruning, support neurogenesis, clear apoptotic cells and cellular debris, and secrete trophic factors that sustain neuronal viability and plasticity [2–6]. Microglia also influence axonal myelination by regulating oligodendrocyte progenitor cell proliferation, differentiation and survival, and by clearing apoptotic oligodendrocytes and myelin debris, thereby facilitating proper myelin formation and remodeling [7–9]. Among their many roles, microglia further modulate vascular structure and function, collaborating with other cells of the neurovascular unit, including endothelial cells, pericytes, and astrocytes, to regulate the

Marie-Ève Tremblay (ed.), *Ultrastructural Neuroscience Methods*, Neuromethods, vol. 238,
https://doi.org/10.1007/978-1-0716-5456-9_7,

blood–brain barrier and cerebral microcirculation [10]. Electron microscopy is a unique tool for studying, at nanoscale resolution, many microglial functions that depend on their direct contacts with other cells in the CNS.

1.2 Microglial Heterogeneity

Microglia are now recognized to exist in a continuum of states characterized by distinct transcriptomic, proteomic, morphological, and functional profiles [11]. Microglia are highly adaptable to their CNS environment, and these states reflect their dynamic nature. Many of these states were first described using single-cell RNA sequencing, such as the disease-associated microglia (DAMs) initially associated with Alzheimer's disease [12]. To provide insights into the roles of different microglial states, across contexts of health and disease and throughout the lifespan, a battery of methodologies can be employed, including electron microscopy. Additionally, in recent years, significant progress has been made in understanding sex differences in brain function and immunity [13, 14]. Microglia exhibit several sex-dependent characteristics, notably in their density, distribution, morphology, molecular profile, and ultrastructure [14, 15].

The use of advanced electron microscopy techniques, including 2D and 3D approaches, enables direct visualization of microglial cell bodies and ramified processes, along with their organelles, such as phagocytic inclusions and contacts with other parenchymal elements, including synapses. Ultrastructural signatures of cellular stress and aging can also be unravelled in microglia, thus providing insights into their health and activity [16, 17]. The aim of this chapter is to provide a tool to recognize and describe microglia at a nanoscale resolution, along with their interactions with other cell types and cellular structures. Using an open-source image-processing software, we will describe the analysis of microglial organelles under homeostatic and distressed conditions. We will also explain how to recognize other cell types and study their interactions with microglial soma and processes.

2 How to Recognize Microglia from the Electron Microscopy Angle?

Electron microscopy can elucidate ultrastructural features that cannot be resolved using light microscopy. Microglia can be identified either with or without immunostaining [17, 18].

Staining reduces the likelihood of misidentifying microglia as other ultrastructurally similar cells, such as oligodendrocytes. Staining for ionized calcium-binding adaptor molecule (Iba)1, revealed with peroxidase 3,3'-diaminobenzidine (DAB), which produces an electron-dense precipitate, or with immunogold conjugated to streptavidin or secondary antibodies, is commonly used to identify microglia [19–21]. The extent of electron-dense precipitates can vary, sometimes altering the visibility of organelles such as phagocytic inclusions [22]. The decision to perform immunostaining is at the discretion of the researcher. Whether stained or unstained, microglia are readily recognized at the ultrastructural level by their size and morphology. They have small cell bodies and a high nuclear-to-cytoplasmic ratio, and various other distinctive features [16] described in detail below.

2.1 Microglial Nucleus and Processes

Nuclear Heterochromatin Pattern and Shape

Microglial cell bodies are easily recognizable by their characteristic "cheetah pattern" of heterochromatin within the nucleus. Although disease conditions can disrupt this nuclear chromatin pattern, as in dark microglia, a microglial ultrastructural state identified using electron microscopy, this pattern is a reliable marker of microglial cells in homeostasis [23]. The microglial nucleus is relatively small compared with other cell types (e.g., oligodendrocytes, astrocytes, and neurons) and often triangular in shape [24, 25].

The microglial nucleus exhibits a sparse pattern of electron-dense heterochromatin and electron-lucent nucleoplasm, forming black-and-gray globular structures. Clumps of peripheral heterochromatin typically accumulate along the edges of the nuclear envelope, producing a darker periphery [16].

Microglial Cytoplasm, Cell Body and Processes

The microglial cytoplasm is generally electron-dense, making the cell appear darker than neurons or astrocytes. It is the smallest glial cell in the CNS parenchyma [16].

Typical microglial cell bodies display a thin enclosure of cytoplasm around their nucleus in a single ultrathin section. These generally comprise numerous mitochondria, long stretches of endoplasmic reticulum (ER) cisternae, frequent lipidic inclusions (e.g., lipid bodies/droplets, and lysosomes). Unlike oligodendrocytes, which can be similar in appearance to microglia, the organelles in microglial cells are distributed throughout the cytoplasm, whereas in oligodendrocytes, they are often concentrated on one side. Microglial cell bodies not only contact synaptic elements, such as axon terminals, dendrites and spines, and perisynaptic astrocytic processes, but also occupy satellite positions (i.e., with their cell body directly touching the cell body of other cells) relative to other glial and neuronal cells. The microglial organelles and contacts will be further discussed in detail in the next sections [16, 17, 26–28].

Microglial processes are long, ramified, and thin; they also have irregular contours and are rarely seen protruding out from the cell body in typical/homeostatic microglia when examined in a single ultrathin section. Similar to the cell body, microglial processes contain dense cytoplasm and house multiple vesicles, including endosomes and phagocytic inclusions. They frequently interact with synaptic elements, such as axon terminals, dendrites and spines, and perisynaptic astrocytic processes, and can encircle part of the synaptic clefts. Additionally, their processes can be associated with pockets of extracellular space that contain degraded elements such as membranes, suggestive of extracellular degradation or digestive exophagy [16, 17, 26–28].

Figure 1 shows examples of a microglial cell body, highlighting unique features of the heterochromatin pattern and cytoplasmic organelles.

2.2 Microglial Organelles

Microglial Endoplasmic Reticulum

In homeostatic conditions, microglial ER appears as narrow flattened cisternae that are sparsely distributed throughout the perinuclear cytoplasm [17] (Fig. 1A, a). When compared to those of oligodendrocytes, microglia's ER cisternae appear longer. Microglial ER becomes dilated under pathological contexts, such as in mouse models of Alzheimer's disease [23], Parkinson's disease [29], Huntington's disease [28], and chronic social defeat stress [30]. This feature can

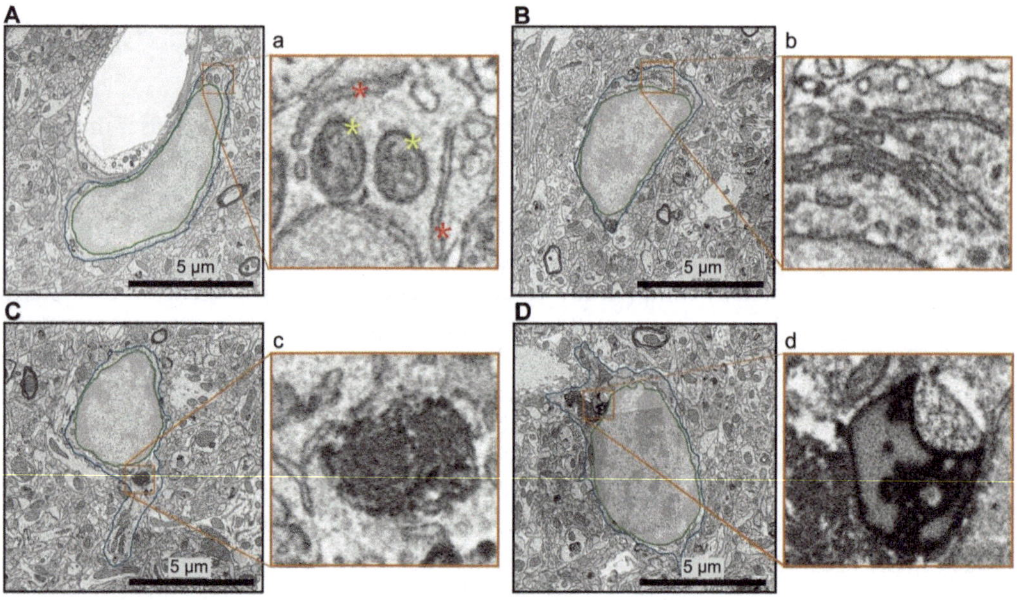

Fig. 1 Microglial organelles under electron microscopy. Panels show pictures of Iba1 immunopositive (+)/ immunonegative (−) microglia from the ventral hippocampus CA1 *stratum radiatum* of adult male wild-type mice. (**A**) Microglia recognized by their characteristic nuclear pattern close to a blood vessel, (**a**) Inset [2×] showing mitochondria (yellow asterisk) and endoplasmic reticulum stretches (red asterisk). (**B**) Microglia recognized by their characteristic nuclear pattern, (**b**) Inset (2×) showing a Golgi apparatus. (**C**) Microglial cell body and part of a process, which contains a lysosome, (**c**) Inset (2×) depicting the lysosome. (**D**) Microglia containing a lysosome and a lipid droplet, (**d**) Inset (2×) showing the lipid droplet in detail. Green outline = nuclear membrane; blue outline = plasma membrane. Scale: 5 µm

be measured by assessing the distance between the two ER cisternae membranes. Dilation is typically defined as a distance greater than 100 nm, indicative of a swollen lumen [25]. These ultrastructural alterations reflect ER stress, which is linked to the accumulation of unfolded or misfolded proteins within this organelle [25, 31]. Under stressful situations, such as chronic social defeat stress in mice, increased contacts between the ER and mitochondria have been described in hippocampal microglia [30].

Microglial Golgi Apparatus

The Golgi apparatus of microglia is commonly observed as a compact set of stacked cisternae and saccules located in the cytoplasm close to the nuclei [17] (Fig. 1B, b). Also, numerous vesicles, probably Golgi-derived, are usually observable close to the stacked cisternae. Similar to the ultrastructural alterations in the ER under stress conditions, the Golgi apparatus can be found dilated, for example, in mouse models of Alzheimer's disease and Parkinson's disease [23, 29]. Golgi stress is associated with disrupted post-translational modifications and trafficking of secretory and membrane proteins [32].

Microglial Mitochondria

In homeostasis, microglial mitochondria are elongated, oval-shaped, and well-structured with inner and outer membranes, and clearly visible cristae [17] (Fig. 1A, a). Mitochondria are found in both the cytoplasm and the processes of microglia, although not all types of ramification contain mitochondria, at least in ultrathin sections [33]. Ultrastructural alterations include distortion of their shape into a donut-like shape or when they become elongated (over 1000 nm), disruptions of the inner and/or outer membranes, swelling or degradation of the cristae (electron-lucent pockets), and the presence of "holes" or inclusions [23, 25]. Mitochondria exhibiting any of these changes are considered to be dystrophic, and their accumulation can be quantified per cell, as a proportion of total mitochondria, or by assessing the fraction of microglia containing at least one altered mitochondrion [25]. Dystrophic microglia are commonly observed in neurodegenerative conditions, including mouse models of Alzheimer's disease [23, 34] and spinal cord injury [35], as well as in human postmortem samples from aged individuals [23].

3 Microglial Lysosomes/Lipofuscin

Lysosomes are single-membrane organelles that degrade and recycle phagocytosed materials. In electron microscopy, their maturation stages can be distinguished by variations in morphology and electron density. Primary lysosomes are small (diameter of 0.3–0.5 μ m), spherical, and display a fine, homogeneous granular content. Secondary lysosomes are larger (diameter 1–2 μm) and irregular in outline, with heterogeneous contents often comprising lipid droplets

(Fig. 1C, c). Tertiary lysosomes, the most mature form, are the largest (diameter of 1.5–2.5 μm) and show irregular shapes with multiple lipid inclusions, frequently juxtaposing lipofuscin granules [18]. Lipofuscin represents an accumulation of undigested material associated with lysosomes and appears electron-dense when stained with osmium tetroxide [25].

4 Microglial Lipid Droplets

In electron microscopy, lipid droplets typically appear rounded with an even, smooth surface (Fig. 1D, d). They can appear in shades of black or gray, or display a dark outer edge enclosing a paler center. These structures often cluster to form lipid bodies and are commonly located near secondary or, more frequently, tertiary lysosomes [25]. In homeostasis, microglia do not present many lipid droplets; however, under inflammatory conditions, such as in aging or neurodegenerative diseases, they become prominent [36, 37].

5 Microglial Phagosomes/Autophagosomes

Consistent with microglia's phagocytic nature, phagosomes and their cargo can be identified within the cytoplasm, providing information on phagocytic activity across different contexts. Phagosomes are identified as single-membrane vesicles formed after the engulfment of extracellular material. This distinguishes them from autophagosomes, which are spherical, double-membrane organelles formed around cytoplasmic cargo [38]. Depending on the sample preparation, the two lipidic membranes of the autophagosome could be visualized as two thin ones (cryofixation) or a thick one (osmium tetroxide) [38]. To study autophagy intermediates, performing immunostaining for microtubule-associated protein light chain 3 (LC3) can help identify different structures involved in the autophagic pathway [38]. Early phagosomes are less electron-dense than lysosomes and contain recognizable material not yet degraded, such as cellular membranes, myelin fragments [35], or synaptic components [1]. Phagosomes can also be distinguished from endosomes: they are usually larger and contain large extracellular cargo, whereas endosomes are smaller and contain smaller cargo [39]. As maturation proceeds, phagosomes fuse with endosomes or lysosomes, forming phagolysosomes [39]. These organelles are more electron-dense and show greater degradation, as evidenced by an increased heterogeneity. They are usually found close to or partially fused with secondary or tertiary lysosomes [35, 38, 39].

5.1 Microglial Cell Body/Processes Contacts

Microglial cell body/process contacts refer to the direct physical interaction between the microglial cell body or processes with different cells or structures of the CNS [40]. In this section, we describe how to identify other cells and structures and how to study microglial interactions with them.

Neurons

Neurons generally show a large, round euchromatic nucleus with small patches of heterochromatin beneath the nuclear membrane, with many visible nuclear pores. They usually possess one or more nucleoli (Fig. 2A). Their soma shows long cisternae of the ER and the Golgi apparatus, abundant mitochondria, neurofilaments, and microtubules [16]. Under electron microscopy, dendrites and axons can be identified, along with synapses. Excitatory synapses, typically glutamatergic in the hippocampus, exhibit asymmetrical junctions with prominent postsynaptic densities and spheroid vesicles (Fig. 2A, a), whereas inhibitory, usually GABAergic synapses, display symmetrical junctions, lack a prominent postsynaptic density, and contain ellipsoidal vesicles [41]. Microglia often contact excitatory synaptic elements: axon terminals (Fig. 2A, b), dendritic spines (Fig. 2A, c), and synaptic clefts (Fig. 2A, a) in the mouse cerebral cortex and hippocampus [1]. Microglia can engulf presynaptic materials, which are observed as membrane-bound structures in the microglial cytoplasm containing 40 nm vesicles or structures reminiscent of juxtaposed presynaptic and postsynaptic elements [4]. In addition, microglia can perform trogocytosis of presynaptic structures, which means nibbling small membrane fragments [42]. Microglia can additionally perform synaptic stripping, through which they physically separate pre- and post-synaptic elements by their intervening processes as a protective mechanism under pro-inflammatory conditions [43] and neurodegeneration [44].

Astrocytes

Astrocytes exhibit stellate cell bodies with branched processes extending into the neuropil [16]. They display pale nuclei with a thin heterochromatin rim and light cytoplasm containing 8–12 nm intermediate filaments [16] (Fig. 2B). Their cytoplasm shows little electron-dense material. In preparations with osmium post-fixation, astrocytic soma and processes are distinguished by lighter mitochondrial membranes than those of neighboring neurons, glia, and vascular cells [16]. Using 3D reconstruction, it was possible to determine their spongiform morphology and their mitochondrial network [45]. Microglia have been reported to contact astrocytic cell bodies and processes [22], including their endfeet that ensheath the vasculature [46] (Fig. 2C) and perisynaptic processes that wrap around synapses [47].

Blood Vessels/Astrocytic Endfeet

Identifying microglial contacts with blood vessels by electron microscopy requires first localizing the vasculature and the blood–brain barrier, defined by endothelial cells joined by tight junctions,

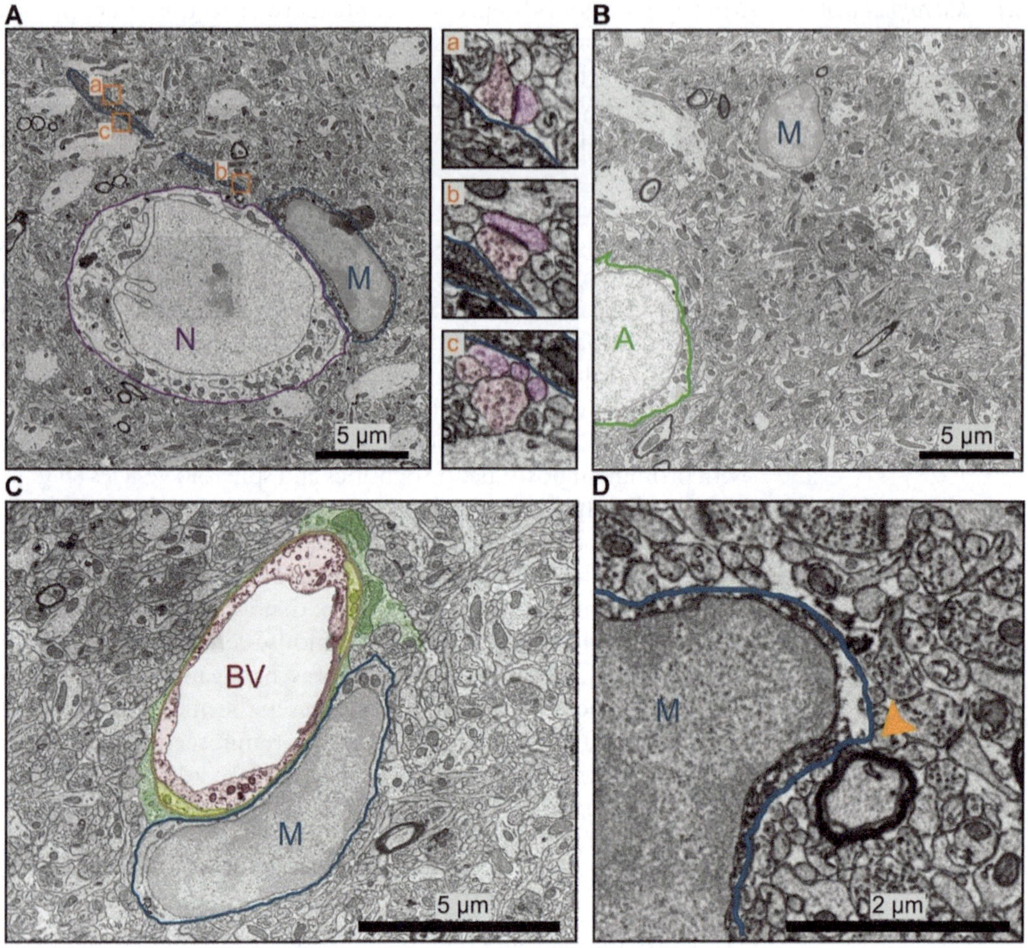

Fig. 2 Identification of the cell types and compartments contacted by microglia. Panels show pictures of Iba1+/− microglia from the ventral hippocampus CA1 *stratum radiatum* of adult male wild-type mice. (**A**) Immunolabeled microglia (Iba1) shown by a dark and spotted cytoplasm (M, outlined in blue) in close contact with a neuronal soma (N, outlined in purple), Insets showing a microglial process touching a synaptic cleft (**a**), a presynaptic element (**b**), and a post-synaptic element (**c**). In pink, the presynapse is highlighted, and the post-synapse is in purple. (**B**) Example of an astrocytic soma showing its nuclear pattern (**A**, outlined in green). (**C**) Microglia (M, outlined in blue) close to a blood vessel (BV). The components of the blood–brain barrier can be identified: endothelial cells in red, pericytes in yellow, and astrocytic endfeet in green. (**D**) Microglia (M, outlined in blue) next to a myelinated axon, touching the myelin sheath (orange arrowhead)

underlain by pericytes, and enclosed within a prominent basement membrane, with astrocytic endfeet [16] (Fig. 2C). Once this structure is observed, the number of microglial processes directly contacting the different cell types that form the blood–brain barrier, as well as the length of these contacts, can be assessed [1]. Microglia–vascular contacts are characterized by direct apposition of the microglial plasma membrane to the vascular basement membrane, typically in regions where the astrocytic endfeet layer is incomplete or absent [48, 49]. It has been shown that microglia of the mouse visual cortex directly contact blood vessels through gaps between adjacent astrocyte endfeet [48]. The contact points are predominantly located near the microglial soma, with their frequency increasing in microglia residing closer to blood vessels [48]. For the contacted cell types, endothelial cells are the most abundant, followed by pericytes [48]. Microglial coverage of blood vessels in the cerebral cortex changes during development [49]. In the early postnatal mouse, microglia associate with large capillaries, mirroring microglia–vascular interactions observed in the developing human brain. As astrocytic endfeet progressively expand and establish continuous coverage along the vasculature, microglial motility and the frequency of vessel-associated contacts markedly decrease [49].

6 Oligodendrocytes

Electron microscopy enables discrimination between oligodendrocyte progenitor cells (OPCs) and oligodendrocytes based on their ultrastructural differences. Mouse OPCs have irregularly shaped nuclei and pale, electron-lucent cytoplasm, which is usually shifted to one side of the cell. They have short ER stretches and small mitochondria [16]. Human OPCs are star-shaped cells with an electron-lucent nucleus containing non-condensed euchromatin [50]. Their soma extends multiple filamentous processes, and their cytoplasm is electron-lucent, rich in short ER, mitochondria, and ribosomes, similar to oligodendrocytes [50]. Mouse oligodendrocytes feature a dark, electron-dense cytoplasm and a nucleus with a heterogeneous chromatin pattern. Their cytoplasm is typically squarish or rectangular, and near the nucleus, they have short, wide ER cisternae and ribosomes [16]. Similarly, in humans, mature oligodendrocytes are small, electron-dense cells scattered throughout the white matter, interspersed among myelinated axons [50]. Their chromatin is highly condensed, often attached to the inner nuclear membrane [50]. The cytoplasm contains short, dilated ER, mitochondria, dense degradation vesicles, and polyribosomes [50]. If immunolabeling against differentiation-stage-specific markers is performed, premature oligodendrocytes can also be differentiated [50]. Oligodendrocytes are the myelinating cells of the CNS, and electron microscopy provides useful information about myelin ultrastructure, including its integrity and compaction (Fig. 2D). Microglia are involved in myelin phagocytosis and remodeling. Microglia contact both OPCs and oligodendrocytes, as well as myelinated axons. Microglia were observed early in mouse development to be phagocytosing OPCs [51] and abnormal myelin [9]. In the corpus callosum of rats, microglial processes were described as contacting the axolemma at nodes of Ranvier, where they were interacting with the paranodal myelin sheath [52]. Microglia were observed adjacent to oligodendrocytes in human postmortem samples of patients with schizophrenia, where they acquired a dystrophic ultrastructure [53]. Microglia were also observed close to abnormal myelin in postmortem samples from patients with multiple sclerosis [54].

7 Space Pockets/Extracellular Digestion

Microglia frequently contact pockets of extracellular space, as first described in the visual cortex of juvenile mice [1]. These pockets, composed of interstitial fluid and diverse extracellular matrix proteins, are observed as electrolucent compartments lacking a delineating membrane under electron microscopy [1]. Macrophages and microglia can also perform digestive exophagy, which has been described using electron microscopy [55]. Digestive exophagy refers to the process by which macrophages and microglia digest large objects, like deposits of lipoproteins, amyloid fibrils or dead cells, by creating a seal with the object to acidify and secrete lysosomal enzymes to hydrolyze the target without taking it up [22, 55, 56].

8 Animals and Materials

Pictures and examples shown are from C57BL/6 J adult male wild-type mice. Experiments were done in accordance with the animal care committee from the University of Victoria (AUP 2020-013). The detailed protocol for sample preparation was previously described [57, 58].

Briefly, at postnatal day (P)110, mice were anesthetized with a mix of ketamine (100 mg/kg)/xylazine (10 mg/kg) and intracardially perfused with 4% paraformaldehyde/0.6% glutaraldehyde. After skull removal, brains were post-fixed with 4% paraformaldehyde for 2 h at 4°C. Coronal brain sections at a 50-μm thickness were cut in ice-cold phosphate-buffered saline (PBS) using a vibratome (VT1200S, Leica Biosystems). Images are representative of the ventral portion of the hippocampus (CA1 *stratum radiatum*, Bregma −2.79 mm to −3.63 mm) selected based on the stereotaxic atlas of Paxinos and Franklin [59].

Sections were immunostained against Iba1 (1:1000, cat# 019-19741, FUJIFILM Wako Chemical). The tissue was enhanced for 60 min with a solution of 4% aqueous osmium tetroxide (cat# 19170, Electron Microscopy Sciences) and 3% ferrocyanide (cat# PFC232.250, BioShop). An aqueous solution made of 1% thiocarbohydrazide (cat# 2231-57-4, Electron Microscopy Sciences) was added for 20 min for contrast. The osmium-enhanced sections underwent dehydration in various concentrations of ethanol (35%–100%), followed by three short washes of propylene oxide. Tissue sections were immersed in Durcupan ACM resin overnight (cat# 44611–44614, Merck KGaA Darmstadt) and then sandwiched between two fluoropolymer sheets (ACLAR®; cat# 50425-25, Electron Microscopy Sciences). After polymerization at 55°C for 3 days, the region of interest was located and excised from the resin-embedded sheets and glued to resin blocks [57, 58].

Ultrathin sections (~75 nm, ARTOS 3D ultramicrotome, Leica Biosystems) were cutout from the blocks and mounted onto a silicon nitride chip, and attached to specimen stubs for EM. A Crossbeam 350 field-emission scanning electron microscope (SEM) with a Gemini column (Zeiss) was operated at 1.4 kV and 1.2 nA to image microglial cell bodies at 5 nm resolution [57, 58].

Our ultrastructural analysis is conducted using QuPath (The University of Edinburgh and The Queen's University of Belfast; v.0.4.4; https://qupath.github.io/), which is available for free. This software enables the creation of annotation classes to manually and easily mark each organelle and/or cell-cell interaction. Additionally, we use ImageJ/FIJI (NIH; v.1.53; https://imagej.net/software/fiji/downloads) to delineate and measure the area of our regions of interest.

9 Methods: 2D Analysis of Microglial Ultrastructure

9.1 A Quick Guide to QuPath: How to Create a Project and Load Your Images

Once QuPath is open, create a new project using the "**Create project**" menu on the left. You will need to select an empty folder where QuPath will save all project files. Make sure to keep all project-related files in this folder. To add images, use the "**Add images**" button in the left menu. This allows you to import the images you want to analyze within this project. It is important to keep the images in their original directory. If you move them, QuPath will prompt you to update the file location. It is important to blind the images to the conditions prior to the analyses.

If no immunostaining is performed, always have someone else confirm the identification of the cells first, as microglia and oligodendrocytes can be somewhat similar. It is important to have the same investigator do the analyses from start to end.

You can reopen your project at any time with the loaded images and saved annotations to continue your work. Simply click the "**Open project**" button in the left menu and select the folder you initially created. To manually save the project at any time, go to "**File**" → "**Save**".

9.2 Creating the Categories

Now that you have created your project and loaded your images, it is time to define the microglial categories in QuPath for your analysis.

First, navigate to **"Annotations,"** then click on the three dots at the edge of the menu. Select **"Add/Remove"** and enter the name of the class you want to analyze in your microglia. This will allow you to categorize organelles and contacts and differentiate microglial ultrastructural states within your project. If you double-click the annotation, you can switch colors of the tags. This might facilitate grouping features. The annotations will be saved within your project; there is no need to redo them every time QuPath is opened.
For microglia, we regularly perform the following analyses:
Contacts:

- Presynaptic element
- Postsynaptic element
- Myelin contact
- Astrocyte cell body
- Neuron cell body
- Oligodendrocyte cell body
- Blood vessel
- Astrocytic endfeet contact
- Iba1 staining of cell body and processes
- Extracellular space
- Extracellular digestion

Organelles:

- Mitochondria
- Elongated mitochondria
- Dystrophic mitochondria
- Dystrophic and elongated mitochondria
- Primary lysosome
- Secondary lysosome
- Tertiary lysosome
- Lipofuscin granules
- Lipid droplet
- ER
- Golgi apparatus
- Dilated ER/Gogi apparatus
- Empty phagosome
- Filled phagosome
- Empty autophagosome
- Filled autophagosome

More features can be added to the analysis as desired. For example, if under a model of Alzheimer's disease pathology, the researcher might add "Dystrophic neurites" or "Amyloid plaques" [26]. Steps 4A and B are shown in Fig. 3.

9.3 Membrane/ Organelle Outlines

For some analyses, it is helpful to measure the plasma membrane, nuclear membrane, and enlarged organelles, as microglia have been shown to enlarge and change their features under stress and pathological conditions. We recommend keeping both QuPath and ImageJ open with the same image simultaneously to facilitate accurate measurements. Open ImageJ, go to **Analyze → Set Scale**, and input your scale (we usually use 1 pixel = 5 nm for 5 nm resolution images). Be sure to click "Global" to apply this scale across all measurements. Then, go to **Analyze → Set Measurements** and ensure **Area, Shape Descriptors, and Perimeter** are checked.

Next, zoom in on the microglia to clearly observe its boundaries and structure. In the top menu, select the **Polygon Tool** (third option), or the **Freehand Tool** (fourth option, bean-shaped icon) for a more fluid manual outline. Carefully trace around the plasma membrane and nuclear membrane. This will also help you identify contacts and organelles, since you now have a clear delineation of the cell edges. You can click 'T' on your keyboard to open the **ROI Manager** and rename/save this region of interest.

To assess elongated mitochondria or dilated ER/Golgi apparatus cisternae, we measure their dimensions. Using the **Straight Tool** (fifth option), drag and draw a straight line on any structure to measure it. For mitochondria, we assess membrane length; if it exceeds 1000 nm, the mitochondria are classified as elongated. In this case,

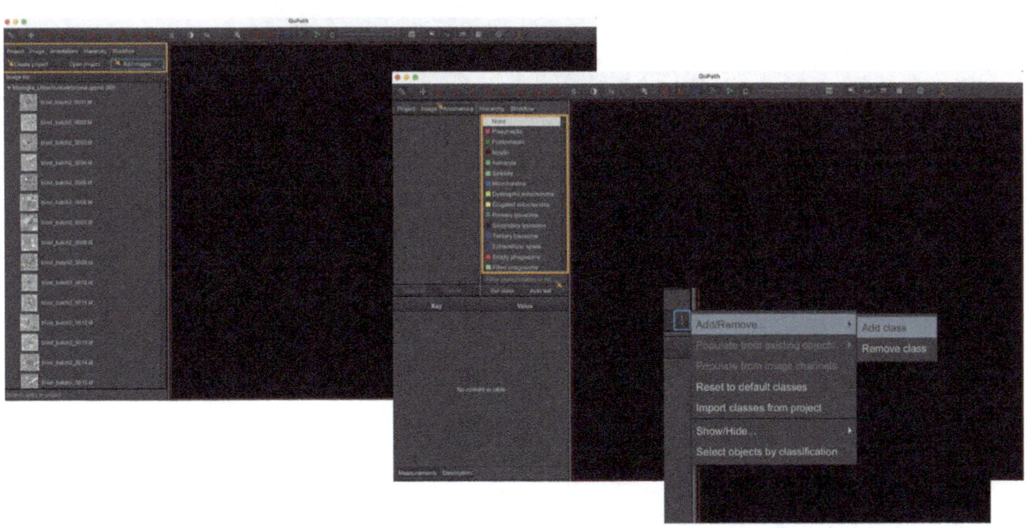

Fig. 3 QuPath workflow. Representation of how to create a new project, add images and create annotations. Created with BioRender.com

an "elongated mitochondria" class should be added in QuPath. For a dilated ER/Golgi apparatus, we measure the intracisternal distance, which should exceed 100 nm to be considered dilated. Other measurements may also be valuable depending on the research objectives. For example, one could measure the size of all mitochondria or quantify extracellular digestion to gain further insights into cellular processes. Hit "M" on your keyboard, and the measurements will show up. Copy and paste the data into a Microsoft Excel sheet and save. Step 4C is shown in Fig. 4.

9.4 Classifying Microglial Contacts and Organelles

Once your project and annotations are saved, you can begin the analysis. Go to **"Project,"** then, in the top menu, click on the **"Annotation**

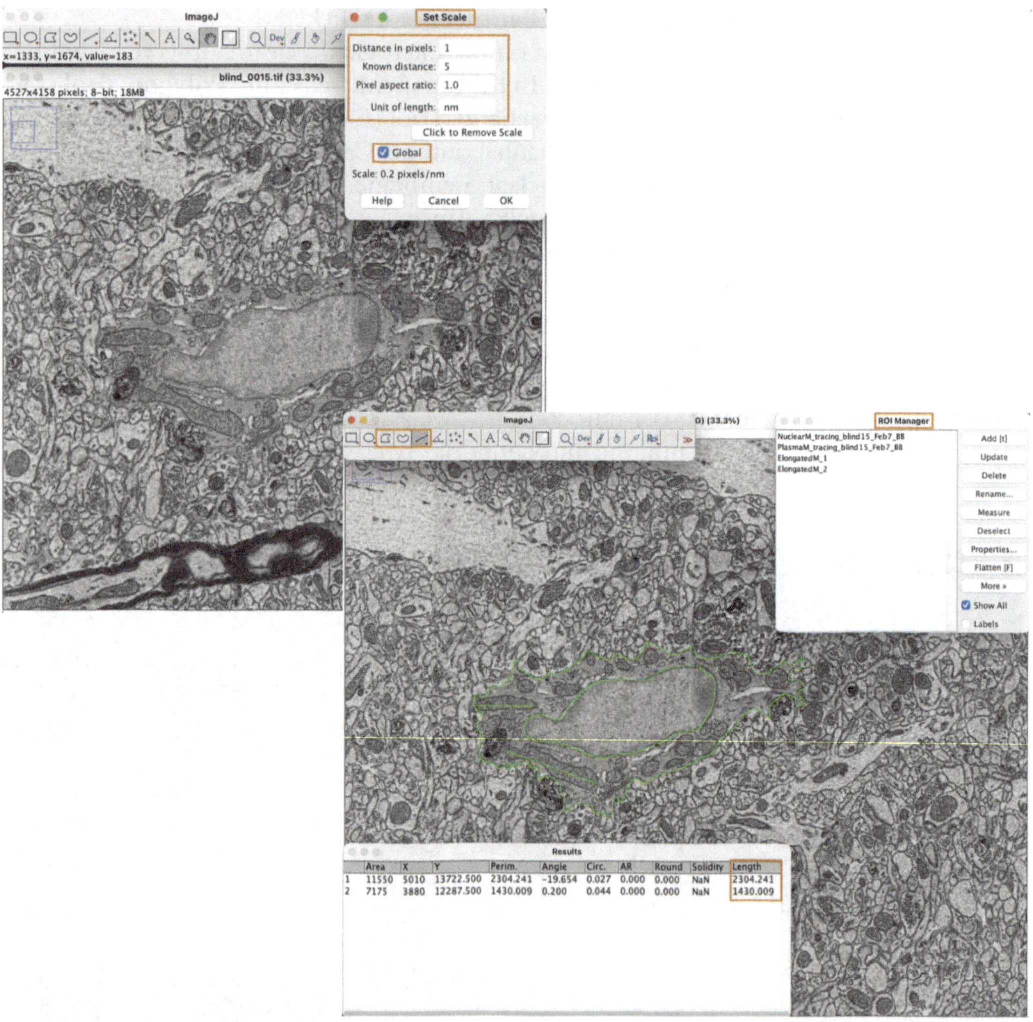

Fig. 4 ImageJ workflow. Representation of how to set the scale, trace, and measure some features of the microglia. The example image is of a randomly selected Iba1- microglia from the ventral hippocampus CA1 *stratum radiatum* of an adult male wild-type mouse. Created with BioRender.com

Counting Tool" (represented by three outlined dots). A small window will appear. Next, click on the three dots button at the edge of this window and select **"Add points for all classes."** This will generate annotations for all the classes you created, allowing you to count and analyze them directly on your images. New classes can also be added during the analysis as you perceive that a feature should also be considered. In the counting Menu, hit "Add" and simply add a new annotation.

After that, all the classes you created will be displayed. You can now select the class you want to start with and click directly on the organelle or contact to mark it. We recommend starting with contacts, followed by organelles. Detailed instructions and examples for identifying microglial organelles and contacts are provided in Sect. 2 above.

Zoom in well on the microglia and add the classes accordingly. You can click and hold the annotation to move it. To erase a marked annotation, hit option "+" click on it. We conducted our analysis on a MacBook; however, QuPath also works well on Windows computers. Note that some commands may differ slightly between operating systems.

Repeat this process for each image in your project. Step 4D is shown in Fig. 5. An example of images with annotations is shown in Fig. 6.

9.5 Exporting QuPath Data

Complete all your analyses before exporting your data. To export your measurements as a text file that can be easily pasted into **Excel**, go to **Measure → Export Measurements**. Click on the arrow and select all available images. Choose the destination folder where the output file will be saved – preferably in the same folder as your project. Under **Export Type**, select **Annotations** to ensure all your marked classes are included. Click **"Populate"** under **Columns to Include** and select:

- **Image**
- **Name**
- **Class**
- **Num Points**

Other export options are also available; review them carefully and select those that best fit your research objectives. The output file is named "measurements.tsv." While the Image option retrieves the image name, the Name and Class options store your annotations. It is important to include both **Name** and **Class** in your export. Of note, if any new annotation was added during the analysis, it will appear under **"Name"** only. Hence, it is crucial to make sure all annotations are properly tracked and categorized. Finally, Num Points shows how many times the annotations appeared in that image.

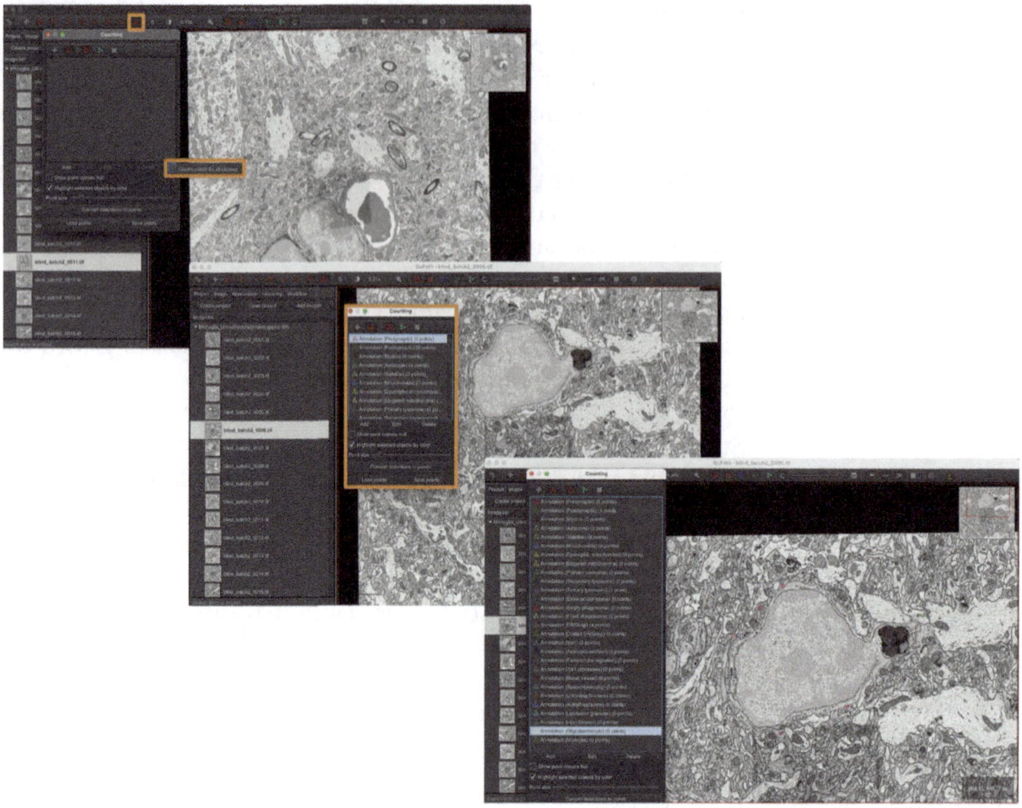

Fig. 5 QuPath counting tool for microglial classification. Representation of how to add the annotations into the counting tool in order to analyze and classify microglia. The panels contain pictures of Iba1+/− microglia from the ventral hippocampus CA1 *stratum radiatum* of adult male wild-type mice. Created with BioRender.com

Select all images and the parameter list and export. Then this information can be further organized and segregated for statistical testing by compiling it into, for example, a Microsoft Excel worksheet (Microsoft Corporation; v16.93.1). Step 4E is shown in Fig. 7.

10 Future Direction: 3D reconstruction and cryofixation

As described in this chapter, analysis of 2D images obtained by electron microscopy provides useful ultrastructural information of microglial organelles and contacts with other cells. However, it is difficult to accurately infer their complete morphology, spatial organization or interaction with various nearby elements. With advances in technology, three-dimensional (3D) electron microscopy, such as serial block-face scanning electron microscopy (SBEM) [60] and focused ion beam-scanning electron microscopy (FIB-SEM), allows the volumetric reconstruction of entire microglial cells and their adjacent environment [17]. This provides precise information on

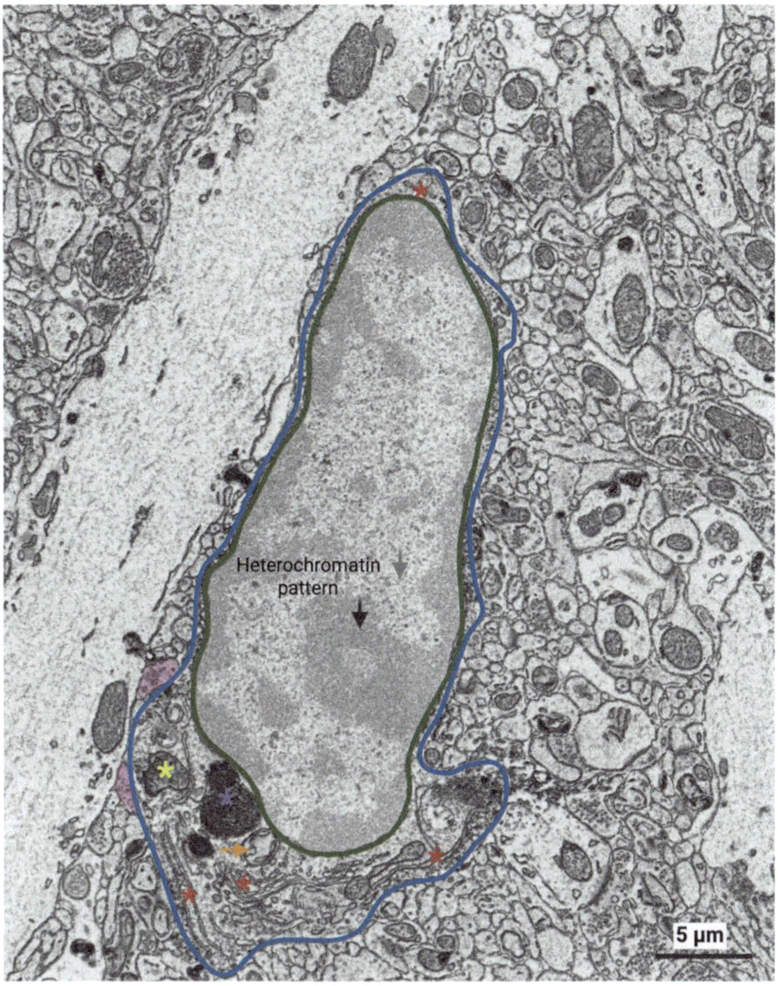

Fig. 6 Microglial cell body seen with a scanning electron microscope at 5 nm resolution. Panel show a picture of a selected Iba1- microglia from the CA1 *stratum radiatum* of the ventral hippocampus of an adult male wild-type mouse. This example highlights the unique "cheetah-pattern" of microglia heterochromatin and some organelles in the electron-dense cytoplasm. Green outline = nuclear membrane; blue outline = plasma membrane; black and gray arrows = globular shapes of dichromatic heterochromatin; orange arrow = filled phagosome; yellow asterisk = mitochondria; red asterisk = endoplasmic reticulum stretches and/or Golgi apparatus; purple asterisk = secondary lysosome. Scale: 5 μm. Created with BioRender.com

their branching and organelle distribution, accurate cell volume, and enables observation of all simultaneous contacts with neighboring cells or elements within a single cell. 3D microglia rendering has already been obtained in the mouse hippocampus [60] and the rat somatosensorial cortex [61]. In terms of studying microglial phagocytic functions, 3D electron microscopy is needed to distinguish

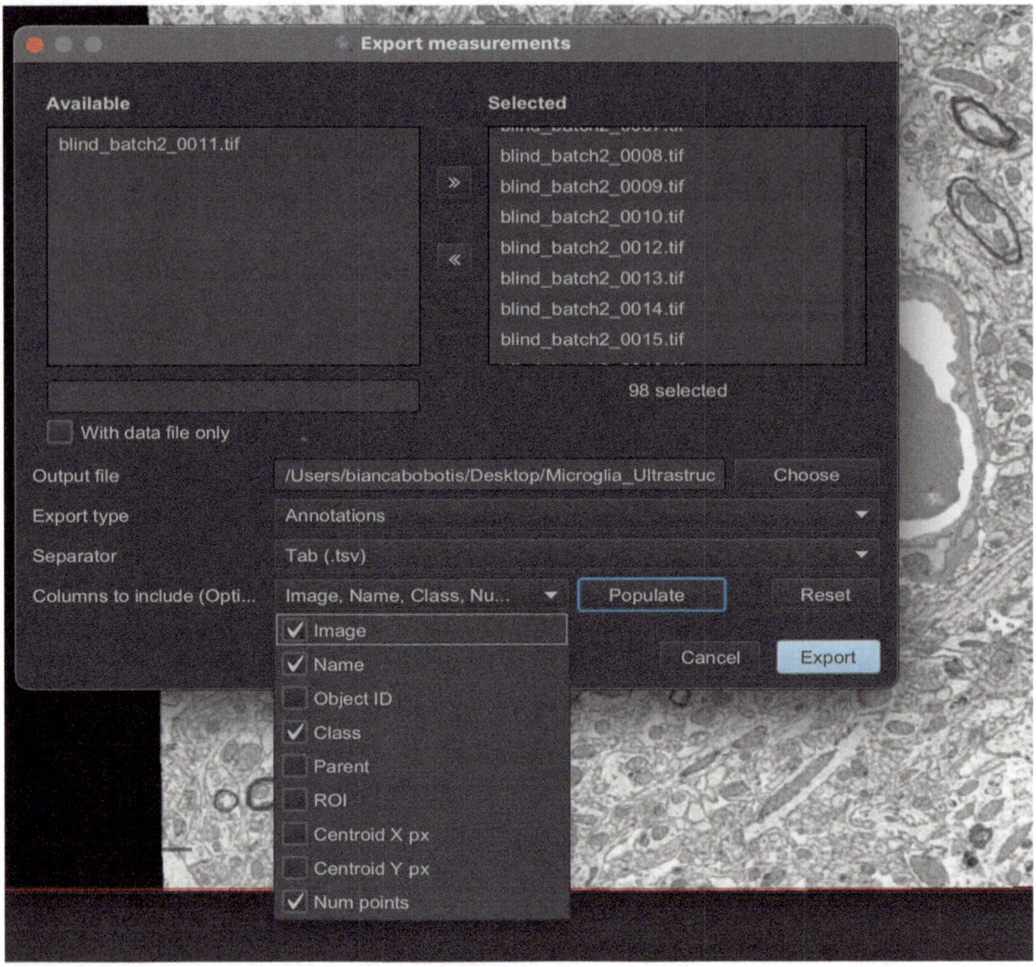

Fig. 7 Exporting data from QuPath. Representation of how to export your counting/annotation data from QuPath. Created with BioRender.com

between partial encirclement, synaptic stripping, phagocytosis, and trogocytosis, among other mechanisms [42].

Last but not least, sample preparation has a huge impact on ultrastructure. The field is moving to cryofixation to prepare the samples to avoid some artifacts that can arise with aldehyde fixation [62]. For example, the latter can affect the study of direct contacts since the extracellular space shrinks, which can lead to overestimation of contacts [62].

Acknowledgments

We acknowledge with respect the Lekwungen-speaking peoples on whose traditional territory the University of Victoria stands and the Songhees, Esquimalt, and WSÁNEC peoples whose historical relationships with the land continue to this day. M-ÈT holds a Tier 1 Canada Research Chair in *Neurobiology of Healthy Cognitive Aging* (CRC-2024-00155). The Tremblay Lab's Zeiss Crossbeam 350 scanning electron microscope was acquired with funding from a Canada Foundation for Innovation John R. Evans Leaders Fund grant (#39965 Laboratory of ultrastructural insights into the neurobiology of aging and cognition). This work was funded by a Discovery Grant from the Natural Sciences and Engineering Research Council of Canada (NSERC) to M-ÈT (RGPIN-2024-06043).

Conflict of Interest The authors have no conflicts of interests to declare that are relevant to the content of this chapter.

References

1. Tremblay M-È, Lowery RL, Majewska AK (2010) Microglial interactions with synapses are modulated by visual experience. PLoS Biol 8:e1000527
2. Paolicelli RC, Bolasco G, Pagani F et al (2011) Synaptic pruning by microglia is necessary for normal brain development. Science 333:1456–1458
3. Sierra A, Encinas JM, Deudero JJP et al (2010) Microglia shape adult hippocampal neurogenesis through apoptosis-coupled phagocytosis. Cell Stem Cell 7:483–495
4. Schafer DP, Lehrman EK, Kautzman AG et al (2012) Microglia sculpt postnatal neural circuits in an activity and complement-dependent manner. Neuron 74:691–705
5. Wiens KR, Wasti N, Ulloa OO et al (2024) Diversity of microglia-derived molecules with neurotrophic properties that support neurons in the central nervous system and other tissues. 29:5525
6. Parkhurst CN, Yang G, Ninan I et al (2013) Microglia promote learning-dependent synapse formation through brain-derived neurotrophic factor. Cell 155:1596–1609
7. Hughes AN, Appel B (2020) Microglia phagocytose myelin sheaths to modify developmental myelination. Nat Neurosci 23:1055–1066
8. Hagemeyer N, Hanft K-M, Akriditou M-A et al (2017) Microglia contribute to normal myelinogenesis and to oligodendrocyte progenitor maintenance during adulthood. Acta Neuropathol 134:441–458
9. Djannatian M, Radha S, Weikert U et al (2023) Myelination generates aberrant ultrastructure that is resolved by microglia. J Cell Biol 222:e202204010
10. Mayer MG, Fischer T (2024) Microglia at the blood brain barrier in health and disease. Front Cell Neurosci 18:1360195
11. Paolicelli RC, Sierra A, Stevens B et al (2022) Microglia states and nomenclature: a field at its crossroads. Neuron 110:3458–3483
12. Keren-Shaul H, Spinrad A, Weiner A et al (2017) A unique microglia type associated with restricting development of Alzheimer's disease. Cell 169:1276–1290.e17
13. Lynch MA (2022) Exploring sex-related differences in microglia may be a game-changer in precision medicine. Front Aging Neurosci 14:868448. https://doi.org/10.3389/fnagi.2022.868448
14. Bobotis BC, Braniff O, Gargus M et al (2023) Sex differences of microglia in the healthy brain from embryonic development to adulthood and across lifestyle influences. Brain Res Bull 202:110752
15. Han J, Fan Y, Zhou K et al (2021) Uncovering sex differences of rodent microglia. J Neuroinflammation 18: 74
16. Nahirney PC, Tremblay M-E (2021) Brain ultrastructure: putting the pieces together. Front Cell Dev Biol 9:629503
17. Savage JC, Picard K, González-Ibáñez F et al (2018) A brief history of microglial ultrastructure: distinctive features, phenotypes, and

functions discovered over the past 60 years by electron microscopy. Front Immunol 9:803
18. El Hajj H, Savage JC, Bisht K et al (2019) Ultrastructural evidence of microglial heterogeneity in Alzheimer's disease amyloid pathology. J Neuroinflammation 16: 87
19. González Ibanez F, Picard K, Bordeleau M et al (2019) Immunofluorescence staining using IBA1 and TMEM119 for microglial density, morphology and peripheral myeloid cell infiltration analysis in mouse brain. J Vis Exp
20. St-Pierre M-K, Bordeleau M, Tremblay M-È (2019) Visualizing dark microglia. In: Garaschuk O, Verkhratsky A (Eds.)Eds. Microglia: methods and protocols. Springer, New York, NY, 97–110
21. St-Pierre M-K, Šimončičová E, Bögi E et al (2020) Shedding light on the dark side of the microglia. ASN Neuro 12:1759091420925335
22. Shapiro LA, Perez ZD, Foresti ML et al (2009) Morphological and ultrastructural features of Iba1-immunolabeled microglial cells in the hippocampal dentate gyrus. Brain Res 1266:29–36
23. St-Pierre M-K, Carrier M, González Ibáñez F et al (2022) Ultrastructural characterization of dark microglia during aging in a mouse model of Alzheimer's disease pathology and in human post-mortem brain samples. J Neuroinflammation 19: 235
24. Bobotis B, Halvorson T, Carrier M et al (2023) Emerging techniques for the study of microglia: visualization, depletion and fate mapping, biology and life sciences
25. González Ibáñez F, VanderZwaag J, Deslauriers J et al (2024) Ultrastructural features of psychological stress resilience in the brain: a microglial perspective 14:240079
26. St-Pierre M-K, Carrier M, Lau V et al (2022) Investigating microglial ultrastructural alterations and intimate relationships with neuronal stress, dystrophy, and degeneration in mouse models of Alzheimer's disease. Methods Mol Biol 2515:29–58
27. St-Pierre M-K, VanderZwaag J, Loewen S, et al (2022) All roads lead to heterogeneity: the complex involvement of astrocytes and microglia in the pathogenesis of Alzheimer's disease 16
28. Savage JC, St-Pierre M-K, Carrier M et al (2020) Microglial physiological properties and interactions with synapses are altered at presymptomatic stages in a mouse model of Huntington's disease pathology. J Neuroinflammation 17: 98
29. Iovino L, VanderZwaag J, Kaur G et al (2024) Investigation of microglial diversity in a LRRK2 G2019S mouse model of Parkinson's disease. Neurobiol Dis 195:106481
30. Zhang J-R, Shen S-Y, Zhai M-Y et al (2024) Augmented microglial endoplasmic reticulum-mitochondria contacts mediate depression-like behavior in mice induced by chronic social defeat stress. Nat Commun 15: 5199
31. Fernández D, Geisse A, Bernales JI et al (2021) The unfolded protein response in immune cells as an emerging regulator of neuroinflammation. Front Aging Neurosci 13:682633
32. Kim WK, Choi W, Deshar B et al (2023) Golgi stress response: new insights into the pathogenesis and therapeutic targets of human diseases. Mol Cells 46:191–199
33. Pietramale AN, Bame X, Doty ME et al (2024) Mitochondria are absent from microglial processes performing surveillance, chemotaxis, and phagocytic engulfment. bioRxiv 2024.10.15.618505
34. Kim YS, Choi S-H, Kim K-Y et al (2024) AIBP controls TLR4 inflammarafts and mitochondrial dysfunction in a mouse model of Alzheimer's disease. J Neuroinflammation 21: 245
35. St-Pierre M-K, González Ibáñez F, Kroner A et al (2023) Microglia/macrophages are ultrastructurally altered by their proximity to spinal cord injury in adult female mice. J Neuroinflammation 20: 273
36. Farmer BC, Walsh AE, Kluemper JC et al (2020) Lipid droplets in neurodegenerative disorders. Front Neurosci 14
37. Marschallinger J, Iram T, Zardeneta M et al (2020) Lipid-droplet-accumulating microglia represent a dysfunctional and proinflammatory state in the aging brain. Nat Neurosci 23:194–208
38. Biazik J, Kallio K, Gudmundsson SR et al (2025) Discerning autophagy pathway intermediates: transmission electron microscopy techniques. In: Loos B, Klionsky DJ (Eds.) Eds. Autophagy— from molecular mechanisms to flux control in health and disease. Springer Nature Switzerland, Cham, 29–42
39. Hartenstein V, Martinez P (2019) Phagocytosis in cellular defense and nutrition: a food-centered approach to the evolution of macrophages. Cell Tissue Res 377:527–547
40. Wake H, Moorhouse AJ, Jinno S et al (2009) Resting microglia directly monitor the functional state of synapses in vivo and determine the fate of ischemic terminals. J Neurosci 29:3974–3980
41. Stewart MG, Popov VI, Kraev IV et al (2014) Structure and complexity of the synapse and dendritic spine. In: The Synapse. Elsevier, pp 1–20
42. Weinhard L, Bartolomei G, Bolasco G et al (2018) Microglia remodel synapses by presynaptic trogocytosis and spine head filopodia induction. Nat Commun 9:1228

43. Chen Z, Jalabi W, Hu W et al (2014) Microglial displacement of inhibitory synapses provides neuroprotection in the adult brain. Nat Commun 5: 4486
44. Hugh Perry V, O'Connor V (2010) The role of microglia in synaptic stripping and synaptic degeneration: a revised perspective. ASN Neuro 2:e00047
45. Aten S, Kiyoshi CM, Arzola EP et al (2022) Ultrastructural view of astrocyte arborization, astrocyte-astrocyte and astrocyte-synapse contacts, intracellular vesicle-like structures, and mitochondrial network. Prog Neurobiol 213:102264
46. Bellver-Landete V, Bretheau F, Mailhot B et al (2019) Microglia are an essential component of the neuroprotective scar that forms after spinal cord injury. Nat Commun 10: 518
47. Akiyoshi R, Wake H, Kato D et al (2018) Microglia enhance synapse activity to promote local network synchronization. eNeuro 5
48. Morris GP, Foster CG, Sutherland BA et al (2024) Microglia contact cerebral vasculature through gaps between astrocyte endfeet. J Cereb Blood Flow Metab 44:1472–1486
49. Mondo E, Becker SC, Kautzman AG et al (2020) A developmental analysis of juxtavascular microglia dynamics and interactions with the vasculature. J Neurosci 40:6503–6521
50. Ulloa-Navas MJ, Pérez-Borredá P, Morales-Gallel R et al (2021) Ultrastructural characterization of human oligodendrocytes and their progenitor cells by pre-embedding immunogold. Front Neuroanat 15:696376
51. Nemes-Baran AD, White DR, DeSilva TM (2020) Fractalkine-dependent microglial pruning of viable oligodendrocyte progenitor cells regulates myelination. Cell Rep 32:108047
52. Zhang J, Yang X, Zhou Y et al (2019) Direct contacts of microglia on myelin sheath and Ranvier's node in the corpus callosum in rats. J Biomed Res 33:192–200
53. Uranova NA, Vikhreva OV, Rakhmanova VI et al (2018) Ultrastructural pathology of oligodendrocytes adjacent to microglia in prefrontal white matter in schizophrenia. NPJ Schizophr 4:26
54. Prineas JW, Parratt JDE (2021) Multiple sclerosis: microglia, monocytes, and macrophage-mediated demyelination. J Neuropathol Exp Neurol 80:975–996
55. Maxfield FR, Barbosa-Lorenzi VC, Singh RK (2020) Digestive exophagy: phagocyte digestion of objects too large for phagocytosis. 21:6–12. https://pubmed.ncbi.nlm.nih.gov/39644493/
56. Jacquet RG, Ibáñez FG, Picard K, et al (2024) Microglia degrade Alzheimer's amyloid-beta deposits extracellularly via digestive exophagy. Cell Rep 43:115052. https://pubmed.ncbi.nlm.nih.gov/39644493/
57. Bobotis BC, Khakpour M, Braniff O et al (2025) Sex chromosomes and sex hormones differently shape microglial properties during normal physiological conditions in the adult mouse hippocampus. J Neuroinflammation 22: 18
58. St-Pierre M-K, Šimončičová E, Carrier M et al (2023) Microglia in human postmortem brain samples: quantitative ultrastructural analysis of scanning electron microscopy images. Methods Mol Biol 2561:63–85
59. Franklin, KBJ, Paxinos, G, 2013. Paxinos and Franklin's the mouse brain in stereotaxic coordinates. Academic Press, Elsevier, Amsterdam
60. Bolasco G, Weinhard L, Boissonnet T et al (2018) Three-dimensional nanostructure of an intact microglia cell. Front Neuroanat 12
61. Calì C, Agus M, Kare K et al (2019) 3D cellular reconstruction of cortical glia and parenchymal morphometric analysis from Serial Block-Face Electron Microscopy of juvenile rat. Prog Neurobiol 183:101696
62. Korogod N, Petersen CC, Knott GW Ultrastructural analysis of adult mouse neocortex comparing aldehyde perfusion with cryo fixation. eLife 4:e05793

Chapter 8

Reproducible Quantification Method to Differentiate Dark *vs* Intermediate *vs* Typical Appearance of Immune Cells in Electron Microscopy A Priori, Including Dark Microglia

Victor Z. Lau and Marie-Ève Tremblay

Abstract

Electron microscopy (EM) is a key technique that allows imaging at nanoscale resolution, resulting in capturing cellular information unavailable or often-missed at lower resolutions. Among its many advantages, EM has been shown to reveal "dark" states in cells including neurons and microglia, the latter of which have been shown to play increasingly important and fundamental roles in neuroscience in recent decades. Dark cells are overall characterized by their display of an especially noticeable electron-dense (i.e., "dark") nucleoplasm and cytoplasm, and are overrepresented in aging and numerous neurodegenerative diseases including Alzheimer's disease. However, quantitative metrics used to robustly differentiate dark cells regardless of a manual observer or adjacent cell have not been established. Furthermore, "intermediate" states have also been observed representing a transitional state between "typical" electron-dense cells and dark cells. Here, we present a novel protocol detailing how to reliably differentiate dark *vs* intermediate *vs* typical immune cells in EM in a quantifiable, consistent manner using predetermined thresholds (agreed on by unblinded analysis from expert observers). While the method and thresholds assigned particularly are based on previous unblinded datasets, this method can easily be trained on and applied to other dark *vs* intermediate *vs* typical cells in EM, regardless of specific cell type, species, or tissue environment (or lack thereof). We also include limitations of our technique but raise that this quantitative approach provides a more robust and reproducible classification of dark *vs* intermediate *vs* typical (central nervous system) cells over previous qualitative identification methods.

Keywords Dark microglia, Dark cell, Quantitative classification, Electron microscopy

1 Introduction

Electron microscopy (EM), which consists of transmission EM (TEM) and scanning EM (SEM), is a powerful technique used to visualize structures at the nanoscale level [1]. TEM, invented by Ernst Ruska in the early 1930s, involves transmitting electrons through

Supplementary Information The online version contains supplementary material available at https://doi.org/10.1007/978-1-0716-5456-9_8

Marie-Ève Tremblay (ed.), *Ultrastructural Neuroscience Methods*, Neuromethods, vol. 238, https://doi.org/10.1007/978-1-0716-5456-9_8,

a thin specimen to create an image. SEM, contrarily, developed in the late 1930s, utilizes scanning a focused electron beam over the surface of a specimen; SEM then measures backscattered and secondary electrons to generate an image [1]. Both TEM and SEM have significantly contributed to advancing multiple scientific fields, including biology, materials science, and nanotechnology.

In biology, visualizing nanoscale (i.e., ultrastructural) features of cells and their environments provide unique strengths in collecting simultaneous data on (i) cell/environmental features not visible or resolved at lower-resolution microscopy, and (ii) multiple features in parallel (e.g., multiple organelle and nucleus appearance, organelle stress) [1]. "Dark" cells displaying an electron-dense cytoplasm and nucleoplasm have been one such unique phenomenon observed in EM, giving them a "dark" appearance in TEM (or via inversed SEM images) [2–5]. Asides from their distinct dark cytoplasm and nucleoplasm, dark cells have been found to display markers of increased cellular stress (e.g., endoplasmic reticulum and Golgi complex dilations, mitochondrial ultrastructure length and morphology alterations, increased glycogen granules) [5–7]. While specific functions remain to be elucidated in dark cells, these associated traits implicate that dark cells are minimally dysfunctional or attempting to buffer damage (e.g., oxidative stress due to increased exposure to reactive oxygen and nitrogen species) [3]; further discerning their functions serves as an important research topic.

Dark cells have notably been observed in EM across both healthy and pathological environments, and as a phenotype across multiple cell types in the central nervous system (CNS). In animal models, astrocytes, neurons, oligodendrocytes, microglia, infiltrating macrophages, and (peri-)vascular immune cells all have been observed presenting dark states under EM [6,8,9]. *Versus* "typical" (i.e., non-dark) cells observed in EM, dark neurons have been found as over-represented in ischemia, epilepsy, and neurodegenerative disease conditions/disease-like models [2]; dark microglia are present in post-mortem samples of aged humans and patients with dementia, and have been found as significantly increased proportions over postnatal developmental, aged, chronic stress, high-fat diet, PolyI:C injections, and Alzheimer's disease-like conditions in mice [7,10–13]. Finally, dark astrocytes were also recently found in aged Alzheimer's disease-like mouse models and aged patient samples without overt dementia diagnosis [5].

While significant progress has been made towards characterizing dark cells in numerous conditions, especially within the CNS, there are minimal studies investigating and characterizing dark cells in other tissue systems or in-vitro conditions. Enriched markers and mechanistic rationales for dark cell formation still need to be robustly confirmed—a hot area of ongoing research. Finally, further complexifying dark cell identification is that microglia can be found in "intermediate" states, or with a cytoplasm and more-noticeable nucleoplasm with a representative "darkness" appearances falling between typical and dark microglia [14]. While intermediate states for other cell types have not been reported in the literature to our awareness, the likely underlying causes of dark appearances (i.e., association with increased cellular stress) reasonably argues for the existence of intermediate of numerous other cell types. Intermediate cells likely have varying levels of cellular stress between those of typical *vs* dark cells [3], but remain much more difficult outside of qualitative approaches to categorize (compared to either typical or dark cells). Altogether, this likely implies that typical to "dark" cell appearances likely belie a continuum, rather than truly-separatable discrete states.

To further reveal function(s) of both intermediate and dark cells in both healthier and disease environments, enhanced rigorous analysis would provide greater sensitivity and discrimination towards identifying, and confidently identifying traits and functions enriched in typical *vs* intermediate *vs* dark cell states. Identifying dark cells has traditionally been performed qualitatively by observers blind to the experimental conditions [4], and while quantitative methods have begun to be employed for identifying these cell states in EM [15], we sought to employ an improved universal, easily-trainable method that could apply to single cells with minimum bias and requirements to conduct, and determine darkness in cells.

Here, we present a simple quantitative, and reliable method using ImageJ that is robust against bias in (i) measuring region of interest (ROI) pixel intensities within the same cell, (ii) calculating ratios out of the measured ROIs, and (iii) assign the resulting ratio to one of three predetermined threshold values to determine if a cell is typical, intermediate or dark. As the same cell is used for all analysis components, our presented method is robust to differences across images in terms of initial imaging conditions (e.g., brightness, contrast) and local environments (i.e., robust analysis regardless of the presence/absence of surrounding tissue and cells). By predetermining the ROI area shape and size range as well detailed below, we find our method to be robust against possible cherry-picking. Finally, due to our focus and previous research on dark microglia, particular threshold values are included here to specifically apply to dark immune cells [including dark microglia and (peri-)vascular immune cells]. However, determining threshold values a priori can easily be done with unblinded data of other cell populations observed in EM.

2 Materials

ImageJ or Fiji

Spreadsheet software (e.g., see Supplementary Figure 1).

The authors assume that the reader is working with pregenerated EM images; for acquisition of EM images and processing tissue for EM quality, detailed protocols can be found elsewhere [16,17] (*see* **Note 1**).

3 Methods

1. Open the EM image on ImageJ or Fiji. Ensure that the EM image and data fulfill the appropriate staining conditions (*see* **Note 1**).
2. Analyze → Set measurements → Tick for the following: "Area," "Standard deviation," "Skewness," "Mean (gray value)," "Median (gray value)" (*see* **Note 2**). Press OK to set measurements after (Fig. 1).
3. For measuring the nucleus ROIs from any given cell (*see* **Note 3**), begin by employing the magnifying tool to zoom into a cell to trace (or the keys "+" or "–" to more easily zoom

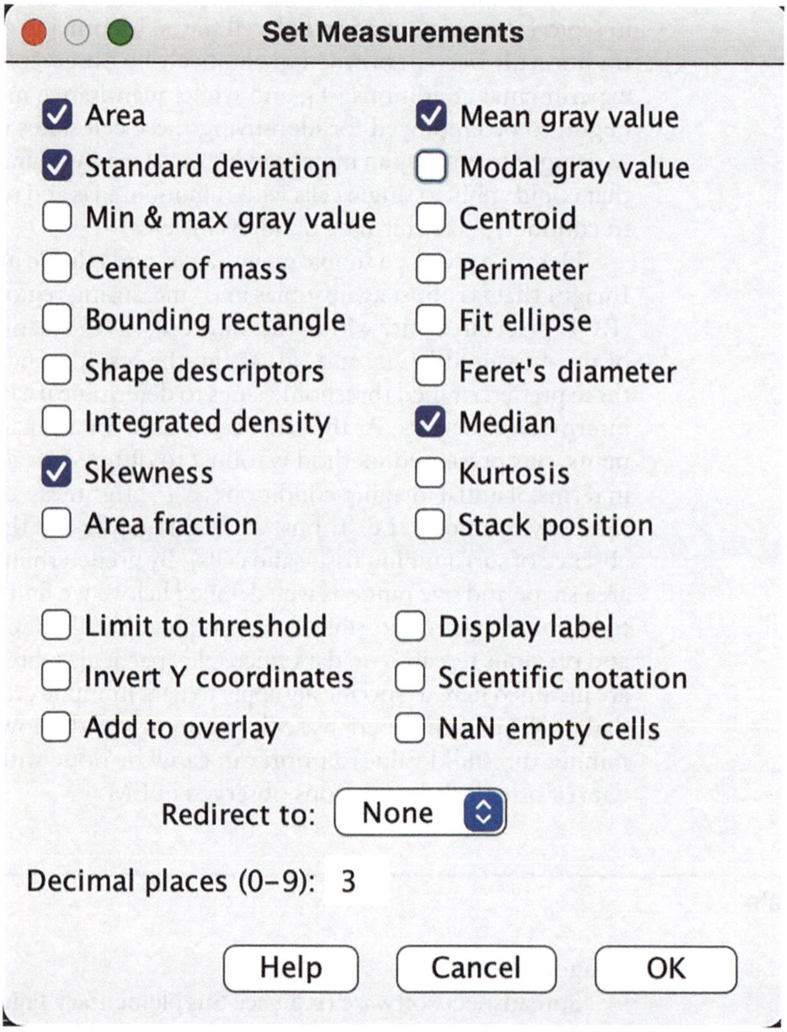

Fig. 1 Setting measurements on ImageJ. While other measurements can also be added, the following, checked measurement types are simplest and sufficient in performing this procedure

in or out to a cell). Switch to the "oval" (or elliptical) selection tool for tracing a ROI in the cell nucleus (*see* **Note 4**).

4. Using the oval selection tool, measure ROI areas between 10,000 and 20,000 for 1 ROI for any "darkest" region within the cell nucleus, and 1 ROI for any "brightest" region within the cell nucleus (*see* **Notes 5–7**) (Figs. 2 and 3).

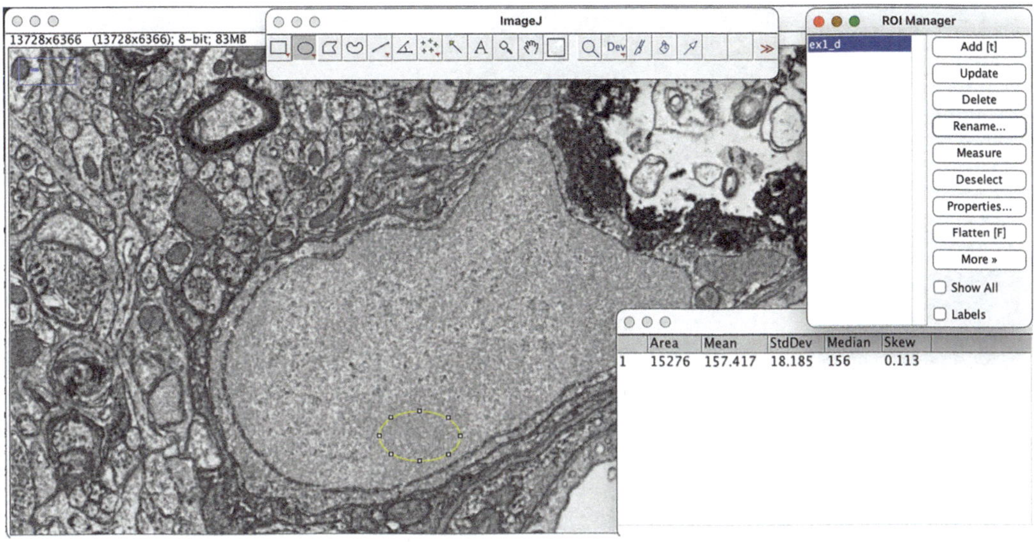

Fig. 2 Example "darkest" ROI selection on Image J. A darker patch of heterochromatin is used here to select the darkest ROI for later analysis, within a likely-"typical" microglia. As skew is <1 and >−1, analysis is expected to be performed using the darkest ROI mean and not median

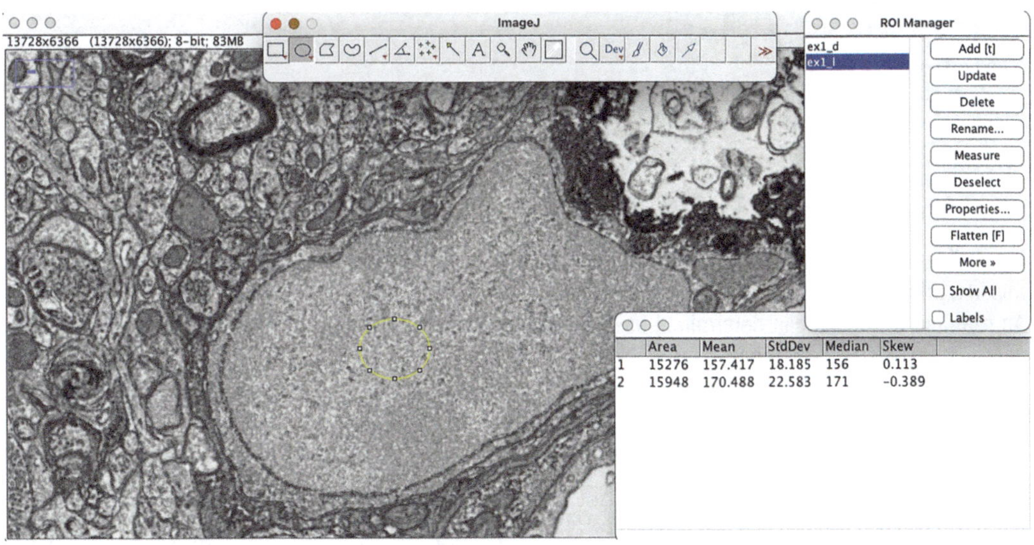

Fig. 3 Example "lightest" ROI selection on Image J. A random patch of euchromatin is used here to select the lightest ROI for later analysis, using the same likely-"typical" microglia. As skew is also not <1 and >−1, analysis should also employ the lightest ROI mean and not median. Finally, examples of extremely dark cells should have "lightest" ROIs that look similarly in appearance and measurement to the "darkest" ROI

Darkest, (Ignore)	Area	Mean	Std Dev	Median	Skewness	Light, (Ignore)	Area	Mean	Std Dev	Median	Skewness		Dark/Light Mean	Dark/Light Median	Assigned Category
1	18546	148.115	21.169	150	-0.264	2	11026	149.862	18.237	150	-0.118		0.988342609	1	dark
3	11392	159.105	19.253	159	0.173	4	18372	174.842	24.513	177	-0.439		0.909993022	0.898305085	typical
5	14100	154.48	19.787	153	0.131	6	11472	171.425	22.343	171	-0.368		0.901152107	0.894736842	typical
7	15736	161.111	19.513	159	0.194	8	19088	176.558	24.734	180	-0.47		0.912510337	0.883333333	typical
9	18220	170.139	24.442	171	-0.198	10	12568	179.78	27.14	180	-0.385		0.946373345	0.95	intermediate
11	19749	168.697	25.159	168	-0.187	12	11684	180.586	27.65	183	-0.43		0.934164332	0.918032787	typical

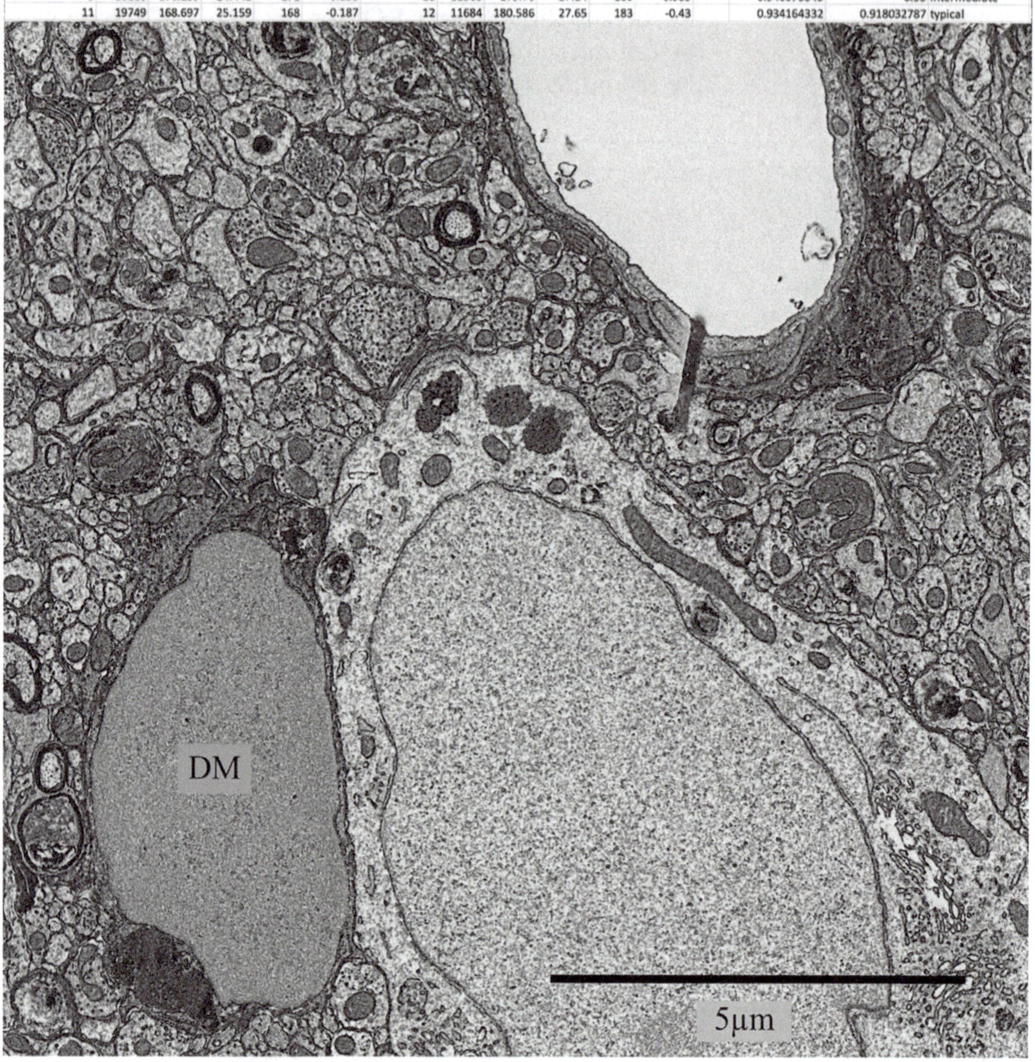

Fig. 4 Example spreadsheet and classification of a dark microglia. Cells should be based on thresholds determined a priori for differentiating typical *vs* intermediate *vs* dark appearances, as discussed in the manuscript. An example of a microglia determined as dark (labeled "DM") via the example spreadsheet (highlighted in yellow) is included for reference

5. After completing the traced ROI, perform Analyze → Measure (or Cmd + M on Mac). Copy and paste measurements from both the darkest ROI and brightest ROI (via the "Results" window) into a row (1 row/measured cell) in a spreadsheet software (e.g., Supplementary Figure 1).

6. On the ImageJ/Fiji menu, select Analyze → Tools →ROI Manager. For reproducibility and tracing data, add the updated measured ROI into ROI manager (*see* **Note 8**).
7. In the Excel Template Sheet (Supplementary Figure 1), calculate the ratio of darkest/lightest ROI means and medians (*see* **Note 9**).
8. Assign cell to one of the following categories, based on the calculated darkest/lightest ROI mean (or alternatively darkest/lightest ROI median, if skewness is found to be <-1 or >1) (*see* **Notes 10–12**) (Fig. 4): Dark microglia/immune cell (DM) $\geq$0.97, 0.95 $\leq$Intermediate microglia/immune cell $<$ 0.97, Typical microglia $\leq$0.95.
9. Save the ROI by clicking More > "Save..." within the ROI manager (*see* **Note 13**).

4 Notes

1. When determining value thresholds (via training with unblinded datasets), all preliminary ROIs should be collected on SEM images without immunostaining specific for the nucleus. Cytoplasmic staining is fine, as such staining does not alter gray values measured within the cell nucleus. Extra contrast staining (performed with uranyl acetate, Mayer's hematoxylin, etc.) can be performed and should be compatible with this method, but it is recommended to be trained on this method first with unblinded data to determine threshold values.
2. EM images used for this technique underlie images typically generated for TEM but can also be performed with inverted gray values on a SEM-generated image (inversion can done with SEM software).
3. This method compares pixel intensities only within the cell nucleus; is not suitable for identifying cells when nuclear staining is involved. The method also does not account for how much heterochromatin is found within any given cell, but rather just the difference between typical/intermediate/dark "appearance" of the cell overall. Additionally, the observer must be blinded to the experimental conditions when doing this analysis.
4. Due to the measurement method (i.e., the oval selection tool and minimum area selected), some cells may be classified as intermediate even when presenting lighter patches of chromatin; this can occur if the patches are very scarce, and most of the cell nucleus is darker in nature. Elliptical

is chosen over freehand selection to allow for more robust sampling of a ROI from the cell, as doing an oval selection minimizes cherry-picking of certain pixels to be analyzed (*vs* freehand or polygonal form selections, which can be manipulated into avoiding darker/lighter dot variation that naturally arises out of the EM processing and imaging methodology). Rectangular selection has a harder time capturing relevant ROIs without capturing unwanted information (e.g., nuclear membrane), and does not offer a more robust ROI sampling method *vs* oval selection. Rectangular selection may be used if the researcher is consistent across situations and works within an acceptable ROI size (determined empirically), but no advantages can be found by the authors here to use rectangular over oval selection.

5. For additional precision, the following are defined here as: "Darkest" = A qualitative region that appears darkest to the measurer, or at least not less dark than other sampled ROIs within the same cell. "Lightest" = A qualitative region that appears brightest to the measurer, or at least not less bright than other sampled ROIs. "Darkest" and "lightest" ROIs must be determined subjectively by the measurer, as no other tested methods were found to sensitively differentiate between dark/intermediate/typical cells.
6. The presented methods here (oval selection tool, area between 10,000 and 20,000, etc.) are used to improve robustness in detecting and minimizing bias caused by cherry-picking ROIs. Area >10,000 was selected to prevent ROIs from being too small and cherry-picking selections for lightest or darkest ROIs; no previous datasets used to obtain this method found areas <10,000 as necessary to sample said ROIs. Area <20,000 was selected to prevent ROIs from overlapping in cells with small nuclei; areas between 10,000 and 20,000 were found in training datasets to sufficiently single out darkest/lightest ROIs. Finally, while possible to perform, taking multiple darkest/lightest ROIs with unblinded datasets (with the same area selection) was not found to change overall classification of cells (for typical *vs* intermediate *vs* dark appearance).
7. In case exceptions (as this may be subject to increased bias), selecting an area <10,000 may be used for cells (a) with a small total nucleus area <60,000, and (b) lacking sufficient dark chromatin patches, which can be applicable for certain cell types including neurons and astrocytes. Although this quantitative approach may be abused into cherry-picking data, other methods (such as quantifying gray values from the whole nucleus) were not found to sensitively discriminate dark *vs* intermediate *vs* typical microglia.

8. Save all ROI drawn so data can be reproducibly traced, and save the ROI on the ROI manager before drawing a next ROI (to prevent losing the first ROI drawn). It is recommended to rename the saved ROI for a particular cell within each image for future accessibility in tracing data, with some reproducible naming convention. One example could include renaming ROI to "Filename_1", where 1 is i) the cell found most to the left, and as 1a, 1b, etc. if ii) the highest cell if 2+ cells share the same x-axis position. For instance: Name Filename_(number)d for the darkest ROI, and Filename_(number)l for the brightest ROI.
9. A template Excel document has been attached as supplementary material (Supplementary Figure 1), to streamline inputting values easily (+ leaving additional columns for calculations) from ImageJ/Fiji.
10. If skewness level is <-1 or >1, use median to calculate the formula instead of mean (as the mean is likely quite skewed as a measure of central tendency, under high skew).
11. Particular threshold values here for typical/intermediate/dark microglia or immune cells were generated using previous datasets but can be recalibrated upon other unblinded datasets people are using (per training on tissue not used for blind analysis, data analysis, etc.). Calibrating particular threshold values via unblinded datasets should be done for other cell types, or in cases where the reader is unsure of particular threshold values.
12. These threshold values are meant to provide "discrete" bounds between difference cell state appearances; in reality, the darkness of a cell truly exists across a gradient of typical into dark cells (i.e., arbitrary cutoffs are made to differentiate cells into multiple categories to assess for significant differences associated with typical *vs* intermediate *vs* dark classifications). The data used to generate and affirm these cutoffs can be provided, if the reader is further interested.
13. It is easier and more time-efficient to save multiple ROIs first, via selecting all (Cmd + A on Mac) before clicking saving ROIs. Saving ROIs into a single.zip file can be done per animal or per condition, for whichever standardized system works best in saving and retrieving data.

Acknowledgments

As part of the University of Victoria, we acknowledge and respect the lək̓ʷəŋən peoples peoples on whose traditional territory the university stands and the Songhees, Esquimalt and W̱SÁNEĆ peoples whose historical relationships with the land continue to this day. V.L. is supported by master and graduate award scholarships from the Faculty of Graduate Studies, University of Victoria. M.E.T. is a Tier I Canada Research Chair in *Neurobiology of Healthy Cognitive Aging* and a College Member of the Royal Society of Canada. We would like to thank Dr. Nicole Templeman and especially Dr. Craig Brown for suggesting the idea for quantitative methods of differentiating dark cell appearances, and Jared VanderZwaag and Mohammadparsa Khakpour for helping to generate data used in optimizing the methods presented in this manuscript. Finally, EM images were generated via a Crossbeam 350 scanning electron microscope acquired with funding from a Canada Foundation for Innovation John R. Evans Leaders Fund grant (#39965 Laboratory of ultrastructural insights into the neurobiology of aging and cognition).

Author Contributions V.Z.L. and M.E.T. contributed to writing and improving the manuscript.

Competing Interests The authors declare no competing interests.

Electronic Supplementary Material

Supplementary File (xlsx 10 kb)

References

1. Carrier M, Robert MÈ, González Ibáñez F, Desjardins M, Tremblay MÈ (2020) Imaging the neuroimmune dynamics across space and time. Front Neurosci 14:903. https://www.frontiersin.org/journals/neuroscience/articles/10.3389/fnins.2020.00903.Accessed 27 Feb 2024 [Online]
2. Ahmadpour S, Behrad A, Vega IF (2019) Dark neurons: a protective mechanism or a mode of death. J Med Histol 3(2):125–131. https://doi.org/10.21608/jmh.2020.40221.1081
3. Bisht K, Sharma K, Lacoste B, Tremblay M-È (2016) Dark microglia: why are they dark?. Commun Integr Biol 9(6):e1230575. https://doi.org/10.1080/19420889.2016.1230575
4. St-Pierre M-K, Bordeleau M, Tremblay M-È (2019) Visualizing dark microglia. Methods Mol Biol 2034:97–110. https://doi.org/10.1007/978-1-4939-9658-2_8
5. St-Pierre M-K et al. (2023) Astrocytes display ultrastructural alterations and heterogeneity in the hippocampus of aged APP-PS1 mice and human post-mortem brain samples. J Neuroinflammation 20: 73. https://doi.org/10.1186/s12974-023-02752-7
6. St-Pierre MK, Šimončičová E, Bögi E, Tremblay MÈ (2020) Shedding light on the dark side of the microglia, ASN Neuro 12:1759091420925335. https://doi.org/10.1177/1759091420925335.
7. St-Pierre M-K et al. (2022) Ultrastructural characterization of dark microglia during aging in a mouse model of Alzheimer's disease pathology and in human post-mortem brain samples. J Neuroinflamm 19(1):235. https://doi.org/10.1186/s12974-022-02595-8
8. Nahirney PC, Tremblay ME (2021) Brain ultrastructure: putting the pieces together. Front Cell Dev Biol 9:629503. https://doi.org/10.3389/fcell.2021.629503
9. Peters A, Folger C (2013) A website entitled 'the fine structure of the aging brain'. J Comp Neurol 521(6):1203–1206. https://doi.org/10.1002/cne.23280
10. Bordeleau M et al. (2020) Microglial and peripheral immune priming is partially sexually dimorphic in adolescent mouse offspring exposed to maternal high-fat diet. J Neuroinflammation 17(1):264. https://doi.org/10.1186/s12974-020-01914-1
11. Bisht K et al. (2016) Dark microglia: a new phenotype predominantly associated with pathological states. GLIA 64(5):826–839. https://doi.org/10.1002/glia.22966

12. Hui CW et al (2018) Prenatal immune challenge in mice leads to partly sex-dependent behavioral, microglial, and molecular abnormalities associated with Schizophrenia. Front Mol Neurosci 11:13. https://www.frontiersin.org/articles/10.3389/fnmol.2018.00013. Accessed 27 Feb 2024 [Online]
13. Flury A et al (2024) Integrated stress response associated with dark microglia promotes microglial lipogenesis and contributes to neurodegeneration. bioRxiv. 10.1101/2024.03.04.582965. Accessed 08 Mar 2024
14. El Hajj H et al. (2019) Ultrastructural evidence of microglial heterogeneity in Alzheimer's disease amyloid pathology. J Neuroinflammation 16(1):87. https://doi.org/10.1186/s12974-019-1473-9
15. Iovino L et al (2023) Investigation of microglial diversity in a mouse model of Parkinson's disease pathology. bioRxiv 2023.11.23.567809. https://doi.org/10.1101/2023.11.23.567809. Accessed 23 Nov 2023
16. St-Pierre MK, Carrier M, Lau V, Tremblay MÈ (2022) Investigating microglial ultrastructural alterations and intimate relationships with neuronal stress, dystrophy, and degeneration in mouse models of Alzheimer's disease, in neuronal cell death: methods and protocols. In: Jahani-Asl A (ed) Methods in molecular biology. Springer US, New York, NY, pp 29–58. https://doi.org/10.1007/978-1-0716-2409-8_3
17. St-Pierre MK, Šimončičová E, Carrier M, Tremblay MÈ (2023) Microglia in human postmortem brain samples: quantitative ultrastructural analysis of scanning electron microscopy images, in Alzheimer's disease: methods and protocols. In: Chun J (ed) Methods in molecular biology. Springer US, New York, NY, pp 63–85. https://doi.org/10.1007/978-1-0716-2655-9_4

Check for updates

Chapter 9

The Dynamic Side of Electron Microscopy: How to Use Quantitative Electron Microscopy to Portray the Gold Particle's Dance Style

Andrée Lessard

Abstract

For decades, electron microscopy (EM) has been an essential tool for neuroanatomists, providing high-resolution details of subcellular components in tissues such as those from the neuronal networks. In addition to its traditional role in anatomy, EM techniques can be used to quantify the densities of molecules. Thereby, quantitative EM is contributing to the development of dynamic models of their movements (namely plasticity or trafficking) within neurons or glial cells. The complete procedure is detailed in this chapter, including critical sequences that must be performed carefully to assure a successful result. In order, the step-by-step EM process consists of (1) aortic perfusion with fixatives, (2) incubations with primary antibodies, (3) incubation with secondary antibodies, (4) embedding the samples in EPON resin, and (5) encapsulating, trimming, and ultra-cut sectioning. These descriptions are followed by images of cells with or without immunolabeling to help the reader to recognize the immunogold and immunoperoxidase labeling on electron micrographs. The data analysis section gives tips to quantify the immunolabeling and representing the results graphically. The chapter concludes with examples of schematic representations of subcellular trafficking of receptors induced by pharmacological agents or physiological conditions (drug administration, stress). In Annex 1, all the solutions and buffers are listed with the recipes and indications to prepare them successfully. Annex 2 provides lists of the consumables, reagents, and equipment needed to obtain quantitative EM results.

Keywords Immunohistochemistry, Subcellular plasticity, Quantitative trafficking, Electron microscopy

1 Introduction

The invention of microscopes has been fundamental to the development of modern neuroscience. It allowed Santiago Ramòn y Cajal to create beautiful drawings reproducing the main parts of neurons [1]. Among these instruments, transmission electron microscopes remain neuroanatomist's best friends, since the launch of its first prototype in the 1930s to today.

Marie-Ève Tremblay (ed.), *Ultrastructural Neuroscience Methods*, Neuromethods, vol. 238,
https://doi.org/10.1007/978-1-0716-5456-9_9,

Electron microscopy (EM) is essential for neuroscience projects since it provides the highest resolution of all available microscopes. Its usefulness in all domains of anatomy is unquestionable.

There is, however, a downside to this distinction: EM has been perceived as a very static expertise giving strict anatomical descriptions of cellular profiles such as neurons and their synapses.

In comparison, photonic microscopy evolved toward new technologies enabling "live" track recording of cells' plasticity. These advancements generated lots of excitement in the neuroscience community, a little bit to the detriment of EM, which was still perceived as a "vintage" tool. Nevertheless, quantitative EM has been developed to complete photonic results by adding detailed subcellular plasticity in specialized pre/post-synaptic compartments as well as membranes, synapses, or organelles which are difficult to identify by light microscopy.

Quantitative EM is an immunohistochemical method that can label molecules of interest using secondary antibodies coupled with gold particles, a technique that allows precise localization of said molecules within portions of a cell know as a "profile." Gold particles can be quantified, for example, by their location on the cell's surface vs within the cytoplasm, or their densities can be evaluated in specific subcellular compartments or organelles. The phenotypes of neurons (e.g., dopaminergic, serotoninergic) can also be distinguished using the technique of immunoperoxidase labeling. The identification of a specific phenotype reveals functional insights into potential signaling pathways or physiological processes involved with the molecule of interest. For example, stress-induced nuclear translocation and surface internalization of neurokinin receptors in vasopressinergic neurons of the ventral tegmental area suggest a role for these receptors in signaling pathways involved in gene transcription as well as the cardiovascular response to stress [2].

In addition to the trafficking of selective receptors [3–5], quantitative EM can also help in measuring synaptic densities in in pre/post-synaptic compartments [6].

Alternatively, neuroscientists can choose serial reconstruction of profiles to measure, for example, synapse size or spine densities [7]. In parallel to these new practices in dynamic EM, exciting developments recently emerged to analyze single-particles in protein complexes using cryo-EM and reconstruct a cellular profile in 3D using tomography [8, 9]. Among these new developments, deep learning and artificial intelligence technologies are reaching out the EM field with useful tools that will soon help scientists to detect quickly their profiles of interests [10, 11].

New platforms allowing the combination of standard photonic microscopy and EM are complementing each other. They can provide complete information on cellular and subcellular plasticity of immunolabeled molecules. When combined, they can describe the cellular and subcellular distribution of a specific molecule (e.g., dopamine transporter) in selected brain areas such as the mice extrapyramidal dopaminergic pathways [12].

This chapter describes a step-by-step procedure to highlight trafficking of identified molecules using immunolabeling techniques.

2 Methods

The protocol involves many steps; all of them must be performed carefully. For example, triple washes are important to eliminate a previous chemical solution. Adherence to this procedure will preserve the integrity of the tissue and assure a complete absence of reagent before the next step. Otherwise, you will not be able to get ultrathin sections on the ultramicrotome.

2.1 Perfusion

This method, described in Table 1, allows a fast replacement of blood with a fixative. It preserves the ultrafine structure needed for EM visualization. For proper safety, perfusions are performed under a hood, by two persons wearing adequate protection gear (lab coat, gloves and protective glasses).

2.2 Primary Antibody Incubation

The procedure described in Table 2 can be apply to sections stored in the freezer (in the storage solution) or in freshly perfused brains after the vibratome sectioning. In both cases, sections should be put in 0.1 M PB. The specificity of the primary antibody (1st A°) is key for the success of immunolabeling. Make sure your primary antiserum is specific to your target, otherwise you could end up with disappointing results.

2.3 Secondary Antibody Incubations

Double-immunolabeling protocols with secondary antibodies (2nd A°s) are described in Table 3 (immunoperoxidase) and Table 4 (immunogold). For single-immunolabeling with gold particles, go straight ahead to Table 4. For single-immunolabeling with peroxidase: follow the procedure in Table 3, wash your section in 0.1 M PB, then go directly to the osmium tetroxide incubation in Table 4. Follow the steps until the end.

2.4 Embedding and Encapsulation

Prepare fresh EPON and mix it for at least 5 min. Then, incubate your sections in 100% EPON for 1 h. Replace EPON and incubate for another 1 hour with 100% EPON to make sure there is no more propylene oxide on your samples.

At this point, you will be ready to embed your sections. You will need a metal tray (any type of tray as long as it is flat and can be put in the oven) and pieces of ACLAR fluorocarbon film. Lay your section flat on a piece of ACLAR, using a wooden stick. There will be an excess of EPON on the tissue at this point. To remove it, use

Table 1
Description of the perfusion procedure

Step	Notes
Prepare paraformaldehyde (PFA) 4% with 2% glutaraldehyde in 0.1 M phosphate buffer (PB) PFA will dissolve in hot water (55–65 °C, don't boil). Add a splash of NaOH to clear the solution. Once dissolved, add half content of 0.2 M PB. Filter the solution (filter paper #3). Adjust the pH at 7.4, store in the fridge.	Can be prepared the day before. Work under a hood. Use a glass Erlenmeyer flask. You will need approximately 500 ml per rat and 300 ml per mouse.
Prepare 0.1 M PB with 0.9% phosphate-buffered saline (PBS)	Can be prepared the day before. Calculate approximately 50 ml of PBS per mouse or 150 ml per rat.
Fill the perfusion tubes One tube is filled with PBS; the other one with the mix of PFA/ glutaraldehyde. Add a regular (rat) or butterfly (mice) needle at the end. Adjust the flow rate at 25 ml/min.	ALWAYS CHECK FOR BUBBLES! A single air bubble can screw up the whole procedure.
Anesthetize the animal	Check for reflexes by pinching the end of the tail or between the toes. There should be no response.
Perfusion procedure Open the chest cavity making a horizontal abdominal incision just below the ribcage, cut the diaphragm, and expose the heart. Once the heart is accessible, cut the right ventricle and quickly insert the needle inside the left ventricle. Open the PFA/glutaraldehyde tube and perfuse for about 10 min.	The animal will show spasms.
Post-fixation Delicately remove the brain, post-fix with 4% PFA for 1 h at 4°C in a scintillation vial.	Brain matrices can be used to separate the brain in several blocks.
Vibratome sections 40 microns in 0.1 M PB	Brain sections can be collected in 6-, 12-, or 24-well plates.
Storage Replace the 0.1 M PB with storage solution. Tape the sides of your plate and label it with a marker before storage.	Your sections can be stored at −20 °C for many months, even years.

laboratory wipes and cotton tip applicators to absorb the EPON. Cut the other piece of ACLAR a little smaller than the piece on bottom. Lay it on top, so that your sections with be in between two pieces of ACLAR. Using your thumbs, press the sections from one side to the other, and take out as much excess EPON as possible. This will assure tissue, rather than EPON alone, will be included on the first's ultrathin sections obtained while cutting the tissue on the ultramicrotome.

Table 2
Description of the primary incubation procedure

Step	Notes
Prepare 2 L of 0.1 M PB and 0.1 M Tris saline (TS).	The remaining solutions can be stored at 4 °C for the following days.
Incubation 1% sodium borohydride in 0.1 M PB for 30 min. Incubate sections in crucibles and petri dishes, mix constantly with Pasteur pipet. Trash the solution in the appropriate waste container.	Important step to neutralize reactive aldehydes and prevent background labeling. The solution will make bubbles.
Three washes with 0.1 M PB, 3 min each. Use petri dishes and crucibles. Put the first wash in the sodium borohydride waste container.	Make sure there is no more bubbles after the thirrd wash.
Cryoprotectant incubations, 10 min each in a −80 °C freezer: 100% cryoprotectant solution 70% cryoprotectant solution 50% cryoprotectant solution 30% cryoprotectant solution 100% 0.1 M PB	Use a 24-well plate. The procedure breaks down lipid membranes helping the antisera to penetrate the tissue. Alternatively, the tissue can be quickly plunged into methyl butane followed by liquid nitrogen.
Wash sections in 0.1 M TS (5 min).	Transfer tissue back to crucibles and petri dishes.
Incubation in bovine serum albumin (BSA) 0.5% in 0.1 M TS for at least 30 min.	Blocking agent. Very important step to block nonspecific binding of the antisera.
Prepare the primary antibody (1st A°) solution in 0.1% BSA in 0.1 M TS.	In scintillation vials
Incubation with 1st A°.	Incubate your sections in scintillation vials overnight on a shaker at room temperature. Then place your sections for another 12–24 h in a cold room or fridge (4 °C).

Label your tissue samples embedded between the two ACLAR pieces using tape and marker, then place heavy stainless-steel bars (or something flat that weights 1–3 kg) on top of it to keep the tissue flat. Bake them in an oven at 60 °C for 18–24 h to polymerize the plastics in the EPON.

Once set, remove your sections from one side only of the ACLAR sheet by gently peeling the plastic embedded tissue off the ACLAR sheet, cut your tissue area of interest, and glue it to an EPON block (previously made with embedding molds) using crazy glue. When gluing the tissue to the EPON block, make sure that the tissue is in contact with the EPON block, and the second piece of ACLAR, that is still attached to the tissue, is placed on top. Put the block in oven at 60 °C for another 18–24 h.

Table 3

Description of the secondary incubation procedure with immunoperoxidase

Step	Notes
For immunoperoxidase labeling Prepare in advance: Weight 11 mg or 22 mg of 3′-diaminobenzidine (DAB). Store safe from light at −20 °C.	**Be careful: hazardous chemical** DAB can cause cancer. Always wear mask, lab coat, and gloves. Weight the powder under a plexiglass cover.
Immunoperoxidase protocol	
Three rinses of TS 0.1 M, 3 min each	Transfer the tissue back to crucibles contained in glass petri dishes.
30 min incubation in biotin secondary antibody (2nd A°) 1:400, in vials on a shaker	Make sure to use the same species as the 1st A° for immunoperoxidase labeling. Prepare the ABC solution during this incubation: In a vial, put one drop of A and one drop of B in 2.5 ml of TS 0.1 M.
Three rinses of TS 0.1 M, 3 min each	Transfer the tissue back to crucibles in glass petri. At this point, you can put DAB powder at room temperature and safe from light (11 mg or 22 mg depending on your needs).
30 min incubation in ABC solution	The solution will fix with the biotin 2nd A° and generate a reaction when adding DAB and peroxidase: the immunoperoxidase labeling will be recognizable as a brownish precipitate under the microscope.
Three rinses of TS 0.1 M, 3 min each	Transfer tissue back to crucibles and petri dishes.
Incubation in 3,3′-DAB Dissolve DAB in 50 ml (11 mg) or 100 ml (22 mg) of 0.1M TS. Just before incubation, add 5 μl of 30% peroxide. Incubate for 6 min.	**Be careful: hazardous chemical** DAB can cause cancer. Always wear mask, lab coat and gloves and work under a plexiglass shield

2.5 Ultracutting and Staining

Place your embedded specimen in the specimen holder of your ultramicrotome. Using a razor blade, shape your specimen into a trapezoid shape, which will help in identifying your areas (e.g., dorsal vs ventral) and also to reduce the resistive force on your diamond knife.

Examine your block under a brightfield microscope. You must be careful to (1) cut the top and bottom parts of your trapezoid parallel and (2) cut it small enough to fit in 40× (at least the sides).

When ready, cut your samples into 70-nm-thick sections using a diamond knife and an ultramicrotome. The immunogold particles lay on the surface of the tissue, so it is important to align your knife perfectly to the sample (meaning that your first section is as complete as possible). The ideal interface is when 30%–50% of your section is filled with tissue, the rest with EPON. At that level, you will have nice morphology and immunolabeling.

Table 4
Description of the secondary incubation procedure with immunogold

Step	Notes
Immunogold protocol	The fish gelatin (part of the washing incubation buffer) is difficult to pipet. Warm it with your hands during the rinses of TS, PB, and PBS.
Three rinses of TS 0.1 M, 3 min each	Tissue incubated with DAB: let the sections under the plexiglass shield and put the first rinse in DAB waste. After the first rinse, you can put back your sections on the shaker.
Three rinses of PB 0.1 M, 3 min each	This step is very important, do not skip it! At this point, your sections must adapt to PB (no more TS).
5 min rinse in 0.01 M PBS	This step will prepare your tissue for the washing incubation buffer.
5 min in washing incubation buffer	You can prepare the 2nd A° immunogold solution at this point: in vials, 1:50 (40 μl in 2 ml of washing incubation buffer). Make sure to use compatible species for your firstA.
2 h incubation with the 2nd A° for gold	Defrost the silver intensification kit at this point (very important).
Washing incubation buffer 5 min	Transfer the tissue back to crucibles in glass petri.
5 min rinse in 0.01 M PBS	Preparing your tissue for the glutaraldehyde solution.
Glutaraldehyde 2%, 10 min in glass Petri, put 115 ml of 0.01 M PBS + 10 ml vial of 25% glutaraldehyde, or 40 ml 0.01 M PBS + 10 ml vial of 10% glutaraldehyde. Cover with plastic wrap or parafilm.	**Work under the hood** Glutaraldehyde is a fixative for perfusion, but it will also stabilize your gold particles.
5 min rinse in 0.01 M PBS	Prepare the silver staining kit. This step will increase the size of your gold particles.
Silver kit staining Use a 12-well plate: First row: Citrate buffer Second row: 16 drops A + 16 drops B Third row: Citrate buffer (to stop the reaction) **Incubate for 5–7 min in silver (second row)**	**Be careful with the incubation time** If you wait too long, your sections will become green; irreversibly useless. From now on, manipulate your sections with wooden stick, avoid metallic tools like brushes.
Wash in citrate buffer for 2–3 min Section back in crucibles. After the wash, place your sections carefully flat on porcelain Coors dishes containing **0.1 M PB**	**Do not use 0.01 M PBS**, otherwise holes will appear on your tissue.
Osmium tetroxide 2%, 1 h incubation 2 ml vial of 4% Osmium + 2 ml 0.1M PB Gently replace the 0.1 M PB in porcelain dishes wells with 2% osmium, cover from light.	**Be careful: hazardous chemical** always work under the hood with lab coat and gloves This chemical is staining tissue by adding electron-dense material to lipids; thus, enhancing contrast. If osmium comes into contact with your skin, put quickly some milk on it.

(continued)

Table 4
(continued)

Step	Notes
Three rinses of PB 0.1 M, 3 min each	First rinse of 0.1 M PB in osmium waste.
Dehydration in coor dishes: EtOH 30% 5 min EtOH 50% 5 min EtOH 70% 5 min EtOH 95% 5 min	These steps are important to train your sections to have less water.
EtOH 100% 2×10 min	Transfer section in vial (gentle, your sections will be as fragile as Egyptian papyrus at this point). Prepare EPON, mix it for at least 5 min.
Propylene oxide 2×10 min	Propylene oxide will help clearing out the alcohol, thereby facilitating the infiltration of the embedding solution (EPON).
1:1 Propylene oxide–EPON Overnight at room temperature on a shaker	

Stain your grids with 2% uranyl acetate and Reynold's lead citrate:

- Place drops of uranyl acetate on a parafilm or dental wax, the grid on top of the drop. Incubate for 10 min.
- Rince your grids: pick them with forceps, plunge them five to ten times on small beakers (three beakers are enough) filled with distilled water or previously boiled water. Let them dry on a filter paper in a plastic petri.
- Place drops of Reynold's lead nitrate on a wax plate, covered by a petri and filled with NaOH pellets. Place your grids inside the drop and incubate for 5–10 min.
- Rince your grids: pick them with forceps, plunge them five to ten times on small beakers filled with distilled water or previously boiled water. Let them dry on a filter paper in a plastic petri.

3 Immunolabeling Detection

The following figures are presented to assist in identifying immunolabeling on electron micrographs. In Fig. 1, comparisons of neurons (somata or S and dendrites or D) took with a confocal microscope [1] or transmission EM (2 and 3) are showed. The electron micrograph 2 is showing a somata (S) or cell body, while the image 3 is showing a dendrite (D).

Figure 2 aims to assist the reader in recognizing profiles endowed with immunoperoxidase and/or immunogold labeling. While

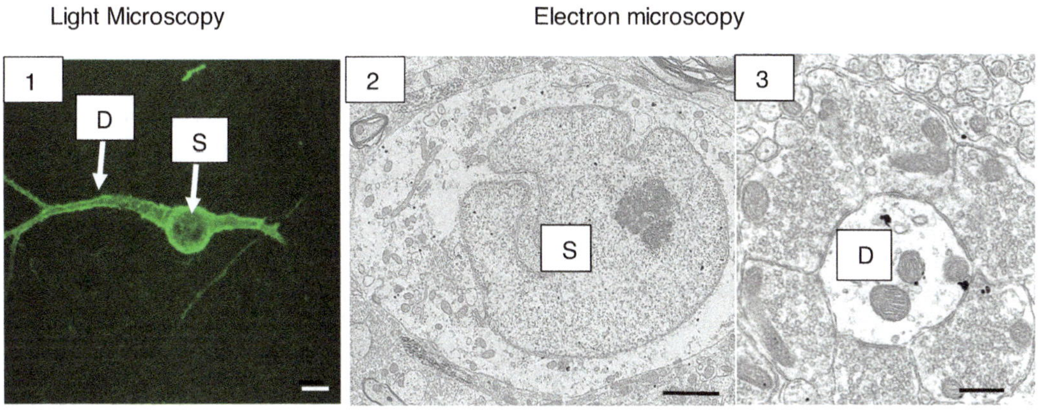

Fig. 1 Confocal (1) and electron micrographs (2, 3) of neurons showing the soma (S) and dendrites (D). In 1, a bi-polar neuron of the rat basolateral amygdala was immunolabeled for the neurokinin-1 receptor. In 2 and 3, neurons of the rat ventral tegmental area were immunolabeled for the neurokin-3 receptor. Scale bars : (**a**) 10 μm, (**b**) 2 μm, and (**c**) 500 nm

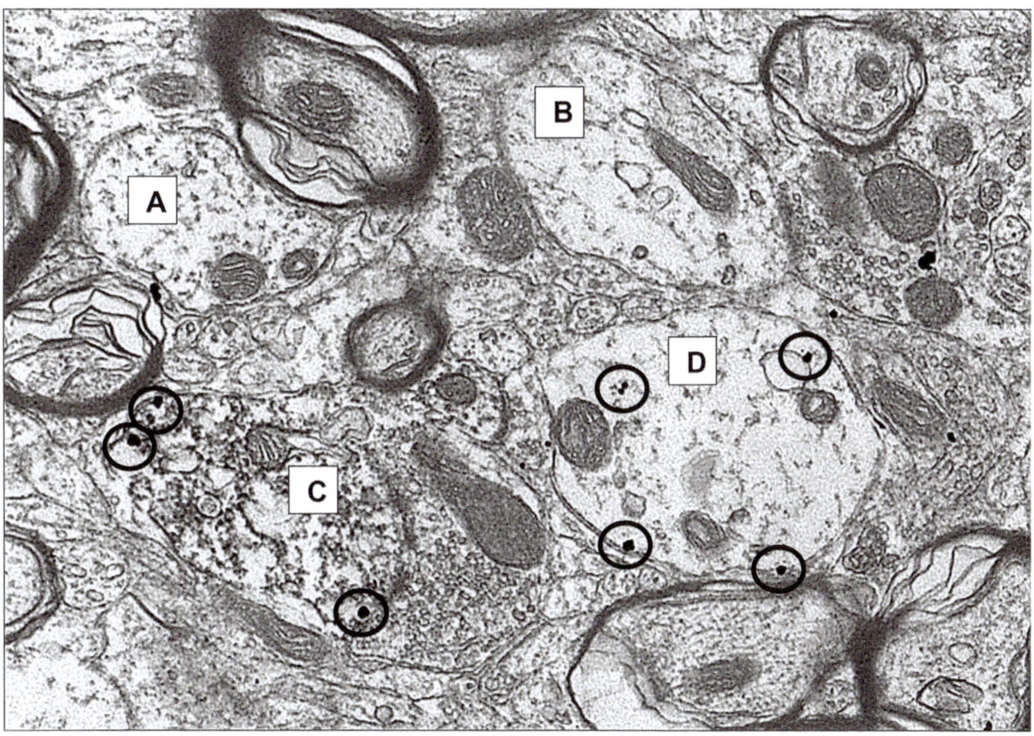

Fig. 2 Electron micrograph showing four different dendrites from the rat ventral tegmental area immunolabeled for tyrosine hydroxylase (marker of dopaminergic neurons) and the neurokinin-3 receptor. Dendrites A and B show no immunolabeling, while dendrite C has both immunoperoxidase (brown precipitate) and immunogold particles (encircled). The dendrite D only presents immunogold particles (encircled). Scale bar: 500 nm

dendrites A and B show no labeling, the dendrite C presents immunoperoxidase (brown precipitate) and gold particles (encircled). The dendrite D only presents gold particles (encircled).

Figure 3 shows the difference between a cell body containing dense brown precipitate (left) and one without it (right) produced by the immunoperoxidase labeling.

The quantitative EM technique allows to distinguish immunogold particles located on the plasma membrane or at a distance from it. Examples of both are shown in Fig. 4, where the dendrite on the left shows a particle touching the plasma membrane (encircled). The dendrite on the right, however, contains a gold particle at a distance from the plasma membrane (encircled). The latter could

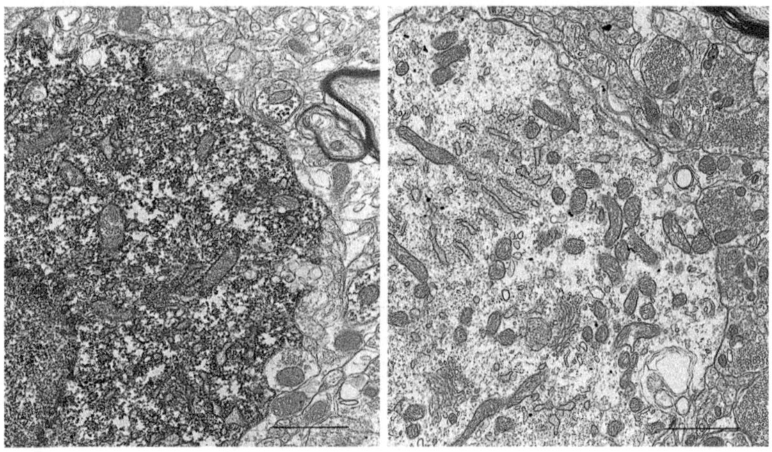

Fig. 3 Electron micrographs showing cell bodies from the rat ventral tegmental area that are immunolabeled (left) or not immunolabeled (right) for tyrosine hydroxylase (marker of dopaminergic neurons). The somata on the left displays the typical brown precipitate produced by immunoperoxidase labeling, which is not present in the soma on the right. Scale bar: 2 μm

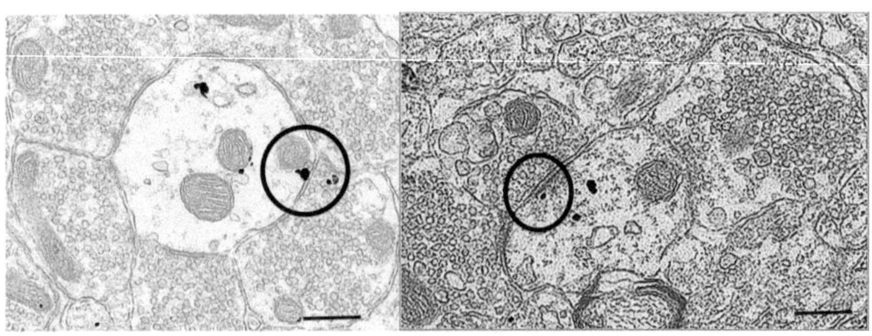

Fig. 4 Electron micrographs of two dendrites of the rat ventral tegmental area immunolabeled for the neurokinin-3 receptor. The immunogold particle encircled on the left is in contact with the plasma membrane, while the one on the right is at a distance from the membrane. Scale bar: 500 nm

be considered on the plasma membrane, depending on the criteria predefined by the investigators. Both dendrites also contain gold particles in the cytoplasm.

4 Data Analysis

Before quantifying your gold particles, make your own rules and stick to it: decide when a gold particle is located on the plasma membrane (at a max of *x* nm from the membrane or touching directly the membrane); cross-diameter lengths of your distal, medium-size, and proximal dendrites. This requires prior examination of your samples without immunolabeling to get a deep understanding of the basic morphology of your brain area and, thus, to determine the relevant criteria to apply before the analysis.

It is tempting to rely on new technologies to identify profiles of interests. Indeed, deep learning technologies are emerging in the EM field [10]. A recent publication reported that immunogold particles can be digitally identified in electron micrographs [11]. However, the software can only predict correct immunolabeling with post-embedding protocols. The latest involve antisera incubations in embedded samples on grids. The present chapter is describing the pre-embedding method. It was chosen over the post-embedding one since it provides higher contrast and the best morphology. The deep learning software Gold Digger, however, cannot yet accurately identify immunolabeling in pre-embedded samples [11].

On the other hand, the deep learning software SynapseNet was developed to compartmentalize neuronal segments such as axon terminals, allowing fast identifications of pre-synaptic profiles along with their synaptic vesicles and mitochondria [13].

These exciting breakthroughs with artificial intelligence (AI) tools will facilitate the work of neuroscientists. Nonetheless, AI results will always need double-checks by senior electron microscopists for validation. To the author's opinion, no AI technology can yet perfectly match the expertise of a trained eye!

Ideally, data analysis should be performed by an investigator blind to the conditions. Selecting the range of proximal, medium-size, and proximal dendrites will decrease the standard error mean and help revealing a clear plasticity on graph bars. Quantification should include at least three samples per conditions, with each sample comprising three or four sections of the selected area, covering a total area of 1000 μm^2 or more. For example, in Fig. 5, the nuclear densities of receptors were calculated in three conditions and represented as number of gold particles per area of 100 μm^2. Plasmalemmal densities can be measured as the ratio of immunogold particles on the plasma membrane per perimeter area of 100 μm [14].

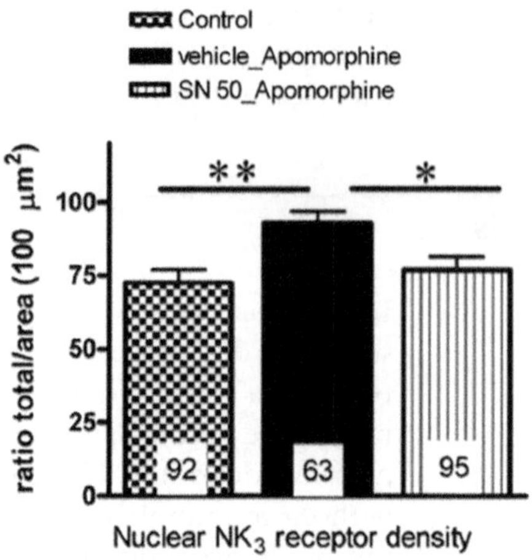

Fig. 5 Ratio of nuclear NK_3 receptor densities in dopaminergic neurons of the rat ventral tegmental area in control conditions, in subjects receiving the drug apomorphine with or without the nuclear blocker SN 50. Vertical bars represent the mean ± standard error mean in the ratio of the total number of gold particles/area of the profile (100 µm²). Numbers in bars represent the total number of profiles in three animals for each condition. Significant differences were determined using one-way ANOVA followed by a Bonferroni post hoc test. $^{*}P<0.05$; $^{**}P<0.01$

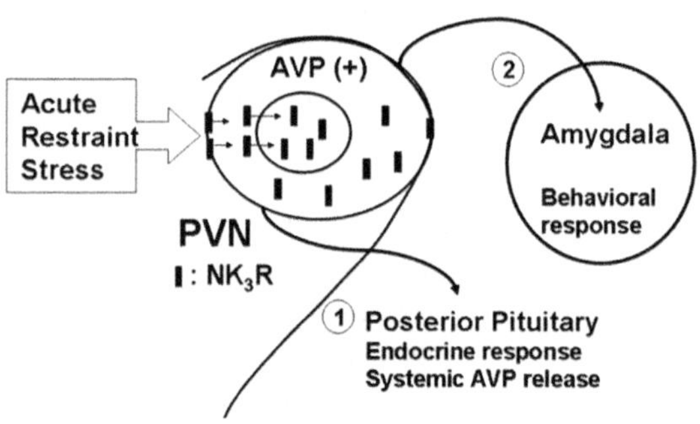

Fig. 6 Stress-induced internalization and nuclear translocation of the neurokinin NK_3 receptor (NK_3R) in vasopressinergic neurons (AVP) of the rat paraventricular nucleus of the hypothalamus (PVN) From [2].

5 Conclusion

Upon carefully following the protocols, users will be able to imagine the dance style of their particles: either quick cha-cha moves toward the synapse, a long waltz from the cytoplasm to the nucleus (nuclear translocation), or groups of particles dancing around in circle from the membrane to the cytoplasm (internalization). These methods will reveal trafficking mechanisms and enable schematics such as those shown in Fig. 6 [2] and Fig. 7 [14]:

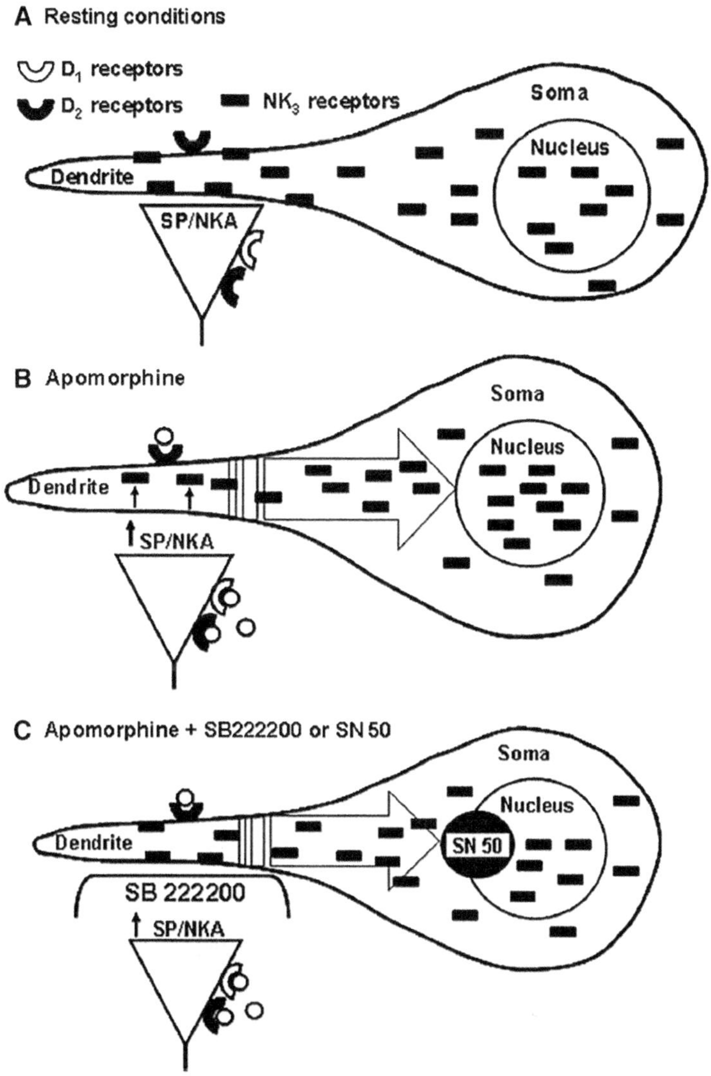

Fig. 7 (**A**) In resting conditions, the neurokinin-3 (NK_3) receptor in the rat ventral tegmental area is distributed in somatodendritic profiles, on the plasma membrane, and in the cytoplasm of dopaminergic and non-dopaminergic profiles. (**B**) Apomorphine, an agonist of the dopaminergic D_1 and D_2 receptors, triggers the release of Substance P (SP) and Neurokinin A (NKA) pre-synaptically, leading to an internalization and nuclear translocation of the NK_3 receptor. (**C**) The specific NK_3 receptor antagonist SB222200 prevents the apomorphine-evoked NK_3 receptor internalization, but not its nuclear translocation. Conversely, the nuclear blocker SN 50 prevents the apomorphine-evoked NK_3 nuclear translocation, but not the internalization From [14].

Annex 1: Main Buffers and Solutions Needed for Quantitative EM

Buffers

0.1 M Phosphate buffer

	1 L	2 L	3 L	4 L
Sodium phosphate dibasic (Sigma) Na_2HPO_4	10.9 g	21.8 g	32.7 g	43.6 g
Sodium phosphate monobasic (Sigma) $NaH_2PO_4\ H_2O$	3.2 g	6.4 g	9.6 g	12.8 g

0.2 M Phosphate buffer

	0.1 L	0.5 L	1 L	2 L	3 L
Sodium phosphate dibasic (Sigma) Na_2HPO_4	2.18 g	10.9 g	21.8 g	43.6 g	65.4 g
Sodium phosphate monobasic (Sigma) $NaH_2PO_4\ H_2O$	0.64 g	3.2 g	6.4 g	12.8 g	19.2 g

0.1 M Tris Saline pH 7.6

	1 L	2 L	3 L	4 L
Trizma HCl (Sigma # T3253)	13.2 g	26.4 g	39.6 g	52.8 g
Trizma Base (Sigma # T1503)	2 g	4 g	6 g	8 g
NaCl	9 g	18 g	27 g	36 g

EPON Resin

Resin	Volume	60cc syringeFill to:	30 cc syringeFill to:
Embed 812 (EMS # 14900)	20 ml	19 ml	9 ml
D.D.S.A. (EMS # 13710)	16 ml	35 ml	17 ml
N.M.A. (EMS # 19000)	8 ml	43 ml	21 ml
B.D.M.A. (EMS # 11400-25)	1.2 ml	1.2 ml (pipetter)	0.6 ml (pipetter)
Total	45.2 ml		

OR (alternative recipe that works as well)

Resin	Recipe 1	Recipe 2	Recipe 3	Recipe 4
Embed 812 (EMS # 14900)	6.1 ml	12.2 ml	18.3 ml	24.4 ml
D.D.S.A. (EMS # 13710)	1.9 ml	3.8 ml	5.7 ml	7.6 ml

N.M.A. (EMS # 19000)	3.9 ml	7.8 ml	11.7 ml	15.6 ml
DMP 30 (EMS # 13600)	3 drops	6 drops	9 drops	12 drops
Total	±12 ml	±24 ml	±36 ml	±48 ml

Storage solution (30% sucrose): for keeping sections at –20 °C

Ethylene glycol	150 ml
0.1 M PB	150 ml
Sucrose	150 g
Add 0.1 M PB until 500 ml	

Cryoprotectant solution (100%)

Put 180 ml 0.05 M PB (90 ml of 0.1 MPB + 90 ml of purified water) in beaker.
Add 50 g of sucrose and 20 ml of glycerol, stir.
Make 100 ml of 70%, 50%, and 30% (add 0.1 MPB 30 ml, 50 ml, 70 ml, respectively).

0.01 M phosphate buffer saline (PBS)

Purelab water	1 L
Sodium phosphate dibasic(Sigma)	1.09 g
Sodium phosphate monobasic(Sigma)	0.32 g
NaCl	9 g
pH = 7.4	

Washing incubation buffer

0.01 M PBS	100 ml
BSA	0.8 g
Gelatin	250 µl
pH = 7.4	

Citrate buffer 0.2 M

Purelab water	125 ml
Sodium citrate	7.36 g
pH = 7.4 [use citric acid (1 M citric acid: 19.2 g in 100 ml water)]	

2% Uranyl acetate

Purelab water	100 ml
Uranyl acetate	2 g
Stir for at least 30 min	
Filter paper; store in amber bottle	

Reynold's lead citrate

Lead nitrate	1.32 g
Mix with 30 ml of Purelab water, add:	
Sodium citrate	1.76 g
Mix it for 1 min (your solution will become milky). Cover it with parafilm and stir it for 30 min. Add:	
NaOH 1 M	8 ml
Stir it until the solution is clear	
Add water until 50 ml	

Annex 2: List of Consumables and Equipment Needed for Quantitative EM

Consumables and Reagents

Glass petris
Crucibles
Porcelain coors dishes
Scintillation vials
Holes wells (6, 12, 24)
Protection (lab coat, gloves, protective glasses, mask)
Brushes
Wooden sticks
Embedding molds
Eppendorfs
Razor blades
Transfer pipettes (glass) and bulbs
Pasteur pipettes
Pipettor and pipette tips
Whatman filter paper #1 and #3
Buffers and solutions in Annex 1 (PB, TS, PBS, citrate, incubation, storage, etc.)
Fixatives (PFA, glutaraldehyde)
Bovine serum albumin (BSA)
Fish gelatin
Sodium borohydride
Osmium tetroxide
DAB
Ethanol
Propylene oxide
Silver intensification kit
ABC kit
Antibodies (primary, secondary)
Peroxide 30%
Grids (Cu, 300 mesh)
Ultrapure water
Tweezers
Scissors
Crazy glue

Uranyl acetate
Lead nitrate
Dental wax
Parafilm
Plastic wrap
Metal tray
Stainless steel bars
Cotton tip applicators
Wipes
Weighting boats
Weighting scoops
Labeling tapes
Timer
NaOH pellets
NaOH solution
HCl solution
Brain matrices

List of Equipment and Glassware

Oven
pH meter
Thermometer
Stirring bars
Glass and plastic beakers
Glass and plastic cylinders
Glass Erlenmeyers
Glass or porcelain funnel
Precision balance
Hotplate stirrer
Rotator
Shaker
Fume hood
Plexiglass cover
Perfusion pump and tubes
Perfusion needles
−80 °C freezer
Fridge or cold room
−20 °C freezer
Ultramicrotome and diamond knife
Microtome or vibratome
Brightfield microscope
Transmission electron microscope

References

1. De Carlos JA (2020) The history of Spanish neuroscience. Anat Rec 303(5):1179–1180
2. Miklos Z, Flynn F, Lessard A (2014) Stress-induced dendritic internalization and nuclear translocation of the neurokinin-3 (NK3) receptor in vasopressinergic profiles of the rat paraventricular nucleus of the hypothalamus. Brain Res 1590:31-44.
3. Ambrose-Lanci L, Peiris N, Unterwald E, Van Bockstaele E (2008) Cocaine

withdrawal-induced trafficking of delta-opiod receptors in rat nucleus accumbens. J. Brain Res 19(1210):92–102
4. Riad M, Watkins K, Doucet E, Hamon M, Descarries L (2001) Agonist-induced internalization of serotonin-1a receptors in the dorsal raphe nucleus (autoreceptors) but not hippocampus (heteroreceptors). J Neurosci 21(21):8378–8386
5. Miklos I, Kovacs K (2002) GABAergic innervation of corticotropin-releasing hormone (CRH)-secreting parvocellular neurons and its plasticity as demonstrated by quantitative immunoelectron microscopy. J Neurosci 4522(02):147–151
6. Nusser Z, Roberts J, Baude A, Richards J, Somogyi P (1995) Relative densities of synaptic and extrasynaptic GABAA receptors on cerebellar granule cells as determined by a quantitative immunogold method. J Neurosci 15(4):2948–2960
7. Ostroff L, Cain C, Bedont J, Monfils M, Ledoux JE (2010) Fear and safety learning differentially affect synapse size and dendritic translation in the lateral amygdala. Proc Natl Acad Sci USA 18(20):9418–9423
8. Barekatain M., Johansson L, Lam J, Chang H, Sadybekov A, Han GW et al (2024) Structural insights into the high basal activity and inverse agonism of the orphan receptor GPR6 implicated in Parkinson's disease. Sci Signal 17(865):eado8741
9. Hylton RK, S. V (2021) Cryo-electron tomography and automatic segmentation of cultured hippocampal neurons. Methods Mol Biol 2215:25–48
10. Ede J (2021) Deep learning in electron microscopy. Mach Learn Sci Technol 2:011004
11. Jerez D, Stuart E, Schmitt K, Guerrero-Given D, Christie J, Hahn N, Smirnov M (2021) A deep learning approach to identify immunogold particles in electron microscopy images. Sci Rep 11(7771):1–9
12. Block E, Nuttle J, Balcita-Pedicino J, Caltagarone J, Watkins S, Sesack S, Sorkin A (2015) Brain region-specific trafficking of the dopamine transporter. J Neurosci 35(37):12845–12858
13. Muth S, Moschref F, Freckmann L, Mutschall S, Hojas-Garcia-Plaza I, Bahr J et al (2025) SynapseNet: deep learning for automatic synapse reconstruction. Mol Biol Cell 1(10):1–19
14. Hether S, Misono K, Lessard A (2013) The neurokinin-3 receptor (NK3R) antagonist SB222200 prevents the apomorphine-evoked surface but not nuclear NK3R redistribution in dopaminergic neurons of the rat ventral tegmental area. Neuroscience 247:12–24

Chapter 10

Multimodal Imaging of Breast-to-Brain-Metastatic Lesions

Sabine Hombach-Klonisch, Danyyl Ippolitov, Emily Fedora and Thomas Klonisch

Abstract

Breast cancer brain metastasis (BCBM) is an increasingly common and devastating stage of disease, and particularly common in human epidermal growth factor receptor (2-positive and triple-negative breast cancer. This is further complicated by the inability of many therapeutic drugs to effectively cross the blood–brain and blood–tumor barriers. This chapter presents a multimodal imaging toolbox spanning clinical magnetic resonance imaging and positron emission tomography, whole-brain histopathology, multiplex immunofluorescence, and transmission electron microscopy to interrogate BCBM biology across spatial scales. This includes non-invasive lesion detection and treatment monitoring, high-content mapping of tumor–microenvironment interactions, and ultrastructural analysis of metastatic breast cancer cells. Together, these approaches provide complementary readouts of tumor burden, brain niche remodeling, and nanoscale cellular adaptations. This enables a more precise preclinical modeling and facilitates translation of mechanistic insights into imaging biomarkers and therapeutic strategies for patients with BCBM.

Keywords Breast cancer brain metastasis (BCBM), Triple negative breast cancer (TNBC), Human epidermal growth factor receptor 2-positive (HER2+) BCBM, Multi-modal imaging, Magnetic resonance imaging (MRI), Immunofluorescence, Immunohistochemistry, Transmission electron microscopy (TEM), Immuno-TEM

1 Introduction to Breast Cancer Brain Metastasis (BCBM)

Investigating breast cancer brain metastasis encompasses the in vivo detection, localization, and monitoring of metastatic tumors, the histological assessment of transcriptional and protein changes in the brain tumor microenvironment, and the ultrastructural analysis of subcellular structural alterations. Here, we describe our established methodology using pre-clinical animal models of breast cancer brain metastasis.

1.1 Clinical Overview and Challenges

Breast cancer brain metastasis (BCBM) represents a devastating progression of metastatic breast cancer, associated with significant

Marie-Ève Tremblay (ed.), *Ultrastructural Neuroscience Methods*, Neuromethods, vol. 238,
https://doi.org/10.1007/978-1-0716-5456-9_10,

morbidity and mortality [1, 2]. As systemic therapies have improved extracranial disease control and prolonged patient survival, the brain has increasingly become a sanctuary site for metastasis [3, 4]. This is primarily due to the protective nature of the blood–brain barrier (BBB), which restricts the entry of many therapeutic agents [5].

The clinical management of BCBM is complicated by the unique biology of the central nervous system (CNS) [6]. Unlike peripheral metastases, brain metastases are shielded from many therapeutics by the BBB and the more permeable but still challenging blood–tumor barrier (BTB) [5]. Furthermore, the brain microenvironment, comprising many cell types including resident astrocytes, microglia, oligodendrocytes, and neurons, actively selects for aggressive tumor phenotypes that can survive and thrive in this hostile nutrient-poor environment [7]. Clinically, BCBM is most frequently observed in human epidermal growth factor receptor (HER)2-positive and triple-negative breast cancer (TNBC) subtypes, which exhibit a distinct tropism for the CNS compared to Luminal (estrogen receptor (ER)+/progesterone receptor (PR)+) subtypes [8, 9].

1.2 Incidence of Brain Metastasis by Primary Tumor

While lung cancer has the highest overall frequency of brain metastasis, breast cancer is the second most common cause (Fig. 1) [10, 11]. The probability of developing brain metastasis varies significantly by primary tumor histology. Key facts in BCBM are:

- HER2+ and TNBC: up to 30%–50% of patients with metastatic HER2+ or TNBC will eventually develop brain metastases, a significantly higher rate than other breast cancer subtypes [8, 9].
- Luminal A/B: this subtype has a lower risk (<10%), though the prevalence of luminal breast cancer means the absolute number of cases remains high [9, 12].

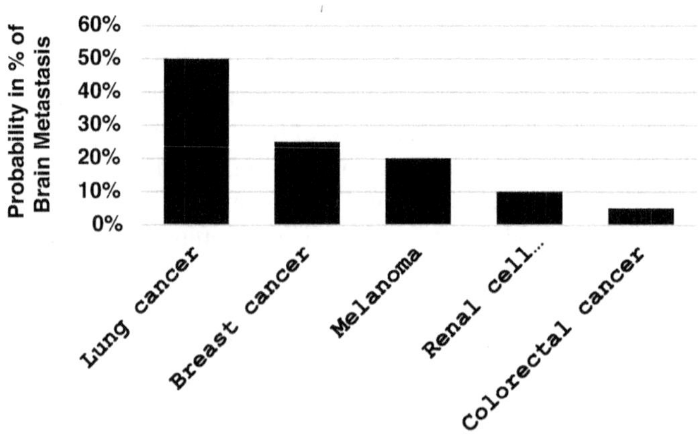

Fig. 1 Incidence of brain metastasis by primary tumor type. The approximate probability of brain metastasis originating from different primary cancers is shown. Data were compiled from multiple cohort studies and reviews [2, 13]

1.3 Animal Models of Breast-to-Brain Metastasis

Studying BCBM requires robust in vivo models that recapitulate the different steps of BCBM: from primary breast cancer (BC) cell clones leaving the primary tumor and gaining access to the blood stream, BC tumor cells circulating in the blood, extravasation from blood vessels, colonization of the brain, to establishing the brain metastatic niche and microenvironmental crosstalk [22]. An ideal model allows for high-resolution imaging while maintaining the relevant tumor biology and host environment (Table 1). For a comprehensive summary of available breast cancer cell models for brain metastasis, we refer to a previous review [23].

1.4 The Metastatic Process and Brain Microenvironment Crosstalk

The colonization of the brain is a multi-step process where tumor cells must survive a hostile and foreign brain environment before establishing new tumor growth [22].

BCBM deploy different survival strategies at the brain metastatic niche (Fig. 2):

Table 1
Syngeneic allograft and xenograft mouse models used in breast-to-brain metastasis (BCBM) research

Model type	Cell line (source)	Phenotype	Primary/key reference(s) for the model
Syngeneic (immune-Competent)	4T1 (BALB/c mouse)	TNBC-like	[14, 15]
Syngeneic (immune-Competent)	TBCP-1 (BALB/c mouse)	HER2+	[16]
Xenograft (human in mouse)	MDA-MB-231-Br	TNBC	[17, 18]
Xenograft (human in mouse)	BT474 / BT474-Br	HER2+ (luminal B)	[19]
Xenograft (human in mouse)	JIMT-1	HER2+ (resistant)	[20]
Patient-derived (PDX)	Patient tumor tissue	Varies	[21]

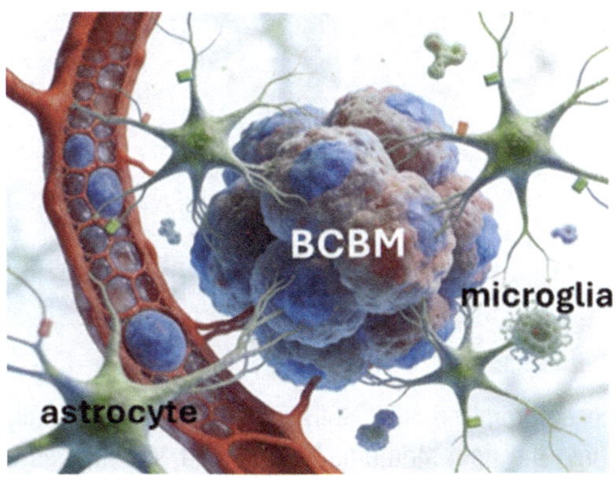

Fig. 2 Schematic illustration of the "neuronal BCBM niche" where HER2+ breast cancer cells (BCBM) establish crosstalk with resident brain cells, including reactive astrocytes and activated microglia, near a blood vessel of the blood-brain barrier. These interactions are critical for tumor survival and proliferation [24, 25]

- *Neuronal mimicry:* BC cells can hijack neuronal signaling pathways for their own benefit [26]. They may upregulate glutamate receptors (like NMDARs, AMPARs) and form pseudo-synapses with neurons, receiving pro-proliferative signals directly from them [27].
- *Astrocyte co-option:* Astrocytes are often the first responders to incoming cancer cells [28]. While initially displaying a defensive reactive glial response, tumor cells can co-opt them [29]. Direct communication through gap junctions made of connexins, like Cx43, allows the transfer of signaling molecules (e.g., cGAMP, Ca^{2+}) that support tumor growth and chemoresistance [30].
- *Functionalized membrane protrusions:* BCBM may also extend nanotubes and cytonemes, long extensions stabilized by an actin cytoskeleton, to facilitate inter-cellular exchange of organelles and molecules, respectively, while also acting as signaling platforms [31, 32].
- *Vascular co-option:* Rather than inducing the growth of new blood vessels (angiogenesis), TNBC cells in particular often migrate along existing brain capillaries [33]. The perivascular space provides a scaffold for invasion and allows for alterations to the BBB [33].

2 Imaging Tools to Study BCBM Lesions in the Brain

2.1 Clinical Imaging Modalities

Magnetic Resonance Imaging (MRI): MRI is the gold standard for the clinical detection of BCBM (Fig. 3).

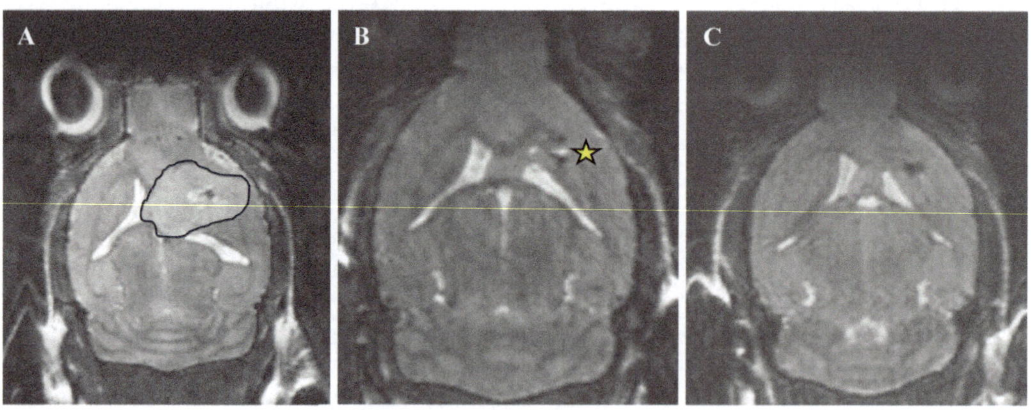

Fig. 3 Immunocompromised NOD-SCID mice were orthotopically xenografted with human HER2+ BCBM cell model BT474 into the right striatum. T2-weighted MRI images of mouse brain were taken two months later. (**A**) Untreated mice had developed large intracranial tumors. (**B**) Massive reduction in tumor mass upon treatment after one week of treatment with the covalent HER1/2/4 inhibitor Poziotinib (2mg/kg). A small cavity (black) and a hyperdense tissue areas (white; star), likely indicating remaining tumor tissue, are seen. (**C**) The void space (black) remains and a further reduction in hyperdense areas is observed upon two weeks of treatment with Poziotinib

- T1-weighted +/− Gadolinium (Gd): This sequence is the most crucial for diagnosis [34, 35]. Metastases appear as hyperintense (bright) enhancing lesions where the contrast agent has leaked through a compromised BTB [34].
- T2/FLAIR: These sequences are sensitive to water content and highlight the tumor and peritumoral edema (swelling) [35, 36]. Fig. 3 shows a large HER2+ BCBM tumor before (A) and after successful treatment (B, C) with the brain-permeable small molecule HER2 inhibitor Poziotinib.

Positron Emission Tomography (PET): ^{18}F-FDG PET utilizes the increased glucose metabolism of tumor cells but is less sensitive for small brain lesions due to high background glucose uptake in normal brain tissue [37]. However, when combined with computed tomography (CT), the hybrid PET/ CT technique has the ability to precisely localize areas of increased FDG accumulation and is widely used in oncology [38].

2.2 BCBM Immunodetection in Brain Tissues

Different BCBM subtypes display distinct morphologies of colonization. TNBC cells tend to grow in an infiltrative pattern, co-opting existing vessels (*"perivascular cuffs"*) (Fig. 4A) [25, 33]. HER2+ BCBM typically forms nodular and well-demarcated sphere-like tumors frequently surrounded by reactive astrocytes (Fig. 4B) [25].

High-resolution immunofluorescence (IF) enables the visualization of critical cell types in the unique brain parenchyma in the vicinity of BCBM tumors. We utilized our previously published multiplex immunofluorescence (mIF) imaging technique by Shanmugam

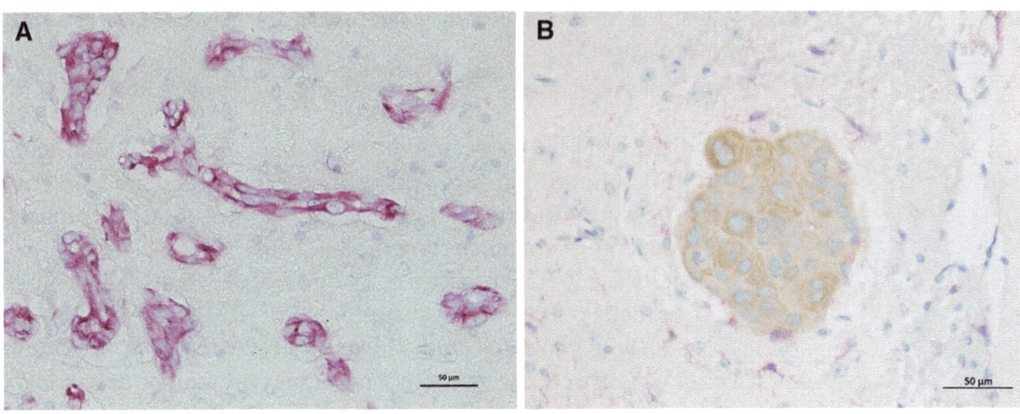

Fig. 4 Morphology of breast cancer brain metastases in mouse brains. TNBC (**A**) and HER2+ (**B**) human breast cancer cells were xenografted by ultrasound-guided intracardial injections for hematogenic colonization of the brain. (**A**) Triple-negative MDA-MB-231/BR (39) brain metastasis colonize the space surrounding blood vessels and grow by co-option. Colorimetric immunohistochemistry (IHC) for vimentin with alkaline phosphatase detection (red precipitate) is shown. (**B**) HER2+ BCBM94 cell model (40) grows spherical tumors. Colorimetric IHC for pan-cytokeratin with horseradish peroxidase (HRP) detection (brown precipitate) and Iba1 microglial marker with alkaline phosphatase detection (red precipitate)

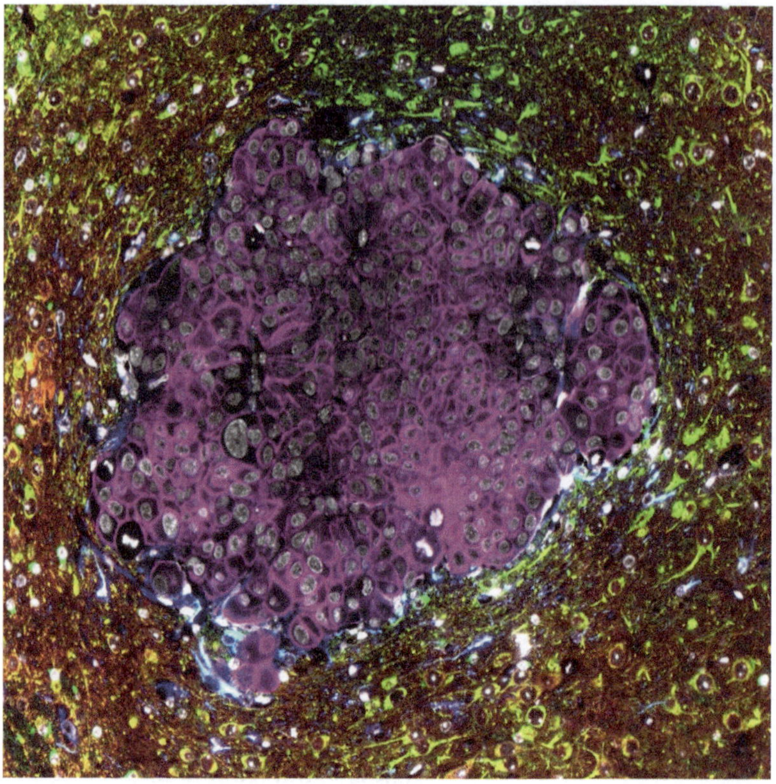

Fig. 5 Composite immunofluorescence image of a HER2+ BCBM94 metastasis in mouse brain. BCBM94 tumor cells show membrane-bound HER2 immunostaining (pink; exposure time [ET]: 20ms). Surrounding the metastasis are reactive microglia cells stained for IBA1 (white; ET: 50ms) and reactive astrocytes visualized with GFAP (blue; ET: 50ms). DISP2 (green; ET: 20ms) and MAP2 (red; ET: 20ms) serve as neuronal markers. Nuclear DAPI stain in grey (ET: 15ms). Magnification: x63

Anandhan et al [41] to localize different brain cell types in the vicinity of BCBM tumors (Fig. 5). This protocol includes details on tissue mIF processing, immunostaining, antibody stripping, autofluorescence bleaching, image alignment, and tiling.

Briefly, we performed three rounds of IF staining on mouse brain tissue sections with xenografted HER2+ BCBM94 tumor lesions (Fig. 5). We selected four antigen targets that provide an overview of three key resident brain cell types and tumor cells at the BCBM lesion. Brain resident cells were immunostained with the microglial marker ionized calcium-binding adaptor molecule 1 (IBA1; 1:500; rabbit polyclonal, Wako, 019-19741, AB_839504), the astrocytic marker glial fibrillary acidic protein (GFAP; 1:100; chicken polyclonal, Aves Labs, GFAP, AB_2313547), the neuronal marker microtubular associated protein 2 (MAP2 [EPR19691]; 1:600; rabbit monoclonal, Abcam, 183830, AB_2895301), and the neuronal marker dispatched RND transporter family member 2 (DISP2; 1:28.5; mouse monoclonal supernatant, DSHB, F66A4B1, AB_2618109). HER2+ BCBM94 tumor cells were immunostained for ErbB2/ HER2 (1:100; rabbit monoclonal, Cell Signaling Technology, 2165, AB_10692490).

The primary antibodies were incubated overnight at 4°C in a humidified chamber. Next, slides were washed with phosphate-buffered saline (PBS) plus 0.05% Tween-20 (PBS-T) 3× 10 min and blocked with 10% normal goat serum (NGS) in PBS-T and 1:50 goat F(ab) anti-mouse IgG in PBS-T prior to incubation for 1 h at RT with the secondary antibodies (1:500) goat-anti-rabbit IgG Alexa Fluor 488 (Thermofisher, A-11034, AB_2576217), goat-anti-chicken IgY Alexa Fluor 568 (Thermofisher, A-11041, AB_2534098), and goat-anti-mouse IgG Alexa Fluor 647 (Thermofisher, A-21235, AB_2535804) in a humidified chamber. The slide was washed in PBS-T 3× 5 min to remove unbound secondary antibody, and autofluorescence quenching solution was applied to the section for 25 min. Upon completion of rounds 1 + 2 of immunostaining, bound antibodies were stripped and followed by immunofluorescence microscopy to confirm successful stripping prior to proceeding with the next round of IF. Finally, all tissue sections received a nuclear DAPI counterstain (1:50,000 DAPI in double-distilled H_2O) for 5 min at RT in a humidified chamber. The slide was washed for 5 min in PBS-T and coverslipped in Fluoromount G prior to imaging. This mIF approach allowed us to use three different rabbit antibodies consecutively on the same tissue section:

Round 1: Primary: IBA1 (rabbit): 1:250—Secondary: goat-anti-rabbit GAR-AF-488nm/ GFAP (chicken): 1:100—Secondary: GAC-AF-568nm/ DISP2 (mouse monoclonal supernatant): 1:28.5 – Secondary: goat-anti-mouse GAM-AF-647nm.

Round 2: MAP2 (rabbit): 1:600—Secondary: GAR-AF-488nm.

Round 3: HER2 (rabbit): 1:100—Secondary: GAR-AF-488nm.

2.3 Transmission Electron Microscopy (TEM) and Immuno-TEM

TEM provides the ultimate imaging tool for subcellular resolution imaging of cellular structures and the immunolocalization of specific antigens in immuno-TEM. Here, we provide complete protocols for the regular and immuno-TEM processing of adherent human HER2+ BCBM94 breast cancer cells. These protocols can be easily adapted to other cell types.

Protocol for Transmission Electron Microscopy (TEM) of Cell Monolayers

We use Permanox™ dishes coated with poly-L-lysine for better cell adherence. Unlike regular cell culture dishes, Permanox™ dishes are resistant to the solvents used for TEM sample processing and compatible with most embedding resins.

- Coat Permanox™ dish (Cat # 70350, EMS, USA) with Poly-L-Lysine (Cat # P4704, Sigma-Aldrich, Germany) as per the manufacturer's protocol to improve cell adherence.

- Plate cells at 200,000–300,000 cells/plate (60 × 15 mm) and culture until 80-90% cell confluency is reached. TEM images show BCBM94 cells, which were cultured in DMEM/ F12 1:1 + 10% fetal bovine serum (FBS) for 7 days.
- Aspirate the media and gently rinse the cells with phosphate-buffered saline, pH 7.0 (PBS) at room temperature (RT) once for 1 min.

  Note: *Avoid pipetting solutions directly onto the cells. Pipette slowly onto the side of a cell culture dish to prevent cell detachment.*

- Add 4–5 mL/ dish of 3% glutaraldehyde (Cat # 16210, EMS, USA) diluted in PBS.
- Fix on a rocking platform set to the slowest setting for 4 h at RT.
- Aspirate the fixative and gently wash with PBS 3× 5 min.

  Note: *Upon fixation, cells can be safely stored in PBS at 4°C for up to 4 weeks until further processing.*

- Aspirate PBS and add 1–2 mL of 1%OsO_4 (Cat # 19150, EMS, USA) diluted in PBS. Ensure that the dish is placed on a flat surface and that the cell monolayer is fully submerged.
- Post-fix for 2 h at RT.
- Aspirate osmium and gently wash the cells with PBS for 1× 5 min.
- Aspirate PBS and dehydrate as follows: 30% ethanol (EtOH)—1× 10 min, 50% EtOH—1× 10 min, 70% EtOH—1× 10 min, 90% EtOH—1× 10 min, 100% EtOH—2× 10 min, 100% methanol—2× 10 min, 100% propylene oxide—2× 10 min (Cat # 20411, EMS, USA). Use 4–5 mL/ dish for each solution.

  Note: *PBS is used for washing after OsO_4 post-fixation; at least the first three changes of EtOH used for dehydration are contaminated with OsO_4 and are considered biohazard waste that must be disposed of appropriately.*

- For resin infiltration and embedding, prepare medium-hard EMbed 812 resin (Cat # 14120, EMS, USA) as per the manufacturer's protocol.
- For infiltration, upon removal of the second change of propylene oxide, immediately add 4–5 mL/ dish of a 1:1 mixture of EMbed812 resin and propylene oxide and incubate for 2 h.
- Remove 1:1 mixture of EMbed 812 resin and propylene oxide, add 4–5 mL/ dish of 100% EMbed 812 resin and incubate overnight.
- Next morning, aspirate the resin, fill the dish to the top (~20 mL) with fresh 100% EMbed 812 resin, and leave for another 24 h at RT.

- For polymerization, remove plate covers and transfer the plates to an oven set to 60°C for 30 h.

  ***Note** : To avoid deformation of the dish bottom during polymerization, place dishes on a flat metal tray.*

- Let the dishes cool down for 30–60 min.
- Peel the dish from the block with pliers and a utility knife.

  ***Note:** Wear safety glasses when separating the dish from the resin block.*

- Rapid submersion ~3–5 s in liquid nitrogen makes the plastic brittle and helps with the separation of the dish from the resin block with embedded monolayer.

  ***Note:** CAVEAT: Prolonged submersion in liquid nitrogen will result in cracking of the block.*

- Cut the block into ~5 × 5 mm rectangular blocks as illustrated in Fig. 6. A jeweler's saw, rotary or oscillating cutting tool with an appropriate disc can be used for cutting.
- The resulting blocks can be used with the standard BEEM capsule ultramicrotomy chucks for further trimming and sectioning.
- Trim the sides of the block to a 45°–60° angle.
- Avoid trimming the block face as it will result in the loss of the embedded monolayer.
- Set the ultramicrotome to 70–80 nm and start sectioning with a semi-thin diamond knife.

  ***Note:** While an ultrathin diamond knife can be used right away, there is an increased risk of damage to this knife because the block surface was not trimmed.*

- Start cutting and perform a step-wise adjustment of the knife rotation and sample tilt angles to obtain a section of the whole block surface, avoiding multiple knife passes.

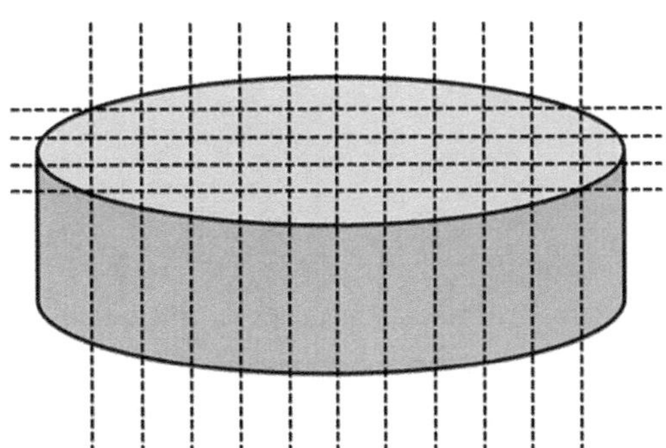

Fig. 6 Schematic diagram depicting the cutting of rectangular-shaped Embed 812 resin blocks containing adherent cells prior to further TEM processing

Note: Based on the type of cell and the way these cells grow, the thickness of the monolayer is typically 5–50 microns. Be aware that the cell monolayer has a finite thickness. One can only obtain a limited number of sections before sectioning completely through the cell monolayer.

- Switch to the ultrathin diamond knife and obtain 70–80 nm sections.
- Mount the sections on 100–200 mesh copper grids.
- Perform contrasting as follows:
- UranyLess (Cat # 22409, EMS, USA) for 1× 2 min followed by 3 × 10 dips in separate containers with double-distilled H_2O.
- Lead citrate (Cat # 22410, EMS, USA) for 1× 1 min followed by 3 × 10 dips in separate containers with double-distilled H_2O.

Note: Upon UranyLess staining, grids should be completely dry before transitioning to the Lead Citrate step to avoid precipitation of the contrasting reagents.

- Once the sections are completely dry (~30 min) after lead citrate staining, the grids are ready for TEM imaging (Fig. 7A, B). We use a JOEL JEM-1400plus TEM, which operates with a lanthanum hexaboride (LaB6) electron source.

Post-embedding Immuno-TEM for HER2 Localization in human BCBM94 cells

- To compensate for cell loss during the processing steps and trimming of the block, fix 2–4 million cells/ sample.
- BCBM94 cells were plated at 500,000 cells/ T75 flask and cultured in DMEM/ F12 1:1 + 10% FBS for 1 week.
- Detach the cells with trypsin, add media, transfer to 15 mL Falcon tubes, and pellet by centrifugation at 300 g for 5 min.

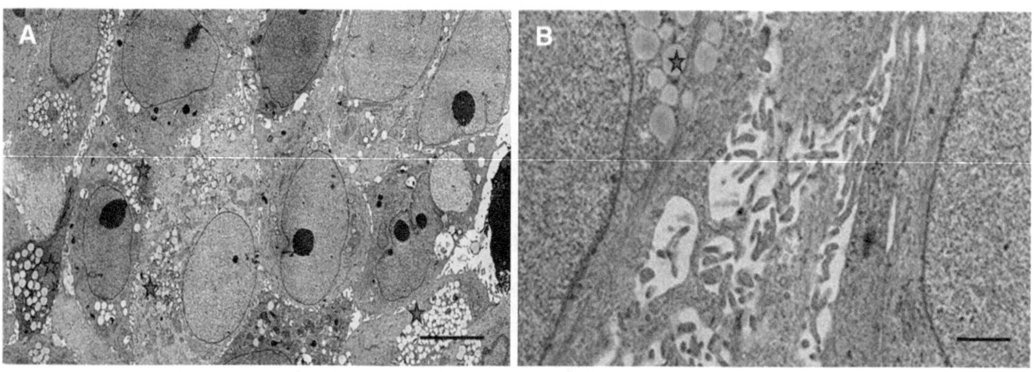

Fig. 7 TEM images of human HER2+ BCBM94 cell monolayers (A, B). BCBM94 breast cancer cells grow in tightly arranged groups (A; x1000) with extensive membrane protrusions of different sizes filling the intercellular space (B; x5000). Other characteristic features of BCBM94 cells are frequent vesiculated structures within the cytoplasm (stars) and prominent electron-dense nucleoli (black) within irregularly shaped nuclei. Scale bar = 1 micrometer

- Wash the pellet 1× 5 min with PBS at RT.
- Resuspend the pellet in 4–5 mL of 4% paraformaldehyde diluted in PBS and fix for 1–4 h. Resuspend the cells by gently tapping the Falcon tube every ~1 h.

  Note: *Prolonged fixation improves ultrastructure but results in loss of immunogenicity. Optimization of fixation time for a particular antibody is recommended.*

- After fixation, the cells are washed for 3× 5 min in PBS at RT. ***Note:*** *Omit OsO_4 post-fixation to preserve immunogenicity.*
- Aspirate PBS and dehydrate as follows: 30% EtOH—1× 10 min, 50% EtOH—1× 10 min, 70% EtOH—1× 10 min, 90%EtOH—1× 10 min, 100% EtOH—2× 10 min.
- For infiltration and embedding, bring medium-hard LR White resin (Cat # 14380, EMS, USA) to RT; then, generate a 1:1 mixture of LR White resin with 100% EtOH.
- Upon removal of the second change of 100% EtOH, immediately resuspend the pellet in 1–2 mL of the 1:1 mixture of LR White resin/ 100% EtOH and incubate for 2 h.
- **Optional:** Incubate in a 2:1 mixture of LR White resin and 100% EtOH for another 2 h.
- Resuspend the pellets in fresh 100% LR White resin and leave overnight.
- Resuspend the pellet in ~50 μL of fresh LR White resin.
- Transfer the cells to gelatin embedding capsules (Cat # 70102, 70103, 70104, EMS, USA) and let the cells fully settle (~30–60 min).

  Note: *Having enough cells is critical because of the round-bottom shape of the gelatin capsules. Ensure that you have a ~2 mm thick cell pellet at the bottom of the capsule. Part of the pellet will be lost during the block trimming step.*

- Slowly fill the bottom part of the gelatin embedding capsule with fresh LR White without disturbing the pellet.
- Close the capsules—a small amount of resin will leak, sealing the bottom and the upper parts of the capsule, and creating an air-free environment is required for polymerization.
- Leave the capsules at RT overnight.
- Transfer the capsules to an oven set to 60°C for 30 h for polymerization.
- Let the capsules cool down (~30 min) and then place them in double-distilled H_2O for 30 min–1 h. Gelatin will soften and can be easily removed with a tissue paper.

- The resulting blocks can be clamped in the standard BEEM capsule ultramicrotomy chucks for further trimming and sectioning.
- Trim the tip of the block to achieve ~2 × 2 mm flat block face suitable for sectioning.
- Trim the sides to a 15° angle.
- Coat nickel or gold 100–200 mesh grids with Coat-Quick "G" pen (Cat # 70624, EMS, USA) as per the manufacturer's protocol. This coating improves the adherence of tissue sections onto the grids and prevents section detachment during multiple incubation and washing steps during the immunostaining part of the protocol.

  Note: *Avoid using copper grids for immuno-TEM applications, as copper can oxidize in buffers, producing visible precipitate.*

- Cut 100 nm sections and mount on the coated grids.
  Note: *Immunostaining can be done on drops on a strip of Parafilm*[R]*, in regular 0.5 mL Eppendorf microtubes, or in caps of flowtubes. The central portion of the flowtube cap perfectly fits one standard-size TEM grid, allowing to perform incubation and washing steps without picking up the grids with tweezers.*
- Prepare PBS washing buffer by adding 0.05% Tween-20 (PBS-T).
- Hydrate the sections in PBS-T washing buffer for 15–30 min.
- **Optional:** Perform heat-induced antigen retrieval by transferring grids in 2 mL Eppendorf tubes (1 grid/tube) containing 1.5 mL of the antigen retrieval buffer of choice (10 mM sodium citrate / citric acid, pH 6.0, or 1 mM EDTA, pH 8.0). Incubate the tubes at 95°C for 30 min with caps open. Take out the tubes and let them cool down to RT. Wash 3× 3 min in PBS-T.
- Block non-specific antibody binding sites with 1% BSA plus 5% normal goat serum (NGS; or any other non-immune serum matching the source of the secondary gold-conjugated antibody) diluted in the PBS-T for 1–2 h.
- Wash 1× 1 min in PBS-T.
- Incubate with primary antibody as per the manufacturer's recommendation. Overnight incubation in a humidified chamber at 4°C is preferable.
- For the demonstrated HER2 staining, sections were incubated at 4°C overnight in a 1:400 dilution of HER2/ErbB2 (29D8) rabbit monoclonal antibody (Cat # 2165, Cell Signaling Technology, USA) diluted in 1% BSA in PBS-T.
- Wash 6× 3 min in PBS-T.
- Dilute your secondary gold-conjugated antibody as per the manufacturer's protocol.

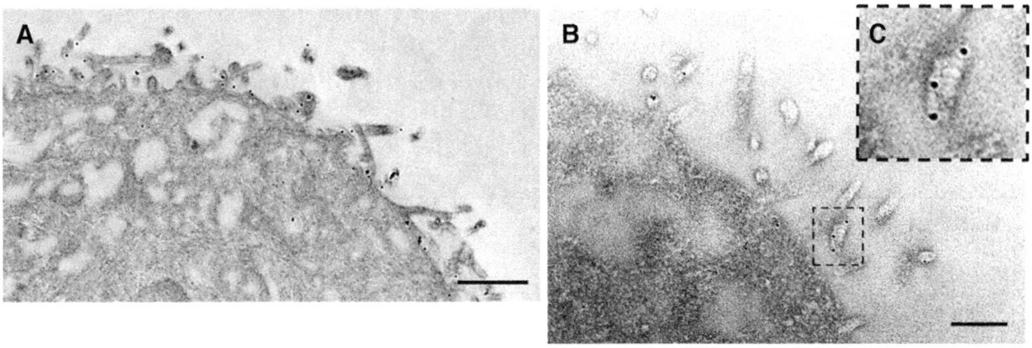

Fig. 8 Immuno-TEM revealed the presence of immunoreactive HER2 localized at cell membrane protrusions of cultured human HER2+ BCBM94 cells as indicated by the 15 nm gold particles (black dots) at the plasma membrane and membrane protrusions. Images were taken at x5000 (**A**) and x10,000 (**B**) magnification. Scale bar = 1 micrometer

- For the demonstrated HER2 staining (Fig. 8A–C), the sections were incubated at RT for 2 h in a 1:100 dilution of 15 nm gold-conjugated goat-anti-rabbit IgG (H&L) (Cat # 815.011, Aurion, Denmark) diluted in 1% BSA in PBS-T.
- Wash 6× 3 min in PBS-T.

  Note: *The number and duration of the washing steps should be adjusted for each primary–secondary antibody combination to reach optimal signal intensity.*

- Post-fix with 4% paraformaldehyde in double-distilled H_2O for 15 min.
- Wash 3× 1 min in PBS-T.
- Once the sections are completely dry (~30 min), check with TEM imaging. If the staining has worked, proceed with the contrasting steps.
- UranyLess (Cat # 22409, EMS, USA) staining for 1× 2 min followed by 3 × 10 dips in separate containers with double-distilled H_2O.
- Lead citrate (Cat # 22410, EMS, USA) staining for 1× 1 min followed by 3 × 10 dips in separate containers with double-distilled H_2O.

  Note: *Let the grids dry completely upon* UranyLess *staining before transitioning to the lead citrate step to avoid precipitation of the contrasting reagents.*
  Note: *Avoiding* OsO_4 *post-fixation results in better immunogenicity, but poorer ultrastructure and contrast. If needed, the duration of staining with UranyLess can be increased to 5–10 min, and for lead citrate to 3–5 min. Decreasing the number of washing steps to 1–2× 10 min will also result in increased contrast.*

- Once the sections are completely dry (~30 min), the grids are ready for imaging (Fig. 8A–C).
  Note: *We have used a JEM-1400plus TEM for immuno-TEM imaging* [13, 24, 39, 40].

Conflict of Interest

The authors have no conflicts of interest to declare that are relevant to the content of this chapter.

References

1. Ivanova M, Porta FM, Giugliano F, Frascarelli C, Sajjadi E, Venetis K et al (2023) Breast cancer with brain metastasis: molecular insights and clinical management. Genes ;14(6):1160
2. Siegel R, Giaquinto AN, Jemal A (2025) Cancer statistics. CA Cancer J Clin 75(1):10–45
3. Kuksis M, Gao Y, Tran W, Hoey C, Kiss A, Komorowski AS et al (2021) The incidence of brain metastases among patients with metastatic breast cancer: a systematic review and meta-analysis. Neuro Oncol 23(6):894–904
4. Valiente M, Ahluwalia MS, Boire A, Brastianos PK, Goldberg SB, Lee EQ et al (2018) The Evolving Landscape of Brain Metastasis. Trends Cancer 4(3):176–196
5. Wang W, He H, Marin-Ramos NI, Zeng S, Swenson SD, Cho HY et al (2021) Enhanced brain delivery and therapeutic activity of trastuzumab after blood-brain barrier opening by NEO100 in mouse models of brain-metastatic breast cancer. Neuro Oncol 23(10):1656–67.
6. Lin X, DeAngelis LM (2015) Treatment of brain metastases. J Clin Oncol 33(30):3475–3484
7. Liu L, He Y, Du H, Tang M, Wang T, Tan J et al (2025) Biological profile of breast cancer brain metastasis. Acta Neuropathol Commun 13(1):78
8. Kennecke H, Yerushalmi R, Woods R, Cheang MC, Voduc D, Speers CH et al (2010) Metastatic behavior of breast cancer subtypes. J Clin Oncol 28(20):3271–3277
9. Kim YJ, Kim JS, Kim IA (2018) Molecular subtype predicts incidence and prognosis of brain metastasis from breast cancer in SEER database. J Cancer Res Clin Oncol 144(9):1803–16.
10. Nayak L, Lee EQ, Wen PY (2012) Epidemiology of brain metastases. Curr Oncol Rep 14(1):48–54
11. Sacks P, Rahman M (2020) Epidemiology of brain metastases. Neurosurg Clin N Am 31(4):481–488
12. Simsek M, Aliyev A, Baydas T, Besiroglu M, Demir T, Shbair AT et al (2022) Breast cancer patients with brain metastases: a cross-sectional study. Breast J 2022:5763810
13. National Cancer Institute (2025) Surveillance, epidemiology, and end results program. National Institutes of Health, Bethesda, MD
14. Pulaski BA, Ostrand-Rosenberg S (2001) Mouse 4T1 breast tumor model. Curr Protoc Immunol 39(1):20.2.1–20.2.16
15. Tao K, Fang M, Alroy J, Sahagian GG (2008) Imagable 4T1 model for the study of late stage breast cancer. BMC Cancer 8: 228
16. Nagpal A, Redvers RP, Ling X, Ayton S, Fuentes M, Tavancheh E et al (2019) Neoadjuvant neratinib promotes ferroptosis and inhibits brain metastasis in a novel syngeneic model of spontaneous HER2(+ve) breast cancer metastasis. Breast Cancer Res 21(1):94
17. Dun MD, Chalkley RJ, Faulkner S, Keene S, Avery-Kiejda KA, Scott RJ et al (2015) Proteotranscriptomic profiling of 231-BR breast cancer cells: identification of potential biomarkers and therapeutic targets for brain metastasis. Mol Cell Proteom 14(9):2316–2330
18. Yoneda T, Williams PJ, Hiraga T, Niewolna M, Nishimura R (2001) A bone-seeking clone exhibits different biological properties from the MDA-MB-231 parental human breast cancer cells and a brain-seeking clone *in vivo* and *in vitro*. J Bone Miner Res 16(8):1486–1495
19. Conlon NT, Kooijman JJ, van Gerwen SJC, Mulder WR, Zaman GJR, Diala I et al (2021) Comparative analysis of drug response and gene profiling of HER2-targeted tyrosine kinase inhibitors. Br J Cancer 124(7):1249–1259
20. Tanner M, Kapanen AI, Junttila T, Raheem O, Grenman S, Elo J et al (2004) Characterization of a novel cell line established from a patient with Herceptin-resistant breast cancer. Mol Cancer Ther 3(12):1585–1592
21. DeRose YS, Wang G, Lin YC, Bernard PS, Buys SS, Ebbert MT et al (2011) Tumor grafts derived from women with breast cancer authentically reflect tumor pathology, growth,

metastasis and disease outcomes. Nat Med 17(11):1514–1520
22. Wang Y, Ye F, Liang Y, Yang Q (2021) Breast cancer brain metastasis: insight into molecular mechanisms and therapeutic strategies. Br J Cancer 125(8):1056–1067
23. Knier NN, Pellizzari S, Zhou J, Foster PJ, Parsyan A (2022) Preclinical models of brain metastases in breast cancer. Biomedicines; 10(3):667
24. Zhang X, Wang X, Shi S, Guo D (2025) Decoding the mechanisms underlying breast cancer brain metastasis: paving the way for precision therapeutics. Biomark Res 13(1):144
25. Gan S, Macalinao DG, Shahoei SH, Tian L, Jin X, Basnet H et al (2024) Distinct tumor architectures and microenvironments for the initiation of breast cancer metastasis in the brain. Cancer Cell 42(10):1693–1712.e24
26. Zeng Q, Michael IP, Zhang P, Saghafinia S, Knott G, Jiao W et al (2019) Synaptic proximity enables NMDAR signalling to promote brain metastasis. Nature 573(7775):526–531
27. Mondal J, Huse JT (2025) Neurotransmitter power plays: the synaptic communication nexus shaping brain cancer. Acta Neuropathol Commun 13(1):85
28. O'Brien ER, Kersemans V, Tredwell M, Checa B, Serres S, Soto MS et al (2014) Glial activation in the early stages of brain metastasis: TSPO as a diagnostic biomarker. J Nucl Med 55(2):275–280
29. Zhou T, Yan Y, Lin M, Zeng C, Mao X, Zhu Y et al (2025) Astrocyte involvement in brain metastasis: from biological mechanisms to therapeutic strategies. Cancer Metastasis Rev 44(3):60
30. Chen Q, Boire A, Jin X, Valiente M, Er EE, Lopez-Soto A et al (2016) Carcinoma-astrocyte gap junctions promote brain metastasis by cGAMP transfer. Nature 533(7604):493–498
31. Osswald M, Jung E, Sahm F, Solecki G, Venkataramani V, Blaes J et al (2015) Brain tumour cells interconnect to a functional and resistant network. Nature 528(7580):93–98
32. Hall ET, Dillard ME, Cleverdon ER, Zhang Y, Daly CA, Ansari SS et al (2024) Cytoneme signaling provides essential contributions to mammalian tissue patterning. Cell 187(2):276–93 e23
33. Kienast Y, von Baumgarten L, Fuhrmann M, Klinkert WE, Goldbrunner R, Herms J et al (2010) Real-time imaging reveals the single steps of brain metastasis formation. Nat Med 16(1):116–122
34. Taha BR, Wu DJ, Sabal LT, Kollitz M, Sloan L, Shoo BA et al (2025) An integrated dataset of metastatic breast cancer to the brain with imaging, radiomics, and tumor genetics. Sci Data 12(1):1851
35. Buchner JA, Peeken JC, Etzel L, Ezhov I, Mayinger M, Christ SM et al (2023) Identifying core MRI sequences for reliable automatic brain metastasis segmentation. Radiother Oncol 188:109901
36. Heyn C, Bishop J, Moody AR, Kang T, Wong E, Howard P et al (2025) Gadolinium-enhanced T2 FLAIR is an imaging biomarker of radiation necrosis and tumor progression in patients with brain metastases. AJNR Am J Neuroradiol 46(1):129–35
37. Guglielmi R, Andreisek G, Halpern BS (2022) (18)F FDG imaging - response criteria in tumors. Eur J Radiol 147:110054
38. Pauwels EK, Coumou AW, Kostkiewicz M, Kairemo K (2013) [(1)(8)F]fluoro-2-deoxy-d-glucose positron emission tomography/computed tomography imaging in oncology: initial staging and evaluation of cancer therapy. Med Princ Pract 22(5):427–437
39. Palmieri D, Smith QR, Lockman PR, Bronder J, Gril B, Chambers AF et al (2006) Brain metastases of breast cancer. Breast Dis 26:139–147
40. Ippolitov D, Lin YH, Spence J, Glogowska A, Thanasupawat T, Beiko J et al (2025) The brain-penetrant Pan-ErbB inhibitor poziotinib effectively targets HER2+ breast cancer brain metastases. Cancer Res 85(8):1514–1529
41. Shanmugam Anandhan S, Spence J, Begum F, Kaur N, Henderson D, Hombach-Klonisch S et al (2025) A customized affordable multiplexed immunofluorescence method visualizes early changes in the mouse brain microenvironment upon laser cytoreduction. Front Cell Neurosci 19:1553058

Chapter 11

SnapG: An Automated Tool for Myelin g-ratio Measurement with Batch Processing Capability

Yoko Bekku, Stephen Cai, Thomas Xin, Joseph Sall, Frances Kestel, Alice F. Liang and James L. Salzer

Abstract

The g-ratio, calculated as the axon diameter divided by the total myelinated fiber diameter, is widely used to assess the degree of myelination in the central and peripheral nervous systems. Changes in g-ratios accompany demyelinating, hypomyelinating, and remyelinating conditions, and can also result from activity-dependent, adaptive myelination. Although manual tracing from electron micrographs has been the gold standard to measure g-ratios, this process is laborious and limits throughput. To address this challenge, we developed SnapG, an automated and user-friendly software tool to streamline g-ratio analysis from electron micrographs. SnapG implements an end-to-end pipeline for axon and myelin segmentation on EM images. The tool integrates robust segmentation and post-processing steps to extract axon and fiber diameters and generates both per-image and cohort-level summary statistics. SnapG produces g-ratio distributions that agree well with expert manual annotations and established tools across imaging modalities. It reduces user intervention, supports large image and batch processing, and achieves faster processing times while delivering accurate, reproducible measurements. SnapG's automated reporting thereby facilitates high-throughput analyses and improves consistency across experiments. SnapG can accelerate studies of myelin biology and provides a practical alternative to existing tools.

Keywords G-ratio, Myelin, Axon, Electron microscopy, Automated quantification, Image analysis

1 Introduction

Myelin, the multilamellar glial sheath that surrounds axons, emerged during evolution some 425 million years ago in the earliest hinge-jaw fishes [1]. The appearance of myelin made the development of the large, complex nervous systems of vertebrates possible. In addition to its key role in enabling saltatory conduction, myelin is now appreciated to have critical roles in the metabolic support of axons

---

Supplementary Information The online version contains supplementary material available at https://doi.org/10.1007/978-1-0716-5456-9_11

Marie-Ève Tremblay (ed.), *Ultrastructural Neuroscience Methods*, Neuromethods, vol. 238,
https://doi.org/10.1007/978-1-0716-5456-9_11,

and in optimizing circuit function via activity-dependent, adaptive changes [2, 3].

Since Remak's initial description of myelin [4], it has been the subject of intense efforts to characterize its function and assess its structure. By using birefringence of images obtained through optical microscopy, Schmitt and Bear were the first to quantify the thickness of myelin [5]. Their studies defined the g-ratio (initially termed g) as the ratio of the axon diameter to the outer diameter of the myelin sheath [6]. The g-ratio has since become the standard metric of myelin thickness, with values of ~0.67 in the peripheral nervous system (PNS) and ~0.77 in the central nervous system (CNS). These values are generally conserved across species and for different PNS nerves or different white matter tracts in the CNS. The measured values in the PNS agree well with an optimal g-ratio of ~0.6 that supports maximal conduction speed as theoretically derived by Rushton [7]. The higher g-ratios in the CNS may reflect additional constraints, including intracranial volume conservation [8]. Alterations in g-ratios are common in neurological disorders accompanied by hypomyelination, demyelination +/− remyelination, such as Charcot–Marie–Tooth disease and multiple sclerosis, underscoring the importance of accurate g-ratio measurements. In addition, g-ratios can also be modulated normally, for example, reflecting increases in myelin thickness during activity-dependent plasticity [9, 10].

Despite its broad utility as a measure of myelin thickness, g-ratio quantification remains a time-consuming process that traditionally requires extensive manual tracing of axon and myelin profiles from electron micrographs. To address the challenges of manual g-ratio quantification, several automated tools have been developed in recent years to improve efficiency, accuracy, and data consistency; examples include MyelTracer, AxonSeg, and AxonDeepSeg [11–13]. These tools are compatible with either electron micrographs or optical microscopic images, depending on the software. They operate in a semi-automated manner. AxonDeepSeg, a deep learning-based axon and myelin segmentation tool, produces highly accurate results across various species and works with both SEM and TEM images [12]. However, it typically requires powerful hardware and longer processing times to achieve a high level of accuracy. MyelTracer, on the other hand, relies on traditional image processing techniques and lower-resolution images, but requires supervision by the user during segmentation [13]. This limits the range of usable images and makes large-scale batch processing difficult.

To further enhance efficiency and reduce user burden, there is a remaining need for fully automated software that eliminates preprocessing requirements, supports unrestricted image sizes, and offers a user-friendly interface. Here, we describe SnapG, an automated tool we have developed for myelin g-ratio analysis from electron micrographs.

2 Materials and Methods

2.1 Mice

C57BL/6 mouse line was used in this study. Mice were sacrificed at 3 months of age. All experiments with mice were performed in compliance with the relevant policies and institutional guidelines issued by the New York University School of Medicine Institutional Animal Care and Use Committee.

2.2 Electron Microscopy

Mice were transcardially perfused with 4% PFA, 2.5% glutaraldehyde, and 0.1 M sucrose in 0.1 M cacodylate buffer (CB), pH 7.4. After an additional 2 h, sciatic nerves, optic nerves, and whole brains were dissected and immersed in the same fixative solution overnight. The genu of the corpus callosum was dissected by a brain slicer. Fixed tissues were washed three times in 0.1% CB for 15 min each, then post-fixed in 1% osmium tetroxide in CB for 2 h at room temperature. The tissues were dehydrated in a series of ethanol solutions including 30%, 50%, 70%, 85%, 95%, 100%, and 100% for 15 min in each solution at room temperature, then immersed in propylene oxide (PO) (Electron Microscopy Sciences #20412), 1:1 solution mix of PO: Epon without BDMA [46% EMbed 812 (Electron Microscopy Sciences #14900) + 36% DDSA (Electron Microscopy Sciences #13710) + 18% NMA (Electron Microscopy Sciences #19000)] for 1 h, 1:2 of solution mix of PO: Epon without BDMA overnight, and embedded in Epon [44% EMbed 812 + 35% DDSA + 18% NMA + 3% BDMA (Electron Microscopy Sciences #14900)]. Samples were sectioned at 70 nm thickness using Leica UC6 ultramicrotome (Germany) and stained with uranyl acetate and lead citrate in a standard manner. Electron micrographs of random fields were captured at the NYU Langone Microscopy Laboratory on a JEOL JEM-1400 Flash TEM (Japan) and photographed with a Gatan Rio16 camera (Gatan Inc.) or a Talos L120 C TEM (Thermo Fisher Scientific) coupled with a Gatan OneView camera. All chemicals and EM grids are purchased from Electron Microscopy Sciences, Hatfield, PA.

2.3 Software Development

SnapG relies on the Python NumPy and OpenCV libraries and consists of two main steps: preprocessing and information extraction. In the preprocessing step, SnapG reads the input image in grayscale, where each pixel's brightness is represented by an integer value from 0 (black) to 255 (white). The image can then be resized to dimensions smaller than the original to increase processing speed and reduce noise; users may choose the appropriate size based on their computer's performance. To further suppress noise, SnapG applies an average filter using a circular kernel, defined by a user-specified integer radius r, and generated with numpy.ogrid:

```
import numpy as np
import cv2
y, x = np.ogrid[-r:r+1, -r:r+1]
kernel = x**2 + y**2 <= r**2
kernel /= np.sum(kernel)
img_gray= cv2.filter2D (src=img_gray, ddepth=-1,
kernel=kernel)
```

The filtered image is then thresholded to a binary mask using a user-defined *threshold_val*, and small black features that individually take up less than 0.06% of the entire image (either image noise or small organelles) are filled in:

```
_, thresh = cv2.threshold(img_gray, threshold_val,
255, cv2.THRESH_BINARY)
inverted = cv2.bitwise_not(thresh)
contours, _ = cv2.findContours(inverted, cv2.RETR_
TREE, cv2.CHAIN_APPROX_NONE)
for c in contours:
area_proportion = cv2.contourArea(c) /
total_image_area
if area_proportion < 0.0006:
cv2.drawContours(thresh, [c], -1, 255, cv2.FILLED)
```

A dilation operation with a user-defined integer kernel size *dilate_val* is applied, followed by a similar erosion with *erode_val*:

```
dilate_kernel = cv2.getStructuringElement(cv2.
MORPH_RECT, (dilate_val,dilate_val))
dilated = cv2.morphologyEx(thresh, cv2.MORPH_
DILATE, dilate_kernel)
erode_kernel = cv2.getStructuringElement(cv2.MORPH_
RECT, (erode_val,erode_val))
eroded = cv2.morphologyEx(dilated, cv2.MORPH_ERODE,
erode_kernel)
```

Contours—outlines of contiguous groups of white pixels—are detected using cv2.*findContours*. Invalid contours are excluded based on the following criteria: touching the image edges; bounding box area outside of a user-defined range [*min_area*, *max_area*] in pixels^2; or insufficient convexness or circularity, calculated as: convexness= convex_hull_area/contour_area

```
circularity=4π·contour_area/(contour_perimeter)^2.
```

Contours that fail to meet the user-defined limits *min_convexness* and *min_circularity* are excluded. The remaining contours are assumed to represent axon inner edges, and the algorithm proceeds to the information extraction step. To approximate the myelin thickness for each contour, a mask of the contour is created and a distance transform is applied, mapping each pixel to its shortest Euclidean distance from the contour (Fig. S1A–C):

```
contour_mask = np.zeros_like(thresh)
cv2.drawContours(contour_mask, [current_contour],
-1, 255, thickness=cv2.FILLED)
distance_transform = cv2.distanceTransform(255 -
contour_mask, distanceType=cv2.DIST_L2, maskSize=5)
```

An exclusion mask and an outline mask are created. The exclusion mask is the same as the preprocessed image, but it excludes the current contour. The outline mask contains the edges of the exclusion mask, derived using cv2.*Canny*:

```
exclusion_mask = (contour_mask == 0) & (eroded ==
255)
outline_mask = cv2.Canny(exclusion_mask,0,0)
```

The outline mask is applied to the distance transform (Fig. S1D–F):

```
distances = distance_transform[outline_mask > 0]
distances = distances[distances > 0]
```

To extract myelin thickness, the *n* closest nonzero pixels to the contour are sampled, where *n* is twice the number of points in the current contour. From the resulting distribution of distances, the value at the *thickness_percentile* percentile is chosen as the representative myelin thickness in pixels, where *thickness_percentile* is a user-defined parameter (Fig. S1G):

```
n = 2 * len(current_contour)
smallest_n = np.partition(distances, n - 1)[:n]
thickness = np.percentile(smallest,
thickness_percentile)
```

The axon's inner radius is approximated as:

```
contour_perimeter = cv2.arcLength(inner_contour[0],
closed=True)
radius=(contour_perimeter)/2π
```

The g ratio is then calculated:

```
g_ratio = radius / (radius + thickness)
```

Other measurements, such as inner and outer diameter, are derived from these approximations.

2.4 Quantification of g-ratio in Electron Micrographs

For comparison between manual and SnapG-based g-ratio calculations, we used images originally acquired for a previous study from the laboratory [14] together with the corresponding manually quantified g-ratio data. Briefly, manual g-ratio quantification was performed by calculating diameters from their respective perimeters measured with the Magnetic Lasso tool in Photoshop (Adobe). For comparison between per-image and batch-level g-ratio calculation, images of the P90 carpus callosum were randomly acquired from a section on a grid, ensuring coverage across the entire region. For P90 sciatic nerves, 10 FOVs (1 FOV = 22.56 × 22.56 μm) per mouse were analyzed from two mice. For P90 optic nerves, 10 FOVs (1 FOV = 6.92 × 6.92 μm) per mouse were analyzed from two mice. For P90 corpus callosum, 26 FOVs (1 FOV = 6.78 × 6.78 μm) were analyzed from one mouse. Only axons with a circularity equal to or greater than 0.6 were included in the analysis. SnapG can run on Windows 11 (Intel Core i5-14500, 32 GB RAM, Intel UHD Graphics 770; Intel Core i7-12700H, 16 GB RAM, Intel Iris Xe and NVIDIA GeForce RTX 3060 Laptop GPU; AMD Ryzen 7 8840HS, 16 GB LPDDR5 RAM, AMD Radeon Graphics), macOS devices including the MacBook Air (M1, 2020) and Mac mini (M2, 2023; macOS Sequoia), and Linux Mint v22.2 Cinnamon (AMD Ryzen 7 8840HS, 16 GB LPDDR5 RAM, AMD Radeon Graphics).

2.5 Statistical Analysis

Figure legends contain the number of samples and additional details of the data plotted. Section 3 also describes the number of axons, animals, or FOVs analyzed. All statistical significances were assessed by Student's *t*-test in Excel (Microsoft 365).

2.6 Procedure

To begin, download the SnapG code, which is freely available online at https://github.com/KK201431873/SnapG. Open the terminal in the SnapG folder and run the software.

SnapG analysis consists of three steps:

1. Data Preparation—Parameter Tuning: Adjusting segmentation parameters and thickness percentiles as part of data preparation. This step is responsible for threshold adjustment and contour detection.
2. Reviewing Segmentation Files: Manually checking and correcting the segmentation results.
3. Data Generation: Performing quantification tasks, including g-ratio and myelin-thickness measurements.

Detailed procedures are also available on the SnapG website.

Step 1. Data Preparation—Parameter Tuning

Before performing data analysis, users should define the thresholds that will be used for quantification.

1. Open image file(s) from the File menu (Fig. 1A).
2. Select the image(s) in the file dialog and click "Open."

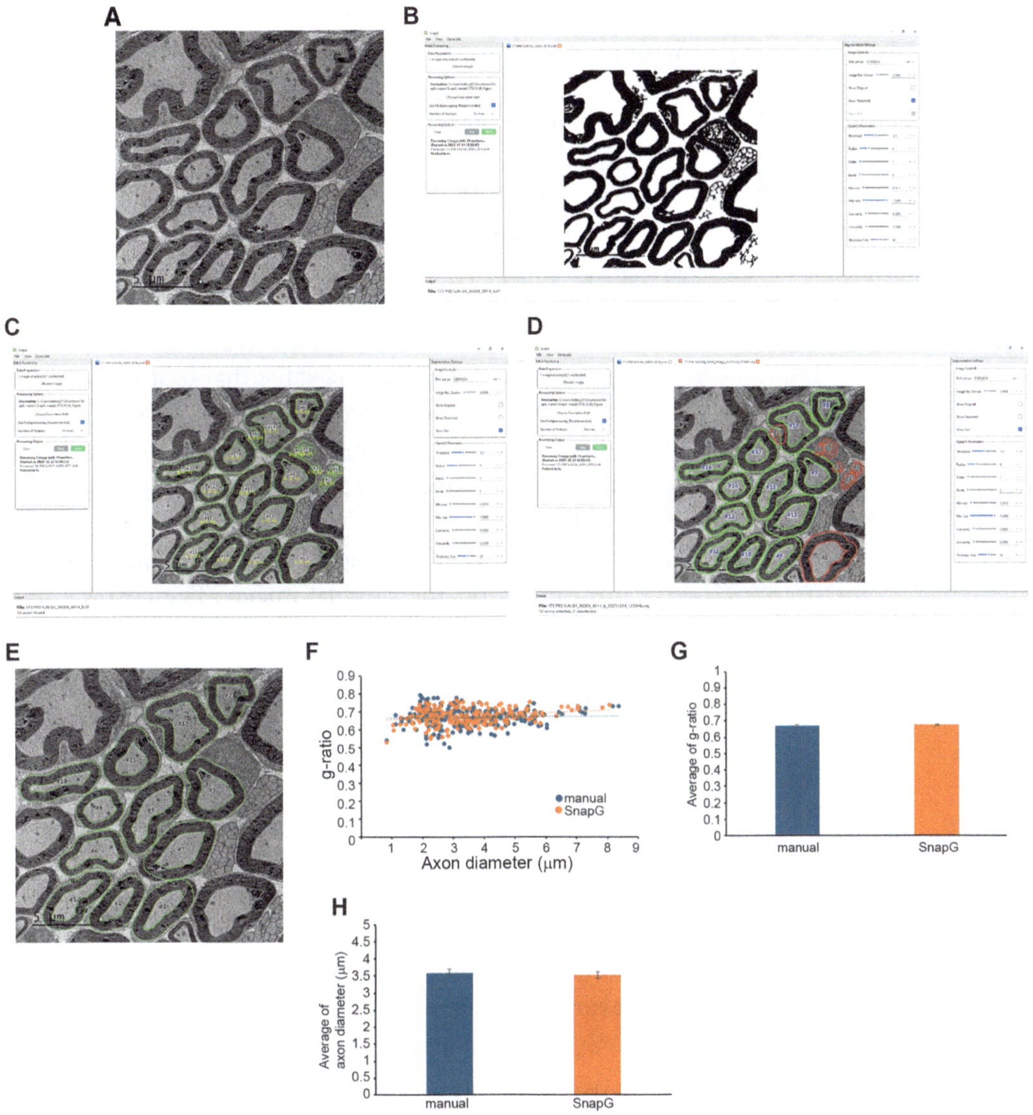

Fig. 1 g-ratio calculated by SnapG versus manual measurements of PNS myelinated axons. (**A**) Electron micrographs of P90 sciatic nerves. Bar, 5 μm. (**B**) The SnapG software interface showing the threshold view with adjusted parameters. (**C**) The outer and inner contours and axon diameters are shown in the processed image. (**D**) Processed image after axons have been deselected. Deselected axons are marked by red lines. (**E**) The final output generated by the code. Axon numbers match those listed in the CSV. (**F**) Comparison of g-ratio values obtained from manual ($n = 201$ axons) and SnapG ($n = 201$ axons) calculations shows both methods yield similar results. (**G**) Bar graph showing the average *g*-ratio calculated manually (0.669 ± 0.004; mean ± SEM) and by SnapG (0.675 ± 0.003). (**H**) Bar graph showing the average axon diameter calculated manually (3.600 ± 0.101 μm; mean ± SEM) and by SnapG (3.529 ± 0.098μm)

3. Tune the segmentation parameters in the "Segmentation Settings" panel.
   a) Set the scale per pixel in "Dist. per pix." You can choose either μm or nm.
   b) Adjust the "Image Res. Divisor" value as needed. We usually use a divisor of 3 to downsample the images, but you may increase it up to 10.
   c) Click "Show Threshold" to display the black-and-white thresholded image. Adjust the myelin-thickness percentile in "Thickness %tile." We generally use 60%, but you may need to determine the value that best fits your samples. Keep this value consistent within the same dataset.
   d) Adjust "Threshold," "Radius," "Dilate," and "Erode" until the black-and-white image looks representative of your sample (Fig. 1B).
   e) Click "Show Text" to display the tracing lines, axon IDs, and g-ratios on the original image. The corresponding data will appear in the output fields (Fig. 1C).
4. (Optional: Batch processing) Open the Batch Processing panel from the View menu, and click "Choose images" under "Data Preparation."
5. Click "Add Images" and select the image(s) for which you tuned the segmentation parameters.
6. Check the "Check Selected" box and click "OK."
7. Click "Choose Destination Path" and select the folder where the segmentation files will be saved.
8. Check "Use Multiprocessing" to allow the computer to use the maximum number of logical processors.
9. Click "Start" in the "Processing Output" section.
10. Check the output folder. It will contain the "xxx.seg" file(s).

Step 2. Reviewing Segmentation Files

While SnapG is designed to avoid selecting myelinated axons that are only partially included in the image, it may occasionally select them. Axons within Remak bundles may also be included, although this is rare. In this step, users can manually filter out any such axons as well as any traced regions that are not axons.

1. Click "Open → Segmentation file(s)" from the File menu and open the "xxx.seg" file(s).
2. Click the axon(s) you want to deselect. The tracing lines will turn red instead of green (Fig. 1D).
3. (Optional: Batch processing) Repeat step 2 for all "xxx.seg" files.

Step 3. Generation of Data

In this step, the segmentation results are converted into a readable format.

1. Click "Segmentation data" from the File menu.
2. Click "Add Files," choose "xxx.seg" file(s), click "Select All," and check the "Check Selected" box.
3. Click "Generate." SnapG will generate the data and open the folder once the data is generated.
4. The output comma-separated values (CSV) includes lists of axon data for each image, with each axon ID corresponding to a number drawn in that image (Fig. 1E).

Save and load setting

SnapG can save the current settings—including threshold values—and load them later.

Saving:

```
Click "Save → Current settings" from the File
menu. The settings will be saved as a "xxxx.snpg"
file.
```

Loading:

```
Click "Open → Settings file" from the File menu,
select the "xxx.snpg" file, and open it.
```

3 Results

3.1 SnapG Is an Automated Tool for Myelin g-ratio Analysis: Setting Threshold and Percentiles

To streamline g-ratio analysis and improve data reliability, we developed SnapG, an open-source, automated, Python-based tool for g-ratio quantification that is compatible with macOS, Linux, and Windows operating systems. As SnapG facilitates batch-wise g-ratio computation, it significantly streamlines the overall analysis workflow. As a threshold-based tool, SnapG achieves high detection rates of myelinated axons and substantially reduces processing time. SnapG thus provides a practical solution for large-scale g-ratio quantification in myelin research.

SnapG is composed of three steps as described in Sect. 2.6. The identification of inner and outer contours is driven by the thresholds defined in the first step. SnapG displays both the threshold image and the detected contours overlaid on the original image. In the processed image, the user can define the pixel-to-nanometer conversion, enabling accurate measurement of inner contour diameters. Additionally, overlaying the contours on the original image allows users to adjust the threshold with precision (Fig. 1A–E).

To quantify myelin thickness, SnapG traces the inner contour and, at each point along it, locates the nearest white pixel outside the contour. Users specify a percentile value for averaging myelin thicknesses, and the distances at that percentile are used as the representative thickness. Percentiles are chosen to reduce the impact of outliers and to provide a more robust estimate of typical myelin thickness. In densely packed regions of the CNS, this procedure may inadvertently capture the inner contours of adjacent myelin sheaths. Consequently, the resulting outer contours overlaid on the original image can appear larger than the true outer boundaries. Selecting a smaller percentile (~30%) helps to limit these enlarged outer contours, facilitating the separation of two closely apposed myelinated axons whose boundaries may otherwise be obscured due to limitations in threshold-based contour detection. In contrast, in images where axons are well separated, such as in the PNS, larger percentiles (60%–70%) may yield slightly more precise myelin thickness measurements.

The CSV output of SnapG provides detailed information, including the total number of axons detected, individual axon IDs, g-ratio, circularity, inner and outer contour diameters (μm or nm), and myelin thickness (μm or nm). Circularity was calculated using the formula 4π (area/perimeter^2); axons with circularity values below 0.6 are typically excluded from g-ratio analyses in our laboratory. However, these axons are retained in the CSV output and are included in the total count of detected axons. This allows for meaningful comparisons of myelinated axon counts per FOV across normal and abnormal conditions.

3.2 SnapG Reproduces Manual g-ratio Quantifications in the PNS

To assess the accuracy of SnapG, we calculated g-ratios with SnapG using electron micrographs of P90 sciatic nerve samples we had previously used in manually measured g-ratios [14], selecting the same 201 myelinated axons for comparison. In the PNS, inner contours were interpreted as corresponding effectively to axon boundaries. First, we compared the 30th to 70th percentiles under the same parameters, including the threshold, to determine which percentile yields a g-ratio most similar to that determined by manual tracing (0.669 ± 0.0044; mean ± SEM). g-ratios calculated using the 30% or 40% were similar but slightly higher (30%: 0.698 ± 0.003, $p < 0.001$; 40%: 0.691 ± 0.003, $p < 0.001$), whereas those calculated using the 60% (0.675 ± 0.003, $p = 0.25$) or 70% (0.667 ± 0.003, $p = 0.64$) were equivalent with no significant differences (Fig. S1H). A scatter plot of g-ratio versus axon diameter is shown in Fig. 1F. There is no significant difference between manual and SnapG measurements at 60% (Fig. 1G). The average axon diameter also showed no significant difference (manual: 3.596 ± 0.1011 μm, SnapG: 3.529 ± 0.098, $p = 0.63$) (Fig. 1H). These results indicate that SnapG is useful for high-throughput quantification in the PNS.

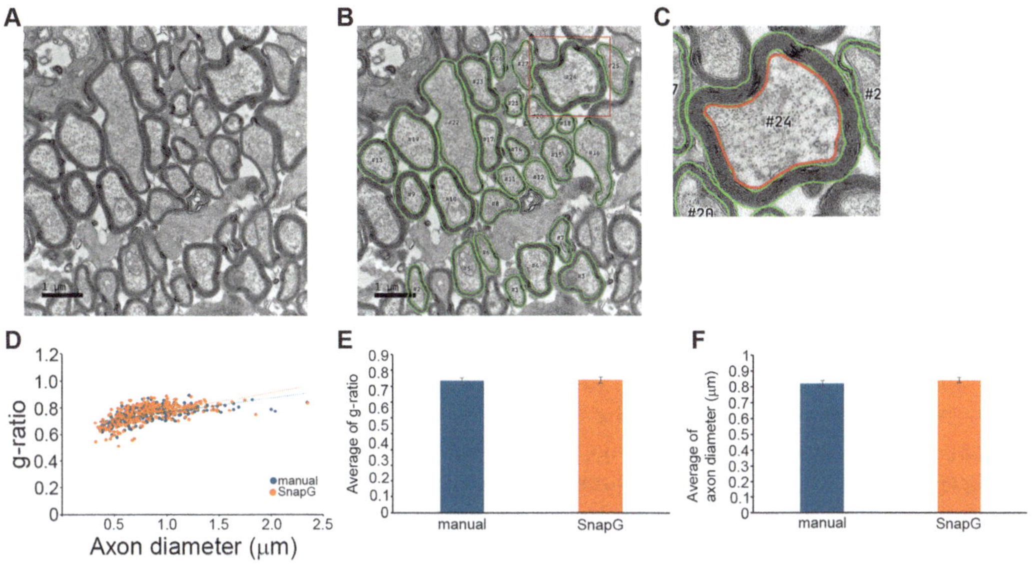

Fig. 2 g-ratio calculated by SnapG versus manual measurements of CNS myelinated axons. (**A**) Electron micrographs of P90 optic nerves. Bar, 1 μm. (**B**) The final output generated by the code. Axon numbers match those listed in the CSV. (**C**) Higher magnification of the red rectangle shown in (**B**). The red line in (**C**) indicates a manually traced axon. Notably, the traced line was drawn to exclude the periaxonal space. (**D**) Comparison of g-ratio values obtained from manual ($n = 245$ axons) and SnapG ($n = 273$ axons) calculations shows both methods yield similar results. (**E**) Bar graph showing the average g-ratio calculated manually (0.748 ± 0.004) and by SnapG (0.757 ± 0.004). (**F**) Bar graph showing the average axon diameter calculated manually (0.835 ± 0.020 μm) and by SnapG (0.839 ± 0.018 μm)

3.3 SnapG Recapitulates Manual g-ratio Measurements in the CNS

To evaluate the accuracy of SnapG in calculating g-ratios in the CNS, we re-analyzed P90 optic nerve samples from our previous study [14], comparing manually obtained measurements with those generated by SnapG. In the CNS, a periaxonal space containing adaxonal myelin is sometimes observed between the axon and its surrounding myelin (Fig. 2A–C). In automated g-ratio quantification using SnapG, axons with a periaxonal space must either be excluded by the user or manually measured using custom contour definitions if inclusion is desired. Most axons were identified based on their inner contours, interpreted as axonal boundaries, while outer contours were used to define the outermost layer of the myelin sheath. In our previous study, axon diameters were measured manually; therefore, to ensure consistency between data sets, axons with periaxonal space were manually measured in this study as well (Fig. 2C). The outer myelin contours were determined using the 30% of the myelin-thickness measurements. Manually measured axons represented 24.61% ± 2.16% of all myelinated axons (18 FOVs; mean ± SEM). The average g-ratio obtained from manual and SnapG measurements showed no significant difference (manual: 0.748 ± 0.003; mean ± SEM; SnapG: 0.757 ± 0.004, $p = 0.132$). Similarly, the average axon diameter did not differ between methods (manual: 0.835 ± 0.020 μm, SnapG: 0.839 ± 0.018, $p = 1.96$). The number of myelinated axons detected (circularity >0.6) was largely consistent between manual and SnapG qualification, with SnapG mode identifying 125.96% ± 7.13% of the axons detected manually (manual: 245 axons; SnapG: 273 axons). These results indicate that SnapG provides reliable large-scale quantification of CNS myelinated axons.

3.4 SnapG Enables Time-Efficient g-ratio Quantification via Batch Processing

Quantification of g-ratios remains a labor-intensive process, primarily due to the need for manual tracing across numerous randomly selected images per animal. This challenge is exacerbated in regions with heterogeneous distributions of myelinated axons, such as the corpus callosum, which necessitate increased sampling to obtain representative measurements. SnapG streamlines this process by enabling either individual or batch-wise image g-ratio calculations. In the individual image mode, each image is thresholded separately. In the batch quantification mode, a series of images captured with consistent brightness and contrast settings are thresholded and quantified together, providing additional time savings. For example, once batch quantification is initiated at the data-generation step, SnapG completes the output generation typically within 1 min, placing all output files into a single folder. This folder contains multiple TIFF images with overlaid contours of selected axons. It also contains a single CSV file summarizing the numbers of myelinated axons detected, individual axon IDs, g-ratio, circularity, inner and outer contour diameters (nm or μm), and myelin thickness (nm or μm). This streamlined format facilitates downstream analysis and enables efficient comparison across datasets or experimental conditions.

We assessed the consistency of SnapG's output between individual and batch quantification modes in the corpus callosum of P90 mice. To compare quantification of individual images vs. the batch-mode processing, we excluded the periaxonal space to consistently define the inner contours as axonal boundaries. We found that the average g-ratios were quite similar between individual and one-batch measurements, although the slight differences were nevertheless significant (individual: 0.819 ± 0.001, mean ± SEM; batch: 0.826 ± 0.001; $p < 0.001$). The average axon diameter showed no significant differences (individual: 0.729 ± 0.007 μm; batch: 0.738 ± 0.008 μm; $p = 0.43$).

Thus, batch processing represents a significant reduction in the time required to analyze g-ratio across multiple animals, albeit with a slight compromise in the accuracy of g-ratio measurements. For the most accurate measurements, individual image analysis is recommended; for the fastest analysis, batch processing is recommended.

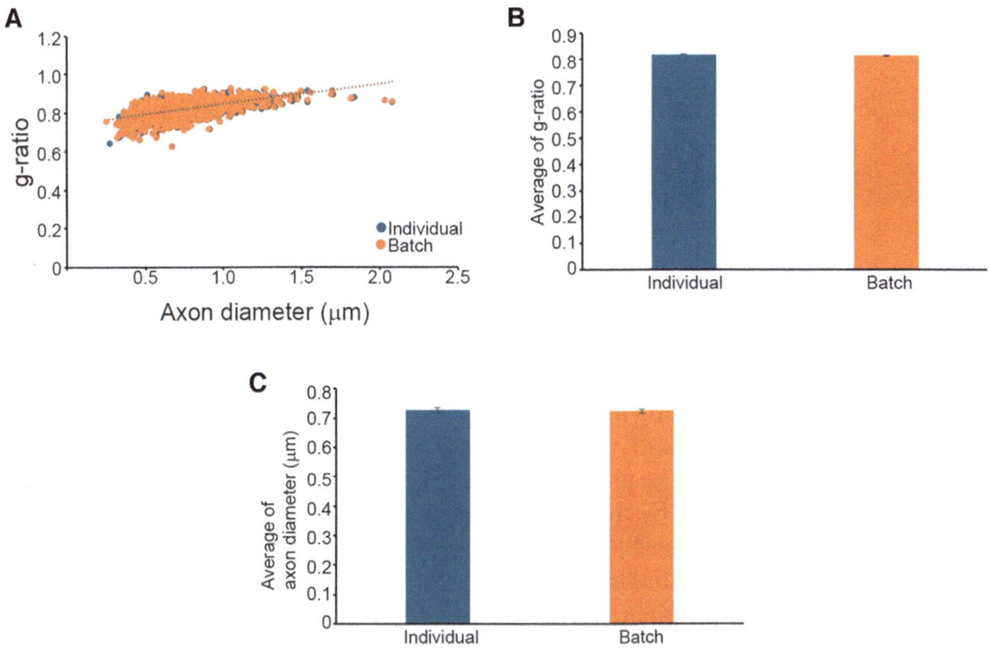

Fig. 3 SnapG enables batch-wise g-ratio calculation equivalent to per-image analysis. (**A**) Comparison of g-ratio values from per-image (Individual; $n = 1027$ axons) and batch-level (Batch; $n = 1047$ axons) calculations in P90 corpus callosum shows both methods yield similar results. (**B**) Bar graph showing average g-ratio calculated at individual image (0.819 ± 0.001) and batch level (0.817 ± 0.001). (**C**) Bar graph showing average axon diameter calculated at individual image (0.729 ± 0.007 μm) and batch level (0.724 ± 0.007 μm)

Finally, we used a hybrid, multibatch strategy of dividing the image series (26 images) into seven subgroups, which were then batch processed separately. In this analysis, there were no significant differences between individual and batch measurements in either the average g-ratio or the average axon diameter (Fig. 3A–C). This latter method can be used to enhance the speed of processing without a sacrifice in accuracy.

4 Discussion

g-ratios have been measured by different methods in multiple studies over the years. Variations in the imaging modalities used (e.g., optical vs. electron microscopy, MRI), in the fixation protocols employed, and the range of axon diameters selected for quantification can all impact values obtained [6, 15]. We developed SnapG to enable accurate and efficient g-ratio quantification from electron micrographs, which are typically used to assess pathological changes in myelinated axons. SnapG incorporates a computationally efficient, unsupervised algorithm for segmenting myelinated axons and myelin sheaths, which we have integrated into an intuitive GUI application. The segmentation algorithm is entirely based on traditional image processing, offering speed and computational efficiency compared to deep-learning methods. The GUI application was also designed to process batches of images and streamline the workflow for manually removing false positives. Although SnapG does not include a built-in feature for manual segmentation of false negatives, it provides a set of tunable parameters that the user can adjust to minimize missed detection.

Among the practical advantages of SnapG are its streamlined batch processing capability and broad image format compatibility, resulting in a substantial reduction in analysis time. SnapG outputs were also highly consistent with g-ratios determined by manual tracing. The precise matching of SnapG to manually measured g-ratios is aided by choosing an appropriate percentile. As noted, SnapG traces the inner contour of each axon and, at every point along this contour, identifies the nearest white pixel located outside it. The user-selected percentile of these distances is then used as the representative myelin thickness. In our experience, the 60th percentile optimized the correspondence between SnapG and manual measurements while minimizing errors in the PNS. The specific percentile to be used in an individual study can be determined empirically, if desired, by comparison to a small set of manually measured profiles that serve as controls.

Despite these advantages, SnapG does have limitations that stem from its reliance on threshold-based detection and contour segmentation, features common to many g-ratio algorithms. While inner contours are indicative of axonal boundaries in the PNS (Fig. 1A), the CNS occasionally exhibits a periaxonal space-containing adaxonal myelin between the axon and its surrounding myelin sheath (Fig. 2B, C). This structural feature complicates automated g-ratio quantification using SnapG. In such cases, users must either exclude axons with periaxonal space from analysis or manually measure them using custom contour definitions if inclusion is desired. In contrast to SnapG, MyelTracer is capable of measuring inner, outer, and axonal diameters, enabling the software to separate the periaxonal space [13]. However, as SnapG does not detect axonal contours automatically, axons must be manually traced when a significant periaxonal space is present. In this study, $24.61\% \pm 2.16\%$ of myelinated axons in P90 optic nerves required manual tracing. While this portion may differ depending on region and age, the full automation of all other steps does enable SnapG to significantly reduce the time required for g-ratio analysis.

A series of images captured under consistent brightness and contrast settings can be quantified simultaneously. However, a single threshold must be applied across all images, which may introduce variability. We evaluated both single-batch processing and multi-batch processing, and users may choose either approach depending on the consistency of their imaging conditions. Batch processing markedly improves the efficiency of g-ratio analysis across multiple animals. Given SnapG's reliance on threshold-based detection, some visually identifiable myelinated axons may fall below the threshold. To obtain accurate counts of myelinated axons per FOV, these undetected axons must be manually added.

In summary, SnapG represents a new method for g-ratio quantification that significantly reduces analysis time with overall strong correlation to manual quantification. SnapG provides a practical and efficient tool for g-ratio quantification, particularly in large-scale studies where high-throughput analysis is required; it is the method of choice in our own laboratory. One limitation shared by this and similar programs that measure g-ratios is the inability to reliably detect and quantify unmyelinated axons and thereby quantify the percentages of myelinated versus unmyelinated axons. This is the result of the relatively low contrast of unmyelinated axons in many EM images. As AI technology continues to advance rapidly, we anticipate that tools capable of detecting unmyelinated axons will become available and serve as an adjunct to SnapG measurements.

Acknowledgments

This research was supported by NIH grant NS043474 and The Duques Family Foundation to JLS. The Microscopy Laboratory is a shared resource partially supported by the Cancer Center Support Grant P30CA016087 at the Laura and Isaac Perlmutter Cancer Center and NIH grant S10OD019974. We used Microsoft Copilot to assist with editing the original draft.

Author Contributions YB and JLS designed the research; YB performed the research and measurements; SC, TX, and FK developed the codes; YB, SC, and TX drafted the paper; YB, SC, JS, AFL, and JLS edited the paper.

Conflict of Interest The authors have no conflicts of interest to declare that are relevant to the content of this chapter.

References

1. Zalc B (2016) The acquisition of myelin: an evolutionary perspective. Brain Res 1641:4–10. https://doi.org/10.1016/j.brainres.2015.09.005
2. Simons M, Gibson EM, Nave KA (2024) Oligodendrocytes: myelination, plasticity, and axonal support. Cold Spring Harb Perspect Biol 16:a041359. https://doi.org/10.1101/cshperspect.a041359
3. Monje M (2018) Myelin plasticity and nervous system function. Annu Rev Neurosci 41:61–76. https://doi.org/10.1146/annurev-neuro-080317-061853

4. Remak R (1839) Über die Ganglien der Herznerven des Menschen und deren physiologische Bedeutung. Wschn f d ges Heilkde 14:149–154
5. Bear RS, Schmitt FO (1937) Optical properties of the axon sheaths of crustacean nerves. J Cell Comp Physiol 9:243–260
6. Gow A (2025) Understanding the myelin g ratio from first principles, its derivation, uses and artifacts. ASN Neuro 17:2445624. https://doi.org/10.1080/17590914.2024.2445624
7. Rushton WA (1951) A theory of the effects of fibre size in medullated nerve. J Physiol 115:101–122. https://doi.org/10.1113/jphysiol.1951.sp004655
8. Chomiak T, Hu B (2009) What is the optimal value of the g-ratio for myelinated fibers in the rat CNS? A theoretical approach. PLoS One 4:e7754. https://doi.org/10.1371/journal.pone.0007754.
9. Fields RD (2015) A new mechanism of nervous system plasticity: activity-dependent myelination. Nat Rev Neurosci 16:756–767. https://doi.org/10.1038/nrn4023
10. Mount CW, Monje M (2017) Wrapped to adapt: experience-dependent myelination. Neuron 95:743–756. https://doi.org/10.1016/j.neuron.2017.07.009
11. Zaimi A, Duval T, Gasecka A, Cote D, Stikov N, Cohen-Adad J (2016) AxonSeg: open source software for axon and myelin segmentation and morphometric analysis. Front Neuroinform 10:37. https://doi.org/10.3389/fninf.2016.00037
12. Zaimi A, Wabartha M, Herman V, Antonsanti PL, Perone CS, Cohen-Adad J (2018) AxonDeepSeg: automatic axon and myelin segmentation from microscopy data using convolutional neural networks. Sci Rep 8: 3816. https://doi.org/10.1038/s41598-018-22181-4
13. Kaiser T, Allen HM, Kwon O, Barak B, Wang J, He Z, Jiang M, Feng G (2021) MyelTracer: a semi-automated software for myelin g-ratio quantification. eNeuro 8:1–9. https://doi.org/10.1523/ENEURO.0558-20.2021
14. Bekku Y, Zotter B, You C, Piehler J, Leonard WJ, Salzer JL (2024) Glia trigger endocytic clearance of axonal proteins to promote rodent myelination. Dev Cell 59:627–644.e10. https://doi.org/10.1016/j.devcel.2024.01.008
15. Fields RD, Dutta DJ (2025) G-ratio commentary—why you've been doing it wrong. ASN Neuro 17:2486962. https://doi.org/10.1080/17590914.2025.2486962

Chapter 12

Ultrastructural Volumetry of Spinal Cord White Matter Mitochondria via 3D Reconstruction from Conventional Transmission Electron Microscopy

Ioannis Pantelidis, Christina Katsigianni, Alexandra Giannaki, Iasonas Dermitzakis, Sofia Gargani, Ioanna Tremi, Sophia Havaki, Maria-Eleni Manthou, Soultana Meditskou and Paschalis Theotokis

Abstract

Transmission electron microscopy (TEM) remains the gold standard for the ultrastructural analysis of myelinated axons and glial compartments in the central nervous system (CNS). However, conventional TEM provides only two-dimensional information, limiting the ability to assess spatial relationships between mitochondria, axons, and myelin. Here, we describe a workflow for the three-dimensional (3D) reconstruction of white-matter structures from serial TEM sections of mouse spinal cord. The protocol includes optimized fixation and embedding for preserving myelin periodicity, serial ultramicrotomy, post-staining, image alignment and preprocessing in Fiji, and 3D rendering in 3D Slicer. This pipeline enables classic quantitative morphometric analyses such as myelin thickness, axonal diameter, and g-ratio, but also mitochondrial quantitative output such as number, density, and extent within reconstructed volumes. Based on the nature of such metrics, the method could infer insight into the interplay between axonal energy metabolism and myelin integrity and can be adapted to diverse models of neurodegeneration or demyelination. By bridging traditional TEM with digital 3D reconstruction, this approach offers a versatile framework for investigating the ultrastructural basis of white-matter physiology and pathology.

Keywords Transmission electron microscopy, TEM, Spinal cord, White matter, Axons, Axonal morphology, Myelin, G-ratio, Mitochondria, Mitochondria dynamics, Volumetry, 3D reconstruction, Segmentation, Myelination, Demyelination, Ultrastructural analyses, Quantification

Supplementary Information The online version contains supplementary material available at https://doi.org/10.1007/978-1-0716-5456-9_12

Marie-Ève Tremblay (ed.), *Ultrastructural Neuroscience Methods*, Neuromethods, vol. 238,
https://doi.org/10.1007/978-1-0716-5456-9_12,

1 Introduction: Ultrastructural Organization and Mitochondrial Dynamics in White Matter

White matter integrity is fundamental for the proper function of the central nervous system [1]. Within the spinal cord, axons are ensheathed by myelin, a multilamellar membrane generated by oligodendrocytes that accelerates impulse conduction and provides trophic support to neurons [2]. The architecture of the white matter reflects the close interaction between axons, glial cells, and vascular elements, forming an integrated system in which even subtle structural alterations can lead to profound functional deficits [1]. Among the subcellular organelles that sustain axonal physiology, mitochondria are of central importance. They provide the energy required for axonal transport, neurotransmission, and ionic homeostasis, and they serve as regulators of calcium buffering and apoptotic signaling [3]. In myelinated axons, mitochondria are strategically positioned along internodal and paranodal regions to satisfy localized metabolic demands [4, 5]. The spatial distribution and morphology of these organelles are tightly coupled to axonal health and myelin stability [6].

A critical aspect of mitochondrial biology is mitochondrial quality control (MQC), a collective term encompassing mitochondrial fission, fusion, biogenesis, transport, and mitophagy [7]. MQC ensures that damaged mitochondria are repaired or removed to preserve energetic competence. Dysregulation of these pathways results in swelling, fragmentation, or depletion of mitochondria within axons, phenomena closely associated with demyelination, axonal degeneration, and disorders, such as multiple sclerosis, leukodystrophies, and spinal cord injury [6, 8, 9]. Recent evidence further demonstrates that the structural remodeling in chronically demyelinated spinal cord tissue can be therapeutically modulated. Intrathecal transplantation of neural precursor cells into the lumbar region enhances oligodendrocyte proliferation in demyelinated hotspots, demonstrating that white matter repair is a dynamic and measurable process [10].

Understanding how mitochondrial architecture and turnover relate to myelin integrity requires imaging strategies that can resolve their 3D organization within the complex white matter microenvironment [11]. Conventional transmission electron microscopy (TEM) provides nanometer-scale resolution but captures only two-dimensional (2D) slices, limiting the visualization of mitochondrial networks and their relationships to axonal membranes and myelin [11]. Three-dimensional (3D) reconstruction from serial TEM sections offers a powerful way to overcome this limitation by enabling volumetric analysis of axons and their associated organelles [12, 13]. This approach reveals spatial features that are otherwise lost in single-section analyses, such as mitochondrial clustering, axon–glia contacts, and variations in myelin thickness along individual fibers [5, 13, 14].

1.1 Rationale for Three-Dimensional Reconstruction Using Transmission Electron Microscopy (TEM)

White matter is a highly organized yet complex tissue where axons, myelin, and glial organelles interact across nanometer scales. Although light microscopy has greatly advanced the study of large-scale organization and molecular labeling, its diffraction-limited resolution prevents detailed visualization of subcellular interfaces that define axon–glia interactions and metabolic coupling [15]. Conventional two-dimensional transmission electron microscopy (TEM) remains the gold standard for ultrastructural imaging but provides only thin snapshots of a continuous three-dimensional (3D) architecture [16, 17]. Because axons, myelin lamellae, and mitochondria extend far beyond the thickness of a single section, spatial relationships between these elements cannot be reliably inferred from isolated images [16].

Serial-section TEM reconstruction overcomes these limitations by combining nanometer resolution with computational volume modeling. Through digital alignment and segmentation of consecutive images, it restores the continuity of cellular and subcellular structures, enabling the direct visualization of axonal trajectories, myelin geometry, and mitochondrial distribution within both axons and glial processes [18, 19]. This volumetric perspective allows the quantification of key parameters, such as axon caliber, g-ratio, myelin periodicity, and mitochondrial density or spatial proximity, offering a comprehensive view of the microanatomy underlying conduction and metabolism [20]. The g-ratio, defined as the ratio of the inner axonal diameter to the total outer diameter including myelin, is a fundamental metric, especially in the context of remyelination. Optimal g-ratio values (~0.6–0.7 in CNS tissue) reflect a balance between conduction speed and metabolic cost. Deviations from this range indicate pathological thinning or hypertrophy of myelin and are widely used in studies of demyelination and remyelination [21].

Applied to the spinal cord white matter, this method provides a unique means to study how ultrastructural organization supports neural conduction, energy homeostasis, and resilience to injury. Mapping mitochondria within the axon–glia unit reveals how energy resources are positioned relative to sites of high ATP demand, including nodes of Ranvier and paranodal domains [5]. Similarly, reconstructing oligodendrocyte processes in 3D allows assessment of their interactions with axons and capillaries, providing insight into metabolic support and myelin remodeling [19, 22]. These features are particularly relevant in disorders involving mitochondrial dysfunction or demyelination, where subcellular disorganization precedes overt histopathology [23]. Methodologically, serial-section TEM bridges the gap between classical electron microscopy and advanced volumetric techniques such as focused ion beam, SEM, or serial block-face SEM, while remaining accessible to most laboratories [24, 25]. When combined with open-source segmentation and rendering software, such as Fiji (ImageJ) and 3D Slicer, this approach yields an integrated, quantitative framework that links structure to function

[22, 26]. It therefore serves both as a research and training tool for exploring mitochondrial quality control and axon–glia ultrastructure in health and disease. For regions requiring higher axial resolution, this workflow can be complemented by tilt-series TEM or electron tomography, which provides three-dimensional information at the level of individual mitochondria and myelin membranes. Integrating these tomographic datasets with serial-section reconstructions further refines the spatial understanding of organelle architecture and axon–glia interactions [27, 28].

2 Materials

2.1 Sample Preparation and Fixation

Thirty female C57BL/6 mice were obtained from the Hellenic Pasteur Institute and housed in the pathogen-free animal facility (SPF; Biosafety level 3) of AHEPA University Hospital (EL 54 BIO29). All animal procedures were conducted in accordance with the European Communities Council Directive 86/609/EEC and the relevant Greek regulations [29]. Reagents used for tissue fixation included paraformaldehyde (PFA), glutaraldehyde, osmium tetroxide (OsO_4), and potassium ferrocyanide ($K_4[Fe(CN)_6]$); for staining, uranyl acetate, lead citrate, phosphate buffer (PB), and sodium acetate buffer; for dehydration, graded ethanol solutions and propylene oxide, and epoxy resin components (Araldite CY212, DDSA with dibutyl phthalate plasticizer and BDMA accelerator) for embedding the tissues. Standard laboratory consumables and 200-mesh copper grids were used.

2.2 Sectioning and Grid Preparation

Resin blocks are trimmed under a high-speed milling system (EM TRIM2, Leica) and sectioned at 70 nm using an EM UC6 ultramicrotome (Leica) with freshly prepared glass knives made out of ultramicrotome glass (406 × 25.4 × 6 mm) using Reichert Knifemaker 705202 (Leica). Semi-thin sections are stained with toluidine blue for orientation and assessment of region quality. Ultrathin serial sections are collected and mounted on 200-mesh copper grids using an eyelash tool or fine anti-capillary tweezers and contrasted with Reynolds' lead citrate.

2.3 TEM and Softwares

Images were acquired using a JEM-1011 (JEOL) transmission electron microscope. Image processing and segmentation were performed using Fiji (an ImageJ distribution), and three-dimensional visualization and quantitative analysis were conducted using 3D Slicer (v5.8.1).

3 Methods

3.1 Sample Preparation and Fixation

Spinal cord white-matter tissue was carefully dissected from perfusion-fixed with glutaraldehyde and PFA (2.5% and 1.5%, respectively) mice and immediately immersed in 0.1 M phosphate buffer (PB, pH 7.4) to maintain physiological osmolarity [17]. Post-fixation is performed in freshly prepared 1% osmium tetroxide (OsO_4) and 1.5%–2.5% potassium ferrocyanide ($K_4[Fe(CN)_6]$) in 0.2 M PB for 2 h at room temperature. The reduced osmium treatment enhances membrane contrast by promoting osmium deposition on lipid bilayers, resulting in improved visualization of myelin sheaths, mitochondria, and other membranous structures during serial sectioning and volume EM workflows [30–32]. Following fixation, samples are rinsed in buffer and stored overnight at 4°C in 0.1 M sodium acetate (SA) to minimize extraction and to improve membrane contrast during subsequent staining [33].

For en bloc staining, tissues are incubated in 1%–2% aqueous uranyl acetate for 30 min in the dark, which enhances electron density and overall signal-to-noise ratio during TEM imaging (*see* **Note 1**) [34]. Dehydration is then performed through a graded ethanol series (25%, 50%, 70%, 95%, and three changes of 100%) for 10 min each, ensuring gradual removal of water while preventing extraction of lipids. Following dehydration, samples are incubated twice for 15 min in propylene oxide (PO) before resin infiltration.

Embedding is carried out using a freshly mixed epoxy resin composed of Araldite CY212 and DDSA (47:53 ratio) with 1 mL BDMA accelerator and 1.5 mL dibutyl phthalate plasticizer per 100 mL resin. Tissue samples are infiltrated overnight in 1:1 mixture of resin and PO, then embedded in flat silicone molds and polymerized at 60°C for 48 h [24, 35]. Once solidified, the resin blocks are stable at room temperature and can be stored indefinitely before sectioning.

For optimal results, ensure that fixatives and embedding reagents are freshly prepared and filtered before use to prevent crystalline artifacts in the final sections (*see* **Note 2**). Consistent fixation and infiltration are critical for preserving axonal ultrastructure, particularly the compactness of myelin and the integrity of mitochondrial membranes, both of which are central to accurate morphometric analysis.

3.2 Ultramicrotomy and Section Collection

Resin blocks containing the embedded spinal cord tissue are first trimmed under a high-speed milling system, EM TRIM2 (Leica), to expose the white-matter region and create a pyramidal block face suitable for ultrathin sectioning. Proper trimming ensures uniform section ribbons and reduces compression artifacts. Semi-thin sections (0.5–1 μm) are cut using a glass knife and stained with 1% toluidine blue in 1% sodium borate to evaluate tissue orientation and identify areas rich in myelinated axons [36] (Fig. 1a). Following identification of the region of interest, ultrathin sections were obtained for TEM analysis.

Ultrathin serial sections approximately 70 nm in thickness are cut with EM UC6 (Leica) ultramicrotome using a freshly prepared glass knife. The knife trough is filled with filtered distilled water to allow sections to float intact and align as ribbons. Sections displaying silver–gold interference color under reflected light typically correspond to the optimal thickness for TEM (*see* **Note 3**). Serial ribbons are collected on 200-mesh copper grids using an eyelash tool mounted on a wooden handle or fine anti-capillary tweezers. Maintaining sequential order is essential for accurate 3D reconstruction; therefore, grids are labeled according to section number.

After collection, grids are dried briefly on filter paper and prepared for post-staining. Consistency in section thickness and alignment greatly improves image registration during digital reconstruction.

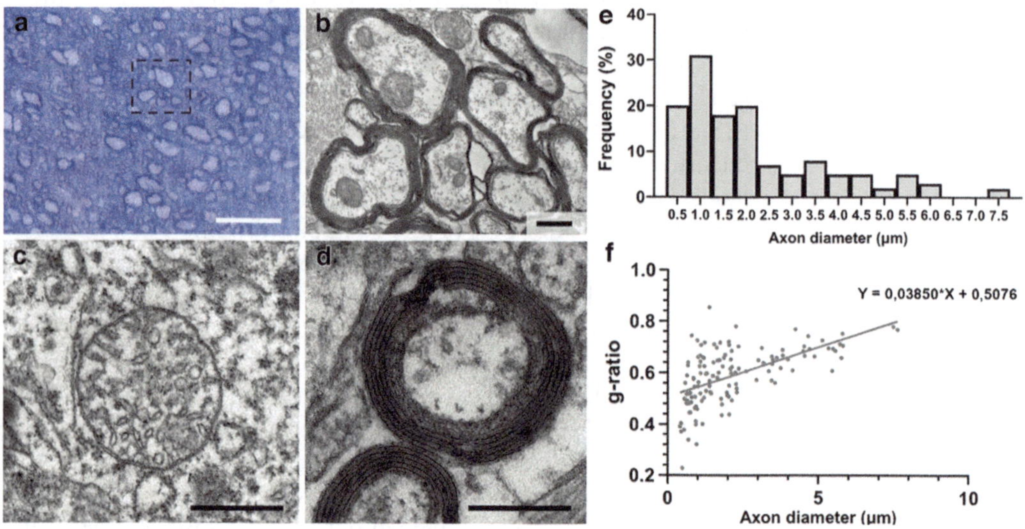

Fig. 1 Classical ultrastructural and morphometric assessment of myelinated axons and mitochondria in lumbar–sacral spinal cord white matter. (**a**) Semithin resin section (0.5 μm, toluidine blue staining) showing the cytoarchitecture of lumbar–sacral white matter. The hatched box indicates the region of interest (ROI) selected for ultrastructural evaluation. Scale bar: 50 μm. (**b**) Representative transverse TEM micrograph showing multiple myelinated axons within the ROI. Scale bar: 1 μm (magnification: 20,000x, voltage: 80 kV). (**c**) High-magnification TEM image of a mitochondrion within the axoplasm, illustrating internal cristae organization and membrane integrity. Scale bar: 0.5 μm. (**d**) High-magnification TEM image demonstrating compact myelin lamellae and axoplasmic organization. Scale bar: 0.5 μm (magnification: 50,000x, voltage: 80 kV). (**e**) Histogram displaying the frequency distribution of axonal diameters within the analyzed white-matter region. (**f**) Scatter plot showing the relationship between axon diameter and g-ratio, including linear regression fit ($Y = 0.03850 \times X + 0.5076$). These classical morphometric parameters provide a structural reference framework for evaluating axonal caliber and myelin thickness prior to volumetric 3D reconstruction. Although the data shown here derive from lumbar–sacral tissue, the workflow is applicable to additional spinal segments, supporting methodological generalizability

Avoid excessive tension on the ribbons, as stretching or folding will introduce distortions that complicate alignment in Fiji or TrakEM2. If section compression is observed, verify the cutting speed and knife clearance angle, or replace the knife edge.

This stage produces a complete serial set of ultrathin sections ready for post-staining and imaging, with well-preserved axonal and mitochondrial profiles that retain the ultrastructural detail necessary for accurate morphometric analysis.

3.3 Post-staining and TEM Imaging

Each grid carrying a serial ultrathin section is contrasted immediately before imaging to enhance membrane visibility and improve overall image contrast. Post-staining is performed by floating the grids, section-side down, on droplets of freshly prepared 0.08 M Reynolds' lead citrate for 10–20 min at room temperature in a CO_2-free humidified chamber [37] (*see* **Note 4**). The CO_2-free environment prevents carbonate precipitation, which otherwise produces granular artifacts on the section surface. After staining, grids are rinsed thoroughly in multiple changes of distilled water and air-dried on filter paper.

For transmission electron microscopy, grids are mounted on a standard single-tilt holder and examined in JEM-1011 (JEOL) TEM operating at 80 kV voltage. A magnification of 20,000x typically provides sufficient resolution to visualize myelin periodicity and mitochondrial cristae [16]. Serial micrographs are acquired using a digital camera at a consistent pixel size of approximately 1 μm/pixel, with a 10%–15% overlap between consecutive fields to ensure continuity for alignment (Fig. 1b, c) (*see* **Note 5**). Exposure settings and beam intensity are kept constant across the series to minimize contrast variation and photodamage [16, 38].

During imaging, it is advisable to start with low-magnification overview scans to document tissue orientation and section order before proceeding to high-magnification acquisition [16]. Images should be saved as uncompressed 8-bit TIFFs and named sequentially (e.g., TEM_001.tif, TEM_002.tif). Regular inspection of section quality, particularly the integrity of myelin sheaths and mitochondrial membranes, helps identify potential fixation or staining inconsistencies early in the workflow [16]. Maintaining stable accelerating voltage and focus across sections ensures uniform brightness and contrast, which are critical for subsequent digital alignment and segmentation [16, 17, 25].

When performed consistently, this imaging procedure yields high-contrast datasets that faithfully preserve the ultrastructure of axons, myelin, and mitochondria, providing the foundation for accurate three-dimensional reconstruction [17, 25].

3.4 Image Preprocessing and Alignment

Digital micrographs obtained from the transmission electron microscope are imported into Fiji (ImageJ) as a sequential image stack [39]. To preserve metadata, images are opened in their original TIFF format and arranged in the correct serial order according to their acquisition numbers. Consistent file naming (e.g., TEM_001.tif, TEM_002.tif) simplifies automated import and prevents misalignment errors. Before any transformation, the entire dataset should be duplicated to maintain an untouched copy of the raw micrographs for archival and quality control purposes.

Preprocessing is applied to enhance the visibility of membranes and minimize background variation across sections [19, 39]. Minor adjustments, such as contrast normalization, histogram equalization, or background subtraction, may be performed when necessary, provided they do not introduce artifacts. Filtering options, including despeckle, Gaussian blur ($\sigma = 0.5$–2), or median filtering, can help remove noise while retaining fine structural detail [19]. All image adjustments are recorded to ensure reproducibility of the dataset [39].

For stack alignment, the MultiStackReg plugin in Fiji is used to assemble and register consecutive sections. Alternatively, alignment can also be performed in TrackEM2 plugin, which supports both rigid and affine transformations [40]. Regardless of the tool used, alignment is guided by fiduciary landmarks, such as blood vessels, axon bundles, or large mitochondria, in order to maintain anatomical continuity [19, 40]. In cases of minor tissue deformation between sections, elastic or affine transformations may be applied locally to correct curvature or stretching. The alignment should always be verified visually by scrolling through the stack to confirm that corresponding structures remain in register along the z-axis [25, 40].

After successful alignment, the dataset is cropped to include only the region of interest (ROI) containing well-preserved white matter. This step reduces file size and improves downstream processing speed [40]. Voxel scaling is defined using the physical calibration obtained from the scale bar on the original TEM micrographs, allowing all measurements to be expressed in micrometers rather than pixels. This transformation ensures correct spacing, when the image stack is imported into the 3D Slicer environment. Proper scaling ensures that quantitative measurements of axon caliber, myelin thickness, or mitochondrial volume remain accurate during 3D rendering [19, 40]. Although the nominal ultrathin section thickness is 70 nm, we assign a z-spacing of 1 μm (0.001 mm) in the final stack. This value corresponds to the approximate distance between consecutive semi-thin levels and is selected to prevent extreme anisotropy in the rendered volume. Using this intermediate spacing improves the recognizability of 3D geometries without altering the in-plane measurements derived from the original TEM images. The choice of z-scaling is therefore a visualization adjustment rather than a representation of the physical cutting thickness. The aligned, cropped, and scaled stack is then saved in TIFF or NRRD format.

Careful alignment and preprocessing are crucial for minimizing reconstruction artifacts. Even slight distortions introduced at this stage will propagate through segmentation and 3D modeling, affecting morphometric accuracy [25]. Therefore, it is recommended to perform alignment iteratively, verifying the continuity of prominent axons or myelin profiles after each correction cycle (*see* **Note 6**) [25, 40].

3.5 Manual Segmentation

Following alignment, individual structures of interest within the white matter are segmented manually to create labeled datasets suitable for 3D reconstruction [19, 40]. Segmentation is performed using Fiji/ImageJ with the ROI Manager tool [40]. Each image in the aligned stack is opened sequentially, and regions corresponding to axoplasm, myelin sheaths, and mitochondria are traced using the Freehand Selection tool. Manual delineation, although time-consuming, ensures accurate recognition of ultrastructural boundaries that automated algorithms often fail to detect in electron micrographs due to variations in staining and contrast [17, 19].

Annotations are saved as Regions of Interest (ROIs) and stored in the ROI Manager for each section. It is important to segment only well-preserved profiles, avoiding regions with folds, tears, or poor contrast that could introduce discontinuities in the reconstructed volume [25]. For mitochondria, contours should closely follow the outer membrane to prevent overestimation of organelle volume [17]. Myelin boundaries should include the compact lamellae but exclude extracellular space (*see* **Note 7**).

Once segmentation of all serial sections is complete, a duplicate of the aligned and scaled stack is generated for each structure type (myelin, axoplasm, and mitochondria). Background is removed, and ROIs corresponding to the structure are filled (intensity value = 255), to produce three binary mask stacks (one per structure). These binary stacks represent each structural class (axoplasm, myelin, mitochondria) as independent 3D volumes [25, 40]. All stacks preserve the voxel metadata inherited from the aligned stack. Then, each binary stack is converted into an integer labeling volume by multiplying the binary masks by 1/255, 2/255, or 3/255, corresponding to myelin, axoplasm, and mitochondria, respectively (background = 0). To visualize that the labeled stacks hold these values, Glasbey LUTs can be applied. The three labeled stacks are then combined, using Fiji's *Image Calculator* (Subtract, XOR, Add) to ensure exclusive voxel labeling. The final multilabel mask (values 0–3) is used for 3D reconstruction. All files are exported in TIFF or NRRD format, which preserves voxel size and slice order for direct import into 3D Slicer.

During segmentation, periodic visual checks along the z-axis are recommended to ensure continuity of each structure between consecutive sections. Misaligned or missing contours can be corrected immediately rather than after volume assembly, saving significant time during reconstruction [17, 25]. Although manual segmentation is labor-intensive, it provides a high degree of precision for morphometric analyses, particularly for quantifying parameters such as g-ratio, mitochondrial density, or inter-organelle distances [17, 19]. Segmentation of axoplasm and myelin boundaries enables quantitative extraction of axonal diameter, myelin thickness, and g-ratio on a per-fiber basis. These measurements are derived from the binary masks prior to smoothing and are therefore directly based on the raw membrane contours [21].

When applied consistently, this step produces a set of geometrically accurate and biologically faithful masks representing the ultrastructure of the spinal cord white matter, ready for 3D rendering and quantitative analysis in 3D Slicer.

3.6 3D Reconstruction and Visualization

The aligned TEM stack is imported into 3D Slicer (v5.8.1 or later) as a scalar volume, and the multilabel mask stack is imported as a label map for volumetric reconstruction [41]. Both image stacks are imported within the Add Data module, as single files to preserve voxel spacing and origin. In the Volume module, the voxel spacing values (x-y-z axes in pixels and mm) are verified and adjusted where necessary to match the voxel metadata defined during Fiji preprocessing. Because true ultrathin section thickness (70 nm) produces a highly compressed z-dimension in volumetric reconstructions, the stack is imported with a z-spacing of 1 μm, the value used during alignment and preprocessing to achieve a proportionally interpretable 3D model [27, 41]. This ensures that all morphometric measurements obtained from the rendered model reflect true physical dimensions.

A new segmentation node is created in the Segmentations module, and the multilabel mask is imported using "Import as LabelMap." This produces three visible segments corresponding to the integer labels (1 = myelin, 2 = axoplasm, 3 = mitochondria). To isolate each anatomical structure individually, the "Split Islands" effect (Segment Editor) is applied to each of the three segments. This operation identifies disconnected components and assigns each of them to a new unique segment (e.g., myelin_1, myelin_2, axoplasm_1, ..., mitochondria_1–7). Each newly generated segment preserves the original voxel-scale geometry and could be further modified with filtering and coloring for better visualization (Fig. 2).

Segmentation refinement is performed using thresholding, interpolation, and smoothing filters to improve surface continuity while preserving fine structural detail [25]. Specifically, after island splitting, segments are refined using multiple Segment Editor effects. Smoothing operations are applied using Gaussian, Median, Joint Smoothing, Curvature Flow, and Anisotropic Diffusion filters, depending on the structure. These filters are used to reduce pixelation artifacts inherent to thin-section TEM data while preserving morphological boundaries. Over-smoothing can distort narrow processes or mitochondrial

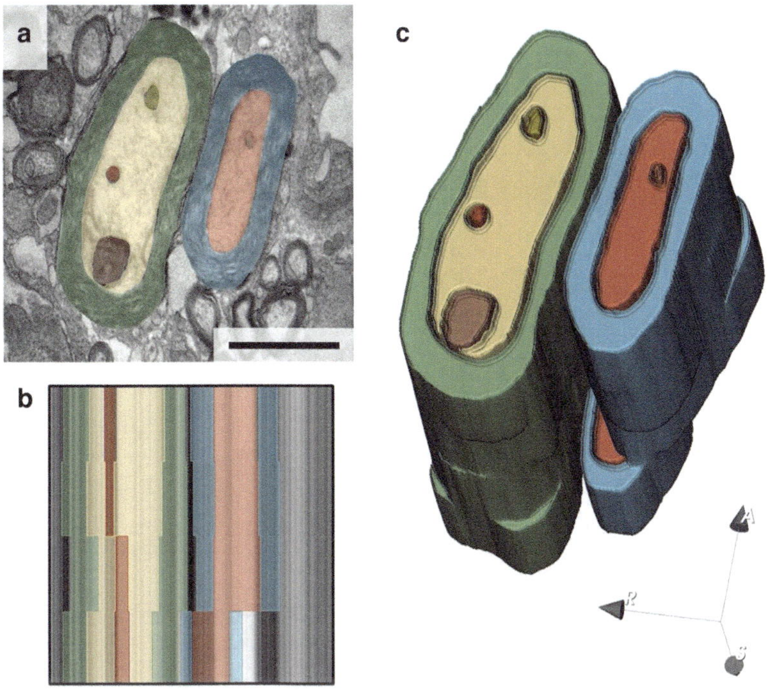

Fig. 2 Validation of segmentation and Split Islands workflow in 3D Slicer. (**a**) Representative TEM slice (20,000x, 80 kV) displayed within the 3D Slicer Axial view, showing the raw micrograph as background and the imported segmentation mask overlaid. Structures (myelin, axoplasm, mitochondria) were imported as a multilabel mask and converted into a Slicer segmentation. The Split Islands module was then applied to each label, automatically separating spatially discontinuous objects into individual segments and assigning distinct colors for visualization (Myelin A = green, Myelin B = blue, Axoplasm A = yellow, Axoplasm B = red, Mitochondrion 1A = brown, Mitochondrion 2A = deep red, Mitochondrion 3A = mustard yellow, Mitochondrion 1B = light brown). Scale bar ≈ 1 μm. The dataset consisted of four serial sections, each 70 nm thick, spaced by 1 μm in the z-dimension. (**b**) Visualization of the segmented structures in the coronal view of 3D Slicer's interface, showing color-encoded segmentation boundaries for each separated object after applying Split Islands. The band-like color structure indicates voxel-wise label assignment across the stack, used here to verify that no object labels unintentionally merge across slices. (**c**) Three-dimensional reconstruction generated from the 3D Slicer 3D view, demonstrating the spatial relationships between myelin, axoplasm, and mitochondria following Split Islands. The iso-surface rendering reflects the final segmented volumes used for downstream quantitative morphometry. Orientation interpreted via axes on the bottom right. Full volumetric and shape descriptor measurements are provided in Tables S1–S3 and Videos S9–S16

outlines, so the smoothing factor should be applied conservatively. The Fill Between Slices function can be used when necessary to close small gaps and discontinuities in the z-axis (interpolation), particularly where sections were missing or damaged. At this stage, visual comparison of the reconstructed volume with the original TEM stack is essential to confirm anatomical fidelity and to correct any misaligned or misplaced features [25, 27].

Once validated, the 3D model can be visualized in the 3D View module (Fig. 3). Individual segments can be toggled on or off to examine specific structures or to highlight their spatial relationships, for example, the position of mitochondria within axonal cylinders or their proximity to the inner myelin membrane. Quantitative measurements were obtained from the final 3D Slicer segmentation.

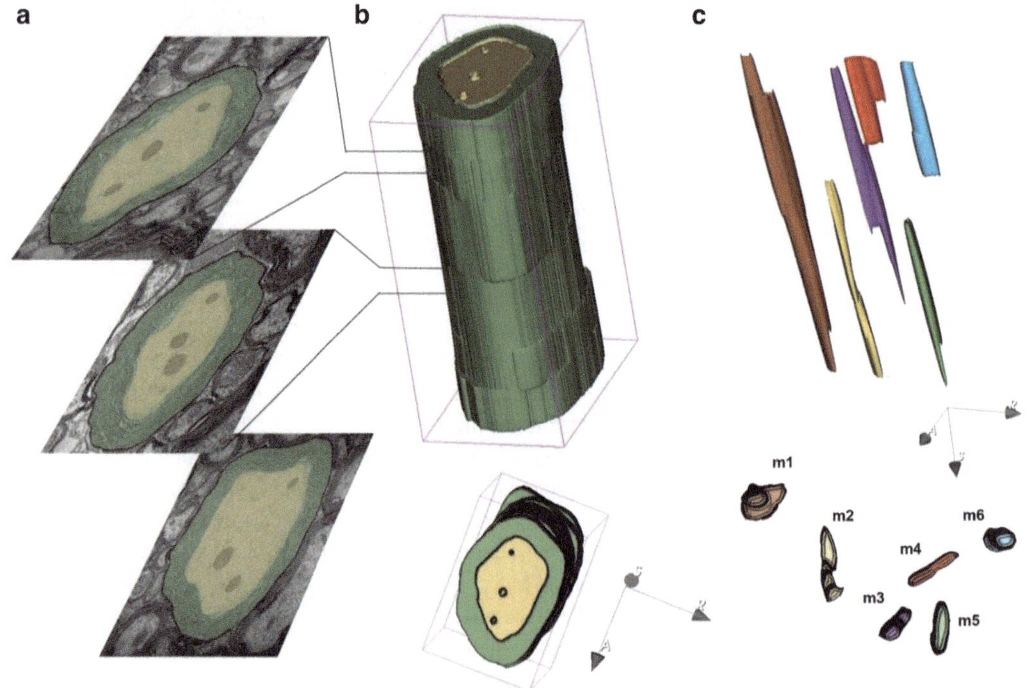

Fig. 3 Pipeline for 3D reconstruction of a myelinated axon from serial TEM sections. (**a**) Original 8-bit grayscale transmission electron microscopy images overlaid with RGB segmentation masks, illustrating manually labeled regions of interest (ROIs). Each color corresponds to a label from the multilabel mask stack: myelin sheath (green), axoplasm (yellow), and mitochondria (brown). Note that the multilabel mask itself is stored as an 8-bit grayscale volume and appears blank prior to LUT color assignment. (**b**) 3D reconstructed axonal model generated from z-stacking of aligned TEM sections and segmentation of ultrastructural compartments. Color representation remains consistent across structures, while mitochondria are displayed with distinct colors following "Split Islands" processing. Parallel and transverse orientations are shown, along with orientation axes and bounding boxes. (**c**) Isolated 3D mitochondrial segments (m1–m6) displaying variability in volume and morphology following successful segmentation separation. Parallel and transverse views are shown, along with orientation axes. This workflow integrates serial-section alignment, manual segmentation, and volumetric surface rendering, providing high-resolution 3D quantification of axonal caliber, myelin geometry, and mitochondrial morphology, metrics not accessible through conventional 2D TEM analysis. Full volumetric and shape descriptor measurements are provided in Tables 1–3 and S1–S8

For each segment, the Segment Statistics module was used to compute volumetric parameters, including segment volume (mm^3), surface area (mm^2), and voxel count. Because myelin and axoplasm represent continuous or semi-continuous compartments rather than discrete organelles, only volume- and surface-related metrics were extracted for these structures.

For mitochondria, which exist as discrete 3D objects, additional shape descriptors were calculated. These included sphericity, elongation ratio, surface-to-volume ratio, Feret diameter, and principal axis lengths. Shape-specific descriptors were applied exclusively to mitochondria segments, as they are not meaningful for non-discrete membrane-bound structures such as myelin sheaths or axoplasm volumes [20, 41]. These data serve as internal quality-control metrics for both biological interpretation and the consistency of image processing [20].

Table 1

Volumetric measurements of myelin and axoplasm using three 3D Slicer representations (LM, SV, and CS)

Structure	Voxel count (LM)	Volume (LM) [mm^3]	Voxel count (SV)	Volume (SV) [mm^3]	Surface area [mm^2]	Volume (CS) [mm^3]
Myelin	1,639,906	4.10E−03	1,639,906	4.10E−03	0.000189295	4.09E−03
Axoplasm	1,670,324	4.18E−03	1,670,324	4.18E−03	0.00010013	4.17E−03

Values represent voxel-based (LM), voxel-corrected (SV), and surface-derived (CS) volumes. Surface-to-volume ratios ensure cross-representation consistency. Volumetry was conducted based on segmentation, illustrated in Fig. 3

Table 2

Volumetric measurements of individual mitochondria (M1–M6) using 3D Slicer's LM, SV, and CS representations

Mitochondrion	Voxel count (LM)	Volume (LM) [mm^3]	Voxel count (SV)	Volume (SV) [mm^3]	Surface area [mm^2]	Volume (CS) [mm^3]
M1	19,743	4.94E−05	19,743	4.94E−05	7.70E−01	4.90E−05
M2	6750	1.69E−06	6750	1.69E−06	3.61E−02	1.66E−06
M3	5875	1.47E−05	5875	1.47E−05	3.91E−01	1.41E−05
M4	5507	1.38E−05	5507	1.38E−05	3.53E−01	1.35E−05
M5	4149	1.04E−05	4149	1.04E−05	2.19E−01	1.00E−05
M6	3060	7.65E−11	3060	7.65e−11	2.15E−02	7.15E−06

These metrics quantify mitochondrial size and surface complexity and allow comparison across organelles. Volumetry was conducted based on segmentation, illustrated in Fig. 3

Table 3

Mitochondria-specific morphometric descriptors: centroid coordinates, Feret diameter, surface area, roundness, flatness, and elongation

Mitochondrion	Centroid_r	Centroid_a	Centroid_s	Feret diameter [mm]	Surface (LM) [mm^2]	Roundness	Flatness	Elongation
M1	0.000676989	0.000861503	0.000164843	0.00700652	3.40E−01	0.889243	129.816	213.584
M2	−0.000438277	−0.000722887	−0.00162844	0.00303236	1.59E−01	0.927824	310.278	736.253
M3	0.00010249	0.000258878	−0.00197166	0.00404711	1.70E−01	0.791488	251.619	13.993
M4	−0.000372707	−0.000308451	0.00133875	0.0050083	1.58E−01	0.816104	206.255	175.599
M5	0.000390309	−0.00121853	0.00224693	0.00201869	9.76E−02	109.394	130.242	122.888
M6	0.000120497	−0.000585288	0.00313333	0.00112714	9.25E−02	0.942068	42.879	398.672

These measurements characterize mitochondrial spatial distribution and shape, providing biologically meaningful information on organelle organization and morphology within the axon. Metrics were not applied to myelin or axoplasm because these structures are continuous and do not represent discrete compact objects. Quantification was conducted based on segmentation, illustrated in Fig. 3

Final visualizations are generated using surface-rendering or volume-rendering modes with adjustable lighting and transparency settings [25]. Screenshots can be exported in high-resolution TIFF or PNG format, and dynamic rotational animations can be produced through the Screen Capture tool to illustrate the 3D organization of axons, myelin, and mitochondria. The resulting models provide a faithful volumetric representation of the spinal-cord white matter, allowing quantitative assessment of mitochondrial quality control and myelin architecture within their native spatial context [25, 27].

4 Results

Quantitative 3D morphometric analysis is performed on individually segmented myelin, axoplasm, and mitochondria using measurements exported from 3D Slicer. For each structure, voxel-based estimates of volume were extracted using three available sampling representations: Label Map (LM), Scaled Volume (SV), and Closed Surface (CS).

The reconstructed segmentation volumes for myelin and axoplasm were 4.09×10^{-3} mm^3 and 4.17×10^{-3} mm^3, respectively, based on closed-surface-derived (CS) volumes (Table 1). These two compartments occupied comparable volumes within the sampled region, consistent with the expected ultrastructural organization of myelinated axons. Mitochondrial segments presented substantially smaller volumes (range: 1.00×10^{-5}–7.15×10^{-6} mm^3), reflecting natural heterogeneity in mitochondrial size within axoplasm (Table 2).

For all segments, volumes derived from the binary label map (LM) and the underlying scalar TEM volume (SV) were nearly identical, with discrepancies $<1\%$. Closed-surface (CS) volumes, computed from triangulated surface meshes, likewise showed minimal deviation from LM/SV values, consistent with polygonal surface smoothing removing stair-step voxel roughness. The convergence of these three measurements confirms that segmentation integrity was maintained during mesh generation and that the geometry of each structure was preserved across representations. Because CS is derived from a surface reconstruction, it is also the most appropriate estimate of physical morphology and is therefore prioritized in the interpretation of results.

Surface area measurements (mm^2) derived from the closed-surface representation were structure-dependent: axoplasm and myelin exhibited large, continuous surfaces with larger external contour complexity of axoplasm relative to myelin confirmed, whereas individual mitochondria showed values proportional to their size and geometry. For instance, mitochondrial segment M1 exhibited a surface area of 0.77 mm^2, while smaller mitochondria (e.g., M6) had ~0.0215 mm^2 (Table 2). The availability of surface area in combination with volume allows subsequent estimation of surface-to-volume ratios for each structure, particularly relevant for organelles such as mitochondria, where membrane-bound processes dominate functionality.

Mitochondria were uniquely assessed with shape-related metrics including Feret diameter, roundness, flatness, and elongation (Table 3). These descriptors provided additional insight into mitochondrial geometry: values ranged from near-spherical to distinctly elongated morphologies, consistent with the natural variability of mitochondrial shape.

Centroid coordinates are additionally recorded for each mitochondrion to describe their spatial distribution relative to the axonal geometry. These metrics describe only positional relationships and do not encode biologically interpretable shape information, but they support potential integration with spatial-distribution analyses in future work, such as clustering or mitochondrial transport inference.

These metrics are not applied to myelin or axoplasm because they are biologically meaningful primarily for compact, self-contained 3D objects. Myelin and axoplasm form extended, continuous structures with large, asymmetric morphologies; thus, principal-axis decomposition or bounding-box fitting would not yield interpretable biological descriptors. In contrast, for mitochondria, these metrics reflect intrinsic 3D organization.

Collectively, these quantitative measures establish a comprehensive 3D ultrastructural profile of myelin, axoplasm, and mitochondria within the imaged white-matter region. The segmentation-based volumes, surface areas, and mitochondrial shape descriptors form the quantitative foundation on which biological implications, such as compartmental proportions, mitochondrial morphology, and ultrastructural organization, are considered in Sect. 5.

5 Discussion

The volumetric quantification performed in the present study provides structural metrics with established biological relevance. Specifically, the calculated axoplasmic and myelin volumes (4.18×10^{-3} mm^3 and 4.10×10^{-3} mm^3, respectively) reflect the caliber and insulating capacity of the analyzed nerve fiber. Axonal volume is strongly associated with conduction velocity and transport efficiency, since increased cross-sectional area lowers internal resistance and supports higher ionic flux and metabolic load [2]. Myelin volume and surface area are similarly linked to saltatory conduction efficiency and energy-efficient action-potential propagation, beyond electrical insulation, oligodendroglia myelin also provides metabolic support to axons, supplying lactate and sustaining energetic homeostasis [2, 42]. Therefore, volumetric balance between axon and myelin, complementary to classical indices, such as the g-ratio, constitutes a meaningful structural proxy for functional potential and white-matter integrity [2].

In addition, the volumetric characterization of individual mitochondria in the present reconstruction (ranging from 1.66×10^{-6} mm^3 to 4.90×10^{-5}), along with shape descriptors, offers biologically relevant insights into axonal metabolic status. Feret diameter represents the maximum caliper distance across the organelle and thus approximates the longest mitochondrial axis, providing an intuitive measure of mitochondrial extension within the axon. Roundness quantifies how closely an organelle approximates a sphere; values approaching unity indicate more isotropic geometry, whereas lower values reflect elongation or irregularity. Flatness and elongation are derived from the principal axes of the segmented object and provide information on anisotropy: flatness increases as an object becomes compressed in one dimension, while elongation increases with elongation along its major axis. Together, these parameters quantify deviations from spherical symmetry and provide biologically meaningful indicators of mitochondrial structural states often associated with fission, fusion, or metabolic demand. Mitochondrial volume and morphology are established indicators of bioenergetic capacity, calcium buffering capability, and resilience to cellular stress. Structural remodeling of axonal mitochondria, such as increases in organelle volume or redistribution along axons, has been shown to occur in response to increased energetic demand and to precede axonal degeneration in demyelinating and neurodegenerative disorders [43, 44]. Furthermore, recent evidence demonstrates that myelination itself shapes mitochondrial localization and accumulation at internodes, underscoring a functional coupling between axon-myelin geometry and metabolic support [14]. Thus, the combined 3D reconstruction and volumetric segmentation approach used here not only captures anatomical organization but also enables biologically meaningful interpretation of axonal function and metabolic competence.

The simultaneous 3D reconstruction of two adjacent myelinated axons within the same microenvironment enabled direct comparison of volumetric properties and mitochondrial profiles across neighboring fibers. This observation highlights the inherent structural and metabolic heterogeneity that can exist within individual white-matter bundles, where axons may differ in caliber, mitochondrial load, and energetic demands despite their spatial proximity [14]. Local heterogeneity of myelin coverage and axonal support has been reported to influence conduction reliability and vulnerability to stress, reinforcing the functional significance of fine-scale volumetric variation [45]. Consequently, volumetric 3D EM does not merely provide morphological detail but serves as a sensitive tool for capturing early microstructural deviations that may precede overt demyelination or axonal degeneration.

Compared to classical electron tomography, which provides nanometer-scale three-dimensional information over very limited volumes, the serial-section TEM reconstruction approach presented here offers a complementary advantage by enabling volumetric analysis across substantially larger tissue extents. While electron tomography excels in resolving fine membrane details within individual organelles, its restricted field of view and technical complexity limit its applicability for comparative analyses of multiple axons, myelin sheaths, and mitochondrial populations within the same microenvironment [27, 28]. In contrast, serial TEM-based 3D reconstruction allows simultaneous assessment of axonal caliber, myelin geometry, and mitochondrial distribution across neighboring fibers, thereby capturing mesoscale ultrastructural organization that is more representative of white-matter architecture [12, 13]. Importantly, this workflow relies on conventional TEM instrumentation and open-source software, making it broadly accessible and scalable for studies requiring quantitative comparisons across conditions, regions, or disease states [38–41]. Thus, the present approach bridges the gap between high-resolution tomographic techniques and tissue-level ultrastructural analysis, providing a practical and biologically informative framework for investigating axon–myelin–mitochondria relationships in health and disease.

Together, these findings underscore the value of integrated axon–myelin–mitochondria volumetry as a structural–metabolic fingerprint of white-matter integrity. Future comparisons between healthy and pathological tissue may determine whether alterations in axon-to-myelin volumetric ratio or mitochondrial surface-to-volume relationship constitute candidate ultrastructural biomarkers of emerging dysfunction and therapeutic response.

6 Notes

1. En bloc staining consistency. Uranyl acetate concentration and incubation time may vary depending on tissue type and thickness. Overstaining leads to increased background, while insufficient staining reduces membrane contrast. It is advisable to protect uranyl acetate from light exposure.
2. Fixative and resin preparation. Both osmium tetroxide and epoxy resin mixtures should be freshly prepared before use. Crystallization of OsO_4 or incomplete resin polymerization can cause granular artifacts that interfere with myelin periodicity measurements. Filtering through a 0.22 μm syringe filter helps remove precipitates.
3. Determining section thickness. Interference colors observed under reflected light are a quick guide to thickness: silver–gold

hues correspond to ~70 nm. Uniform thickness minimizes distortion during 3D reconstruction and quantitative scaling.

4. Lead citrate should be freshly prepared prior to use, as aged solutions can precipitate and generate background artifacts on ultrathin sections.
5. Maintaining imaging consistency. Acquire all images using identical exposure, brightness, and magnification settings. Differences between sessions can introduce artificial variation in contrast and complicate alignment algorithms.
6. Alignment validation. Verify registration quality by scrolling through the stack and focusing on large, easily identifiable structures, such as axon bundles or blood vessels. Misalignment of even one section can propagate errors throughout the volume.
7. Mitochondrial segmentation. Ensure that contours follow the outer membrane, excluding the adjacent axoplasmic matrix. Overestimation of organelle area inflates calculated mitochondrial density and distorts volumetric analysis.

Conflict of Interest

The authors have no conflicts of interest to declare that are relevant to the content of this chapter.

References

1. Kimura I, Hayashi T, Autio JA (2025) Linking neuron-axon-synapse architecture to white matter vasculature using high-resolution multimodal MRI in primate brain. Imaging Neurosci 3:IMAG.a.77
2. Stassart RM, Möbius W, Nave K-A, Edgar JM (2018) The axon-myelin unit in development and degenerative disease. Front Neurosci 12:467
3. Li W, Gui Y, Guo C et al (2025) Molecular mechanisms of mitochondrial quality control. Transl Neurodegener 14:45
4. Ohno N, Kidd GJ, Mahad D, Kiryu-Seo S, Avishai A, Komuro H, Trapp BD (2011) Myelination and axonal electrical activity modulate the distribution and motility of mitochondria at CNS nodes of Ranvier. J Neurosci 31:7249–7258
5. Tosolini AP, Abatecola F, Negro S, Sleigh JN, Schiavo G (2024) The node of Ranvier influences the in vivo axonal transport of mitochondria and signaling endosomes. iScience. https://doi.org/10.1016/j.isci.2024.111158
6. Emamnejad R, Pagnin M, Petratos S (2024) The iron maiden: oligodendroglial metabolic dysfunction in multiple sclerosis and mitochondrial signaling. Neurosci Biobehav Rev 164:105788
7. Ma Y, Song R, Duan C (2025) Mitochondrial quality control and transfer communication in neurological disorders and neuroinflammation. Front Immunol. https://doi.org/10.3389/fimmu.2025.1542369
8. Licht-Mayer S, Campbell GR, Mehta AR et al (2023) Axonal response of mitochondria to demyelination and complex IV activity within demyelinated axons in experimental models of multiple sclerosis. Neuropathol Appl Neurobiol 49:e12851
9. Wang K, Chen X, Liu M et al (2025) Mitochondrial dynamics reveal potential to facilitate axonal regeneration after spinal cord injury. J Transl Med 23: 617
10. Theotokis P, Kesidou E, Mitsiadou D, Petratos S, Damianidou O, Boziki M, Chatzidimitriou A, Grigoriadis N (2021) Lumbar spine

intrathecal transplantation of neural precursor cells promotes oligodendrocyte proliferation in hot spots of chronic demyelination. Brain Pathol 32:e13040
11. Marshall AG, Neikirk K, Stephens DC et al (2023) Serial block face-scanning electron microscopy as a burgeoning technology. Adv Biol 7:e2300139
12. Faitg J, Lacefield C, Davey T, White K, Laws R, Kosmidis S, Reeve AK, Kandel ER, Vincent AE, Picard M (2021) 3D neuronal mitochondrial morphology in axons, dendrites, and somata of the aging mouse hippocampus. Cell Rep 36:109509
13. Turner NL, Macrina T, Bae JA et al (2022) Reconstruction of neocortex: organelles, compartments, cells, circuits, and activity. Cell 185:1082–1100.e24
14. Kole K, Voesenek BJB, Brinia ME, Petersen N, Kole MHP (2022) Parvalbumin basket cell myelination accumulates axonal mitochondria to internodes. Nat Commun 13:7598
15. Ouyang Z, Wang Q, Li X et al (2025) Elucidating subcellular architecture and dynamics at isotropic 100-nm resolution with 4Pi-SIM. Nat Methods 22:335–347
16. Wang Y, Sun B, Shibata B, Guo F (2022) Transmission electron microscopic analysis of myelination in the murine central nervous system. STAR Protoc 3:101304
17. Nahirney PC, Tremblay M-E (2021) Brain ultrastructure: putting the pieces together. Front Cell Dev Biol 9:629503
18. Abdollahzadeh A, Belevich I, Jokitalo E, Tohka J, Sierra A (2019) Automated 3D axonal morphometry of white matter. Sci Rep 9:6084
19. Abdollahzadeh A, Belevich I, Jokitalo E, Sierra A, Tohka J (2021) DeepACSON automated segmentation of white matter in 3D electron microscopy. Commun Biol 4:179
20. Gow A (2025) Demystifying the myelin g ratio: its origin, derivation and interpretation. ASN Neuro 17:2542166
21. Gow A (2025) Understanding the myelin g ratio from first principles, its derivation, uses and artifacts. ASN Neuro 17:2445624
22. Nave K-A, Asadollahi E, Sasmita A (2023) Expanding the function of oligodendrocytes to brain energy metabolism. Curr Opin Neurobiol 83:102782
23. Oost W, Huitema AJ, Kats K, Giepmans BNG, Kooistra SM, Eggen BJL, Baron W (2023) Pathological ultrastructural alterations of myelinated axons in normal appearing white matter in progressive multiple sclerosis. Acta Neuropathol Commun 11:100
24. Cocks E, Taggart M, Rind FC, White K (2018) A guide to analysis and reconstruction of serial block face scanning electron microscopy data. J Microsc 270:217–234
25. Schadt LC, Ruhwedel T, Pape C, Sasmita AO, Steyer AM, Möbius W (2024) Wrapped up: advancements in volume electron microscopy and application in myelin research. Methods Microsc 1:119–136
26. TrakEM2. In: ImageJ Wiki. https://imagej.github.io/plugins/trakem2/index. Accessed 23 Oct 2025
27. Kuan AT, Phan S, Kim K-Y, Mackey M, Kim M, Peltier ST, Ellisman M, Lee W-CA (2024) Scalable electron tomography for connectomics. bioRxiv 2024.06.05.597487
28. Sosinsky GE, Deerinck TJ, Greco R, Buitenhuys CH, Bartol TM, Ellisman MH (2005) Development of a model for microphysiological simulations. Neuroinformatics 3:133–162
29. Theotokis P, Kleopa KA, Touloumi O et al (2015) Connexin43 and connexin47 alterations after neural precursor cells transplantation in experimental autoimmune encephalomyelitis. Glia 63:1772–1783
30. Hua Y, Laserstein P, Helmstaedter M (2015) Large-volume en-bloc staining for electron microscopy-based connectomics. Nat Commun 6:7923
31. Ströh S, Hammerschmith EW, Tank DW, Seung HS, Wanner AA (2022) In situ X-ray-assisted electron microscopy staining for large biological samples. eLife 11:e72147
32. Rivlin PK, Raymond PA (1987) Use of osmium tetroxide-potassium ferricyanide in reconstructing cells from serial ultrathin sections. J Neurosci Methods 20:23–33
33. Deerinck TJ, Shone T, Bushong EA, Ramachandra R, Peltier ST, Ellisman MH (2018) High-performance serial block-face SEM of non-conductive biological samples enabled by focal gas injection-based charge compensation. J Microsc 270:142–149
34. Watson ML (1958) Staining of tissue sections for electron microscopy with heavy metals. J Biophys Biochem Cytol 4:475–478
35. Knott G, Marchman H, Wall D, Lich B (2008) Serial section scanning electron microscopy of adult brain tissue using focused ion beam milling. J Neurosci 28:2959–2964
36. Bozzola JJ, Russell LD (1999) Electron microscopy: principles and techniques for biologists. Jones & Bartlett Learning
37. Reynolds ES (1963) The use of lead citrate at high pH as an electron-opaque stain in electron microscopy. J Cell Biol 17:208–212
38. Tremi I, Havaki S, Georgitsopoulou S, Lagopati N, Georgakilas V, Gorgoulis VG, Georgakilas AG (2021) A guide for using transmission electron microscopy for studying the radiosensitizing effects of gold nanoparticles in vitro. Nanomaterials 11:859
39. Schindelin J, Arganda-Carreras I, Frise E et al (2012) Fiji - an open source platform

for biological image analysis. Nat Methods 9. https://doi.org/10.1038/nmeth.2019

40. Cardona A, Saalfeld S, Schindelin J, Arganda-Carreras I, Preibisch S, Longair M, Tomancak P, Hartenstein V, Douglas RJ (2012) TrakEM2 software for neural circuit reconstruction. PLoS One 7:e38011
41. Rolfe S, Pieper S, Porto A, Diamond K, Winchester J, Shan S, Kirveslahti H, Boyer D, Summers A, Maga AM (2021) SlicerMorph: an open and extensible platform to retrieve, visualize and analyse 3D morphology. Methods Ecol Evol 12:1816–1825
42. Coggan JS, Bittner S, Stiefel KM, Meuth SG, Prescott SA (2015) Physiological dynamics in demyelinating diseases: unraveling complex relationships through computer modeling. Int J Mol Sci 16:21215–21236
43. Kiryu-Seo S, Ohno N, Kidd GJ, Komuro H, Trapp BD (2010) Demyelination increases axonal stationary mitochondrial size and the speed of axonal mitochondrial transport. J Neurosci 30:6658–6666
44. Campbell GR, Mahad DJ (2011) Mitochondria as crucial players in demyelinated axons: lessons from neuropathology and experimental demyelination. Autoimmune Dis 2011:262847
45. Talidou A, Lefebvre J (2025) Spatial heterogeneity in myelin sheathing impacts signaling reliability and susceptibility to injury. eNeuro 12:ENEURO.0402-24.2025

Check for updates

Chapter 13

Postmortem Human Brain Tissue Processing for Transmission Electron Microscopy of Gray Matter Myelin

Bastiaan Lodder and Maarten H. P. Kole

Abstract

Postmortem human brain tissue is invaluable to determine the fundamental architecture of the brain and the biological basis of disease. To resolve the cellular organization at the highest resolution for ultrastructure, down to the atomic level, transmission electron microscopy (TEM) is the gold standard. However, preparing postmortem tissue for TEM is associated with many sequentially performed steps and faces several challenges, such as the preservation of the membrane ultrastructure and cellular organelles. Here, we describe preparation steps for human cortex optimized for the preservation of oligodendroglial myelin sheaths in postmortem tissue.

Keywords Human postmortem, Brain, Transmission electron microscopy (TEM), Resin embedding, Myelin, Axon

1 Introduction

Postmortem human brain tissue is a highly valuable resource to identify the biological mechanisms of disease, stratify individual variability, or determine the fundamental organization of cells and circuits of the brain [1]. Oligodendrocytes, which produce the myelin sheaths in the central nervous system, are involved in various functions, including metabolic support and membrane excitability, and are emerging as key drivers of neurodegenerative and neuropsychiatric diseases [2]. To resolve the integrity of the myelin sheath, composed of multilamellar layering of lipid-rich membranes enveloping axons, transmission electron microscopy (TEM) is the gold standard. Ultrastructural studies using TEM from biopsy or autopsy tissue from patients with multiple sclerosis led to landmark identification of the complex cellular mechanisms of myelin degradation, the occurrence of remyelination, or the role of immune cells in the pathology [3–7], reviewed in [8].

Myelination within the gray matter is, however, highly diverse. Patterns of myelinated axons are anatomically oriented in radial and

Marie-Ève Tremblay (ed.), *Ultrastructural Neuroscience Methods*, Neuromethods, vol. 238,
https://doi.org/10.1007/978-1-0716-5456-9_13,

tangential directions, in part reflecting the wrapping of incoming and subcortically projecting excitatory axons together with locally projecting inhibitory axons [9, 10]. Gray matter myelination is also heterogeneous along the cortical layers, in a brain region-dependent manner, and continues to develop in density until ~50 years of age [11–13].

Examining the myelin ultrastructure, therefore, requires careful selection of anatomically identified regions. While myelin TEM imaging has been reported for brain tissue with long postmortem intervals and prolonged storage [14, 15], the delays between time of death, storage medium, and tissue preparation cause substantial cellular degradation [16]. The passive immersion in large volumes of chemical solutions, typically used to fix or stain human postmortem blocks, hinders effective penetration of chemicals and metals, causing inhomogeneity and tissue instability [17]. In addition, aldehyde fixation itself, required for the large human neocortex, the chemical crosslinking of proteins in the lipid-rich compact myelin sheaths is subjected to artifacts, including membrane splitting and loosening, when compared to high-pressure freezing [18]. These factors causing suboptimal fixation may hinder to discriminate between pathophysiologically relevant changes in the myelin sheath.

Here, we report optimized fixation parameters and TEM procedures focused on the preservation of the myelin sheath ultrastructure in the neocortex of postmortem tissue from the Netherlands Brain Bank (NBB) (https://www.brainbank.nl/ and [19]). We describe the general approach and step-by-step protocols to target specific deeper layers, yielding well-preserved ultrastructural features at the resolution of single myelin lamellae.

2 Materials

2.1 16% PFA Solution

Paraformaldehyde is toxic; all steps should be carried out in a fume cupboard, and appropriate safety protections should be worn.

1. Add 6.4 g of paraformaldehyde (PFA) to a 50 mL Erlenmeyer flask.
2. Preheat 100 mL double-distilled H_2O (ddH_2O) in a 200 mL glass beaker to 60–65°C.
3. Add 32 mL of this heated ddH_2O to the PFA and add a magnetic stir bar.
4. To heat up the solution "au bain-marie," place the 200 mL beaker with the remaining heated ddH_2O on a hotplate stirrer and place the Erlenmeyer flask with PFA inside the 200 mL beaker.
5. Add a thermometer and heat the solution to 60–65°C. It is crucial that the temperature does not exceed 70°C.

6. Pipette 200 µL of 1 M NaOH (approx. 14 drops with a fine Pasteur pipette) to the PFA.
7. After approximately 20 min, or when all the PFA has dissolved, check and adjust the pH to 7.4.
8. Let the solution cool off and adjust the pH again if necessary.
9. Make up to a final volume of 40 mL with ddH_2O.
10. Filter the solution through a 0.2 µm filtration unit with a syringe to remove any undissolved PFA.
11. Store the PFA solution at 4°C for up to 2 weeks.

Materials:

1. Paraformaldehyde (prilled 95%).
2. ddH_2O.
3. 50 mL Erlenmeyer flask.
4. 200 mL beaker.
5. Magnetic stir bar.
6. Thermometer.
7. Hotplate stirrer.
8. pH strips.
9. 1 M NaOH.
10. 0.2 µm filter.
11. Syringe.

2.2 Fixative

4% paraformaldehyde, 2.5% glutaraldehyde in 0.1 M sodium cacodylate buffer.

1. Mix 10 mL of 10% aqueous glutaraldehyde solution with 10 mL of 16% PFA and 20 mL of 0.2 M sodium cacodylate buffer.
2. Store at 4°C for up to 2 weeks.

2.3 Storage Buffer

0.5% paraformaldehyde in 0.1 M sodium cacodylate buffer.

1. Mix 1.56 mL of 16% PFA with 25 mL of 0.2 M sodium cacodylate buffer and 23.44 mL ddH_2O.

2.4 Embed-812 Resin

1. In a fume cupboard, place a disposable plastic cup on a 60°C hotplate stirrer.
2. For a 24 mL solution (sufficient for six tissue blocks), add subsequently the following reagents: 10 mL Embed-812, 8 mL DDSA, 4 mL NMA, and 0.35 mL DMP-30. When adding a new component, wait for the mixture to return to 60°C before adding a new component.
3. Mix with a magnetic stir bar on the lowest speed to prevent the formation of air bubbles.

4. With the DMP-30, the mixture will obtain a red color but returns to yellow once properly mixed. Using a disposable wooden stick (e.g., a coffee stirrer) helps to mix the solution. Clean spills and bottle necks with acetone.

Materials:

1. Embed-812.
2. Dodenyl succinic anhydride (DDSA).
3. Methyl norbornene dicarboxylic anhydride (NMA).
4. DMP-30.
5. Magnetic stir bar.
6. Hotplate stirrer.
7. Disposable cup.
8. Disposable wooden stirrer.

2.5 Resin Sandwich

1. On top of a glass slide, add a piece of Aclar of the same size (Fig. 1).
2. Add a drop of freshly prepared resin to the Aclar.
3. With a toothpick, transfer the sample to the resin drop.
4. Add an additional drop of resin on the sample.
5. Add another piece of Aclar on the sample; be careful not to create air bubbles.

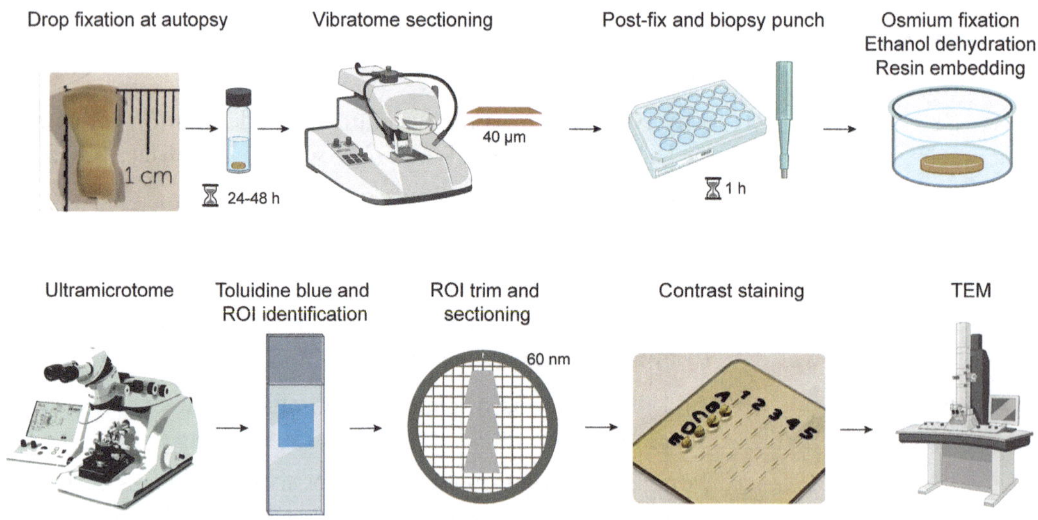

Fig. 1 Schematic overview of the general TEM workflow. Example of a small tissue block of the middle temporal gyrus drop-fixed at autopsy in fixative. Subsequently, thin sections are cut on a vibratome and post-fixed for 1 h. The ROI is collected with a biopsy punch, fixed with osmium, and prepared with resin embedding. Toluidine blue staining is performed on 1 μm ultramicrotome sections to identify and isolate the ROI. 60 nm sections are collected on a grid and stained before imaging with a TEM

6. Add a glass slide on top and add small weights.

Materials:

1. Glass slides.
2. Aclar.
3. Toothpick.
4. Resin.
5. Small weights.

2.6 1% Toluidine Blue

1. Add 1 g toluidine blue and 1 g disodium tetraborate to 100 mL ddH_2O.
2. Filter through a 0.2 μm filter before use.

Materials:

1. Toluidine blue.
2. Disodium tetraborate.
3. ddH_2O.
4. 0.2 μm filter.
5. Syringe.

3 Methods

The following section provides a detailed, step-by-step guide to the depicted TEM workflow (Fig. 2). The method describes the process to optimally preserve human postmortem brain tissue, isolate a specific neocortex region, and how to prepare the tissue for myelin visualization using TEM.

3.1 Human Postmortem Brain Fixation

It is strongly advised to use fixative (see Sect. 2.2) as fresh as possible, although up to 2 weeks old fixative can provide adequate results. In our experience, the most important contributor to the fixative quality is the mixing of glutaraldehyde (GA), and mixing the GA with the PFA should be done shortly before autopsy.

The postmortem delay (PMD) is a critical factor in determining the quality of the sample (refs.). Furthermore, the cerebrospinal fluid pH (typically ranging between 6.0 and 7.0) will impact the sample quality and interacts with PMD [16]. When using EM for comparative analyses, great care should therefore be taken to match the pH and PMD. Here, we show results from autopsies of the NBB, which typically have PMD <7 h and pH between 6.0 and 6.7.

3.2 Drop-Fixation

Immerse-fix a small piece of human brain tissue with at least 1 dimension smaller than 2 mm in 4% PFA + 2.5% GA in 0.1 M sodium cacodylate buffer for 24–48 h at 4°C [16, 20].

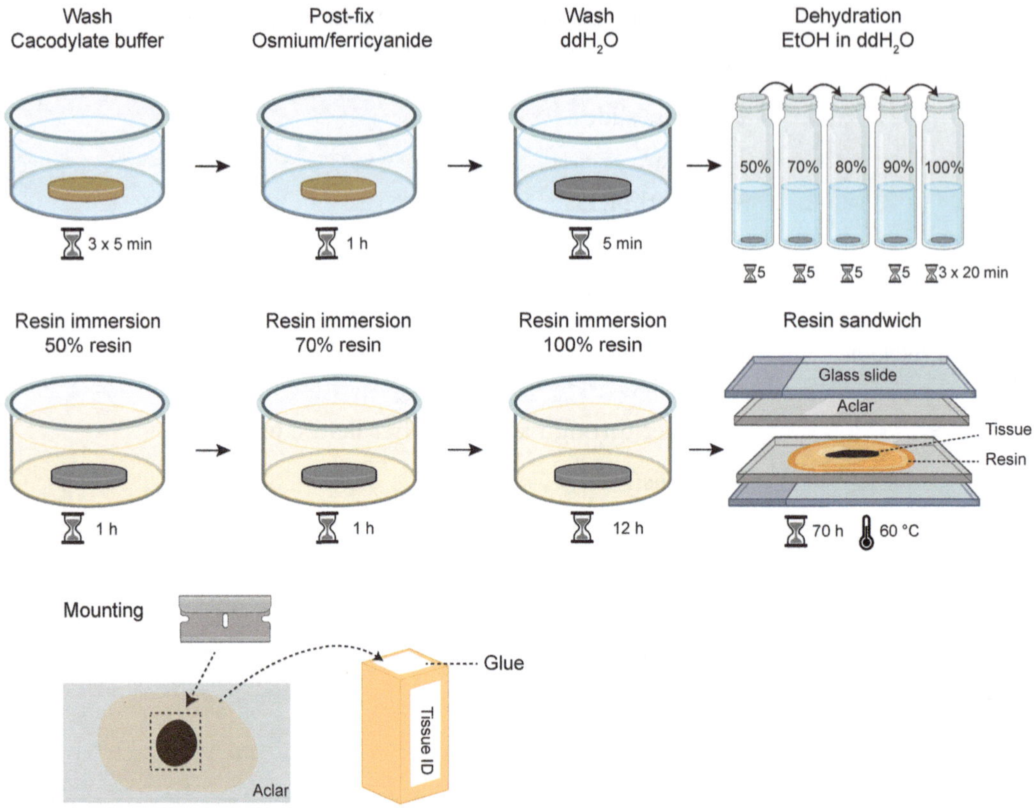

Fig. 2 Schematic overview of osmium fixation and resin embedding workflow. The 40 μm sections are washed and post-fixed with 1% osmium tetroxide and 1.5% (w/v) $K_3Fe(CN)_6$ in 0.1 M sodium cacodylate buffer for 1 h. Thereafter, sections are dehydrated in ethanol and immersed in resin. Sections are polymerized in the sandwich, placed in an oven, and mounted on a resin block

After fixation, immediately continue with vibratome sectioning. Alternatively, the sample can be stored in 0.5% PFA in 0.1 M sodium cacodylate buffer at 4°C for months.

3.3 Vibratome Sectioning and Post-fix

The orientation of the interest region should be considered when sectioning the sample. To image a specific neocortical layer with EM, the vibratome sections should contain all layers (Fig. 3A). During later steps, a specific layer can be isolated and imaged.

1. Wash the fixed brain samples in 0.1 M sodium cacodylate buffer 3× for 5 min.
2. Cut the tissue into smaller blocks using a double-edged, sharp razor blade to avoid compression damage to the sample.
3. Glue the block on the vibratome plate in the correct orientation.
4. Fill the vibratome bath with 0.1 M sodium cacodylate buffer.
5. Cut 40 μm thick slices, collect in a well plate, and post-fix in fixative (see Sect. 2.2) for 1 h.

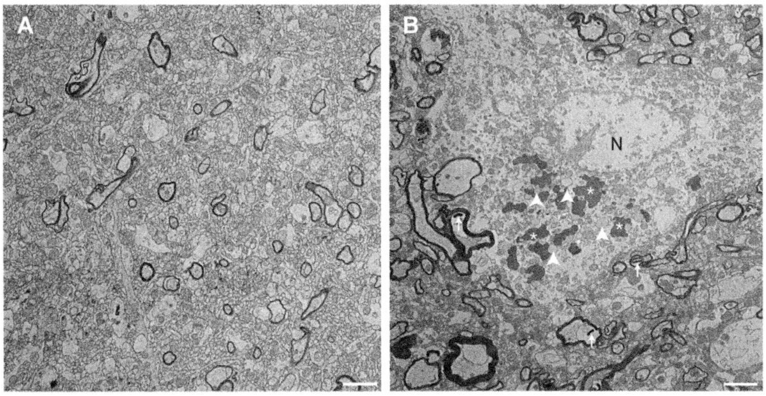

Fig. 3 Low-magnification EM overview of lamina 5. (**A**) Transmission electron micrograph overview (×2600 magnification) covering a 22 × 22 μm area shows the typical cellular organization of gray matter, including dendrites, unmyelinated axon terminals and synapses, as well as darkly stained multilamellar myelin plasma around axons, mostly cross-sectional. Donor age 34 years. (**B**) Layer 5 pyramidal neuron soma with a pale, round nucleus (N), showing ultrastructural features of aging, including lipofuscin granules (asterisks) and closely associated lipid droplets (arrowheads). Outside of the soma plasma membrane, multiple myelinated axons are present, including axons with myelin body inclusions (white arrows). Donor age 53 years. Scale bars, 2 μm

6. Cut out the region of interest (ROI) with an appropriately sized biopsy punch (e.g., a 4 mm punch encompasses all neocortical layers) and store it in a glass vial with storage buffer (see Sect. 2.3) until the osmium fixation.

Materials:

1. Vibratome.
2. 0.1 M sodium cacodylate buffer.
3. Superglue.
4. Double-edged razor blade.
5. Fixative (see Sect. 2.2).
6. Well plate.
7. Biopsy punch.
8. Glass vials.

3.4 Osmium Fixation and Resin Embedding

Osmium tetroxide is highly toxic, and its vapors are irritating to the eyes and respiratory tract. All handling must be performed in a fume hood. Appropriate safety protection should be worn at all times, and care should be taken to ensure proper storage and waste disposal.

To identify the samples during these steps, use an ethanol-resistant marker or a label to mark the glass vials. Vials should be on a shaker in the fume hood to ensure proper penetration. All steps described below are done at room temperature. Freshly made resin (see Sect. 2.4) is advised for these steps.

Resin embedding ensures that cellular structures are stabilized, allowing hardened tissue to be cut at ultrathin (60 nm) sections and electrons to pass through during TEM. However, resin is hydrophobic, so a dehydration step needs to be added to ensure proper penetration of the resin. Before this step, another fixation with osmium tetroxide is performed to preserve structures by immobilizing lipids and to give an electron-dense staining of cellular membranes [21].

1. Wash the samples in 0.1 M sodium cacodylate buffer 3× for 5 min.
2. Fix samples with 1% osmium tetroxide and 1.5% (w/v) potassium ferricyanide in 0.1 M sodium cacodylate buffer for 1 h.
3. Wash briefly with ddH_2O.
4. Dehydrate in increasing ethanol concentrations in ddH_2O for 5 min each; 50%, 70%, 80%, 90%.
5. Dehydrate 3× in 100% ethanol for 20 min each.
6. Prepare fresh resin.
   A rotary mixer inside the fume hood is advised for the next steps to ensure proper penetration. Removing the vial caps for these steps helps the ethanol to evaporate over time.
7. Immerse the sample in 50% resin in ethanol for 1 h.
8. Immerse the sample in 70% resin in ethanol for 1 h.
9. Embed the sample in 100% resin overnight.
10. The next day, prepare fresh resin again and make the resin sandwich as described before (see Sect. 2.5). This ensures the sample remains flat while hardening, making it easier to section with an ultramicrotome. The tissue ID can be written on the glass slide using a pencil.
11. The leftover resin can be used to make mounting blocks in molds. While most molds come with an ID imprinted on the resin, a small paper label can be added to the mold, corresponding to the sample that will be glued to this block (Step 13). Air bubbles in the resin during polymerization can cause sectioning problems. Bubbles can be removed with a wooden toothpick.
12. Polymerize the resin sandwiches and mounting blocks in an oven at 60°C for ~70 h. For safety reasons, all other materials with resin residues should also be polymerized in the oven.
13. Once the resin has polymerized, cut out a square around the sample with a razor blade. Carefully pry the sample with resin off the Aclar and glue it on the resin mounting block. Ensure that the mounting blocks have the correct measurements for the size of the sample. Dry for at least 8 h before sectioning with an ultramicrotome.

Materials:

1. Glass vials.
2. 0.2 M sodium cacodylate buffer.
3. 4% osmium tetroxide aqueous solution.
4. Potassium ferricyanide.
5. ddH_2O.
6. Ethanol.
7. Resin (see Sect. 2.4).
8. Resin sandwich (see Sect. 2.5).
9. Oven.
10. Pencil.
11. Resin mold.
12. Paper ID.
13. Super glue.
14. Razor blade.
15. Wooden toothpick.

3.5 Ultramicrotome Sectioning

When sectioning with an ultramicrotome, it is advised to first collect a thick section of 1 μm and examine it by light microscopy. This allows for quality assessment of the tissue and identification of the ROI. Collecting thick sections should be done with a glass knife, instead of the diamond knife. To get an overview of the sample, a toluidine blue staining can be performed on the section (Fig. 3). This will also help to identify if enough resin and/or glue has been trimmed in front of the sample. Ensure using clean tools when handling samples and grids since these are rapidly contaminated with dust during handling.

1. Place the resin block in the specimen holder of the ultramicrotome's trimming block.
2. Using a fresh double-edged razor blade, trim the resin around the block to create a square encompassing mostly the sample. To facilitate the orientation of the section, trim one edge of the square differently. Additionally, small indentations in the resin on one side of the sample help with the orientation.
3. Move the resin block to the cantilever arm and, using a glass knife, make 1-μm thick sections to trim the resin and/or glue in front of the sample.
4. To check if enough resin has been trimmed and the sample has been reached, a staining can be performed on a section. Add a droplet of water on a glass slide and place the section on the water. Alternatively, a glass knife with a water-filled trough can be used to create floating sections. A trough can

be made with electrical tape [22] or by waxing on premade troughs. The floating section can be picked up with a fine wire loop and transferred to the droplet on the glass slide.

5. Place the glass slide on an 80°C hot plate until the water evaporates and the section adheres to the slide. To help stretch out the floating section, a cotton swab dipped in chloroform can be hovered over the sample. Be careful that it does not touch the sample or droplet.
6. Add a droplet of 1% toluidine blue (see Sect. 2.6) on the section and heat on the hot plate until the droplet edge appears green/yellow.
7. Wash the section gently with water and dry the glass slide on the hot plate.
8. Inspect the section with a light microscope. The tissue in the section will stain blue (Fig. 3). If the section does not contain enough tissue, adjust the knife alignment and trim more resin and/or glue. Repeat these steps until an entire section contains colored tissue.
9. Locate the ROI in the section and use orientation markers to correlate the ROI with the sample block.
10. Using a glass or trim knife, trim the sample block around the ROI into a trapezoidal shape with a maximum diameter of 0.2 mm, but smaller is better.
11. When the sample block is an appropriate size, switch to thin sections. Thin sections can be made with either a glass or diamond knife. Due to the diamond knife's fragility (requiring extra care), it is advised to learn sectioning first with a glass knife.
12. Cut 60-nm-thick floating sections. Sections should appear gray/silver in color when floating in the knife boat. A color reference chart from ultramicrotome and diamond knife manufacturers can be used to estimate the correct thickness.
13. If the sample was properly and parallel-trimmed into a trapezoidal shape, the sections should form a ribbon on the water surface.
14. Mount the sections on a copper grid by holding the grid with forceps and scooping up the sections out of the water boat. Multiple sections can be scooped on one grid. TEM samples often need a thin electron-transparent support film, commonly Formvar, on top of the grid. Grids with support film can be made or purchased. Mount the sample on the support film side of the grid.
15. Dry the grid on filter paper, inspect under the ultramicrotome binoculars for quality, and store in a grid box.

Materials:

1. Ultramicrotome.
2. Glass knife.
3. Diamond knife.
4. 1% toluidine blue (see Sect. 2.6).
5. ddH_2O.
6. Glass slide.
7. Formvar-covered copper grids.
8. Filter paper.
9. Grid box.

3.6 Contrast Staining

Uranyl acetate is highly toxic and mildly radioactive. All handling must be performed in a fume hood, and appropriate safety protection should be worn at all times. Additional care should be taken to ensure proper storage and waste disposal.

An additional stain is often necessary to give more contrast to the sample. A frequently used contrast method consists of a combination of uranyl acetate and lead citrate. The first enhances the contrast of lipids and proteins, and the latter enhances the contrast of proteins and glycogens. Uranyl acetate is light sensitive and should be stored in the dark.

1. Centrifuge the uranyl acetate at 7000 rpm for 10 min.
2. Place the grids on a grid staining pad. Place the pad on a filter paper in a Petri dish.
3. Pipette a droplet of uranyl acetate (3.5% in ddH_2O) against the sample-facing side of each grid and leave for 5 min in the dark.
4. Rinse the grids with a gentle continuous stream of ddH_2O, approx. 50 mL per grid, and dry with filter paper.
5. Carefully add a droplet of NaOH to a corner of the staining pad, opposite the grids. The NaOH will absorb atmospheric CO_2 and prevent its interaction with lead citrate.
6. Add a droplet of lead citrate against each grid for 2 min.
7. Rinse with ddH_2O again as described above and dry the grids with filter paper.

Materials:

1. 3.5% uranyl acetate in ddH_2O.
2. Grid staining pad.
3. Petri dish.
4. Filter paper.
5. ddH_2O.
6. NaOH.
7. Reynolds lead citrate.

4 TEM of Postmortem Human Myelinated Axons

Using a Talos L120C microscope at 120 kV with a Ceta 16 M camera, we collected TEM images of the deeper layers of the neocortex in the inferior temporal gyrus (Fig. 3A, B). Staining with toluidine blue facilitates the identification of lamina 5, which is characterized in the human brain by large and often heterogeneously shaped pyramidal

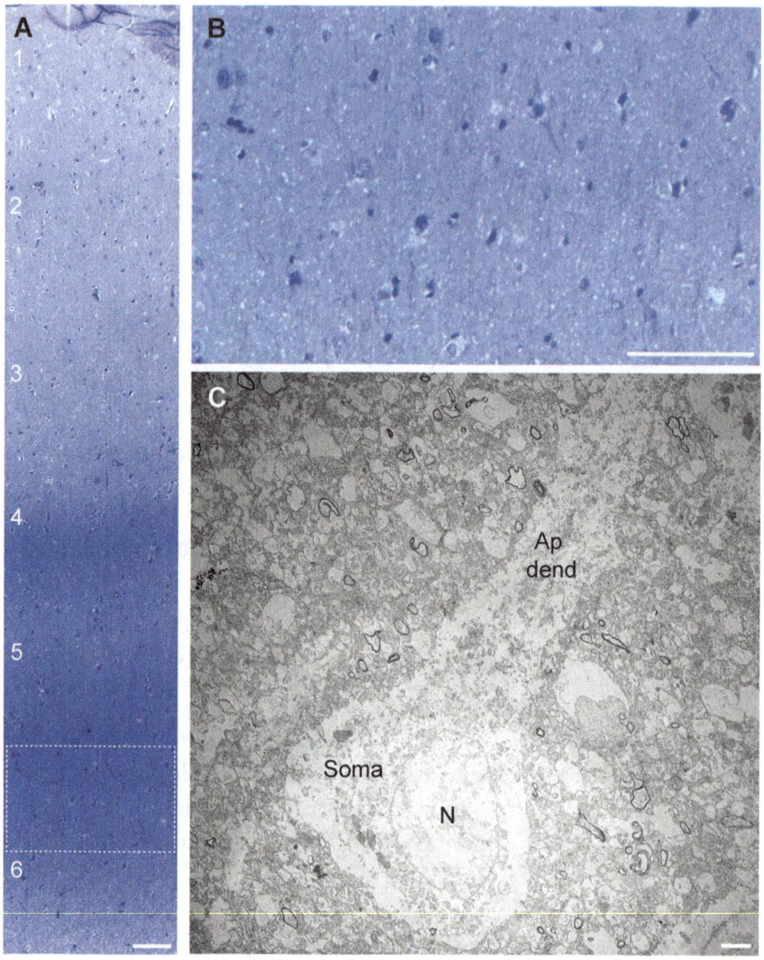

Fig. 4 Toluidine blue-stained section of the cortical layering. (**A**) Thin section stained with toluidine blue, revealing layers 1–6 in the inferior temporal gyrus. Donor age 83 years. Scale bar, 100 μm. (**B**) High magnification of lamina 5, white dotted square indicated in (**A**). Multiple neurons are visible in the staining. Scale bar, 100 μm. (**C**) Electron micrograph of a pyramidal neuron soma in layer 5, characterized by a round nucleus (N) with homogeneously pale karyoplasm and a large diameter dendrite emerging from the apical pole of the perikaryon. At one side of the soma, several dark lipofuscin granules are visible. Scale bar, 2 μm

neuron cell bodies [23–25]. In the temporal cortex, layer 5 neurons are on average located between ~1.5 and 2.5 mm from the pia [26]. TEM imaging within the ROI (Fig. 3B, 2 mm from the pia) shows an example of a large-diameter pyramidal neuron cell body with an apical dendrite, characteristic of an excitatory extratelencephalic-projecting layer 5 pyramidal neuron (Fig. 3C).

The general ultrastructural appearance of the gray matter is variable. Figure 4 shows well-preserved cellular membranes and cytoplasm of dendrites, synaptic terminals, and mitochondria-enriched presynaptic terminals and myelinated axons. Within layer 5, there is a mix of afferent thalamocortical axons, local interneuron axon myelination, and efferent corticocortical-projecting axons from layers 2 and 3 traversing layer 5. Indeed, the myelinated axons

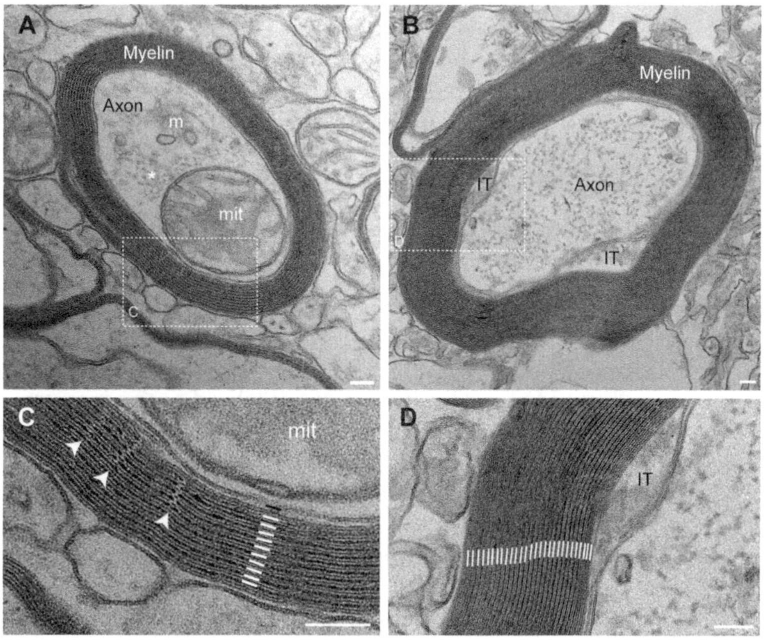

Fig. 5 Electron micrographs of myelin ultrastructure. (**A**) Example TEM image of a 102-year-old donor showing the transverse section of a myelinated axon with well-preserved ultrastructure. The axoplasm contains neurofilaments (asterisk), microtubules (m), and a mitochondrion (mit). Image captured with ×36,000 magnification. (**B**) Example TEM image from a 58-year-old donor of a cross-sectional cut of a myelinated axon with numerous neurofilaments, a thick myelin sheath, and a non-compacted inner tongue (IT) containing cytoplasm. Image captured at ×22,000 magnification. (**C**) Zoom of a region of the compact myelin sheath (white box in A), in which the periodicity of the oligodendrocyte membranes is well visible (white bars align with major dense line), as well as the radial bands of myelin (arrowheads). (**D**) Zoom of the multilamellar layering of the paired plasma membranes. White lines positioned on the major dense line indicated that the sheath comprises 28 myelin lamellae (white bars) and a non-compacted IT. Scale bars, 100 nm

comprised of longitudinal and transverse cuts in the TEM images (Fig. 4). Another pyramidal neuron cell body showed a higher density of irregularly shaped lipofuscin organelles, with a dark appearance due to the content of unsaturated fatty acids, sugars, and metals. To test whether the protocol preserves the myelin sheath ultrastructure, we next imaged at >22,000× magnification. Figure 5 shows two examples of transverse cuts of myelinated axons. The axoplasm contains neurofilaments, microtubules, and mitochondria. The dark appearance of the myelin sheath arises from the spiral wrapping of multiple layers of oligodendrocyte membranes. The multilamellar sheath periodically alternates the tight fusion of the inner cytoplasmic membranes (forming the major dense line) and closely apposed outer leaflets (the intraperiod lines). Further magnification of the sheath shows the clear visibility of the oligodendrocyte lipid membrane bilayers and three radial components, sites where the membranes abut to create a tight junction (Fig. 5B). The outer membranes of the axon and most inner oligodendrocyte, the adaxonal membrane, are separated by a narrow periaxonal space containing extracellular solution. Myelin sheaths are variable in thickness, which depends, in part, on the axon diameter. Figure 5C, D shows a larger axon with a thick sheath of 28 myelin lamellae and the presence of an inner tongue containing oligodendrocyte cytoplasm.

Acknowledgments

The authors are indebted to the support of the Electron Microscopy Center Amsterdam (EMCA). Figures 1 and 2 were created in part using BioRender. Lodder, B. (2026) and https://BioRender.com/ygl r274.

Funding This publication was supported by a grant from the Friends Foundation of the Netherlands Institute of Neuroscience ("Stichting Hersenvrienden") and is part of the project Institute for Chemical Neuroscience (iCNS) with file number 024.006.009, a Gravitation 2023 project of the Dutch Research Council (NWO), which is financed by the Ministry of Education, Culture and Science (OCW) of the Netherlands.

References

1. Murray ME, Smith C, Menon V, Keene CD, Lein E, Hawrylycz M, Aguzzi A, Benedetti B, Brose K, Caetano-Anolles K, Sillero MIC, Crary JF, Jager PLD, Faustin A, Flanagan ME, Gokce O, Grant SGN, Grinberg LT, Gutman DA, Hillman EMC, Huang Z, Irwin DJ, Jones DT, Kapasi A, Karch CM, Kukull WT, Lashley T, Lee EB, Lehner T, Parkkinen L, Pedersen M, Pritchett D, Rutledge MH, Schneider JA, Seeley WW, Shepherd CE, Spires-Jones TL, Steen JA, Sutherland M, Vickovic S, Zhang B, Stewart DJ, Keiser MJ, Vogel JW, Dugger BN, Phatnani H (2025) Accelerating biomedical discoveries in brain health through transformative neuropathology of aging and neurodegeneration.

Neuron 113:3703–3721. https://doi.org/10.1016/j.neuron.2025.06.014

2. Nave K-A, Ehrenreich H (2014) Myelination and oligodendrocyte functions in psychiatric diseases. JAMA Psychiatry 71:582–584. https://doi.org/10.1001/jamapsychiatry.2014.189
3. Prineas J (1975) Pathology of the early lesion in multiple sclerosis. Hum Pathol 6:531–554. https://doi.org/10.1016/s0046-8177(75)80040-2
4. Prineas JW, Connell F (1978) The fine structure of chronically active multiple sclerosis plaques. Neurology 28:68–75. https://doi.org/10.1212/wnl.28.9_part_2.68
5. Raine CS, Hummelgard A, Swanson E, Bornstein MB (1973) Multiple sclerosis: serum-induced demyelination in vitro. A light and electron microscope study. J Neurol Sci 20:127–148. https://doi.org/10.1016/0022-510x(73)90026-9
6. Rodriguez M, Scheithauer B (1994) Ultrastructure of multiple sclerosis. Ultrastruct Pathol 18:3–13. https://doi.org/10.3109/01913129409016267
7. Bosch AMR, Hümmert S, Steyer A, Ruhwedel T, Hamann J, Smolders J, Nave K, Stadelmann C, Kole MHP, Möbius W, Huitinga I (2023) Ultrastructural axon–myelin unit alterations in multiple sclerosis correlate with inflammation. Ann Neurol 93:856–870. https://doi.org/10.1002/ana.26585
8. Oost W, Meilof JF, Baron W (2025) Multiple sclerosis: what have we learned and can we still learn from electron microscopy. Cell Mol Life Sci 82:172. https://doi.org/10.1007/s00018-025-05690-0
9. Micheva KD, Wolman D, Mensh BD, Pax E, Buchanan J, Smith SJ, Bock DD, Slutsky I (2016) A large fraction of neocortical myelin ensheathes axons of local inhibitory neurons. eLife 5:e15784. https://doi.org/10.7554/elife.15784
10. Kole K, Voesenek BJB, Brinia ME, Petersen N, Kole MHP (2022) Parvalbumin basket cell myelination accumulates axonal mitochondria to internodes. Nat Commun 13:7598. https://doi.org/10.1038/s41467-022-35350-x
11. Nieuwenhuys R (2013) The myeloarchitectonic studies on the human cerebral cortex of the Vogt-Vogt school, and their significance for the interpretation of functional neuroimaging data. Brain Struct Funct 218:303–352. https://doi.org/10.1007/s00429-012-0460-z
12. Timmler S, Simons M (2019) Grey matter myelination. Glia 67:2063–2070. https://doi.org/10.1002/glia.23614
13. Faria O, de, Pivonkova H, Varga B, Timmler S, Evans KA, Káradóttir RT (2021) Periods of synchronized myelin changes shape brain function and plasticity. Nat Neurosci 24:1508–1521. https://doi.org/10.1038/s41593-021-00917-2
14. Liu X-B, Schumann CM (2014) Optimization of electron microscopy for human brains with long-term fixation and fixed-frozen sections. Acta Neuropathol Commun 2: 42. https://doi.org/10.1186/2051-5960-2-42
15. Sele M, Wernitznig S, Lipovšek S, Radulović S, Haybaeck J, Birkl-Toeglhofer AM, Wodlej C, Kleinegger F, Sygulla S, Leoni M, Ropele S, Leitinger G (2019) Optimization of ultrastructural preservation of human brain for transmission electron microscopy after long post-mortem intervals. Acta Neuropathol Commun 7:144. https://doi.org/10.1186/s40478-019-0794-3
16. Glausier JR, Konanur A, Lewis DA (2018) Factors affecting ultrastructural quality in the prefrontal cortex of the postmortem human brain. J Histochem Cytochem 67:185–202. https://doi.org/10.1369/0022155418819481
17. Lewis AJ, Genoud C, Pont M, Berg WD, van de, Frank S, Stahlberg H, Shahmoradian SH, Al-Amoudi A (2019) Imaging of post-mortem human brain tissue using electron and X-ray microscopy. Curr Opin Struct Biol 58:138–148. https://doi.org/10.1016/j.sbi.2019.06.003
18. Möbius W, Nave K-A, Werner HB (2016) Electron microscopy of myelin: structure preservation by high-pressure freezing. Brain Res 1641:92–100. https://doi.org/10.1016/j.brainres.2016.02.027
19. Rademaker MC, Lange GM, de, Palmen SJMC (2018) Chapter 1 The Netherlands Brain Bank for Psychiatry. Handb Clin Neurol 150:3–16. https://doi.org/10.1016/b978-0-444-63639-3.00001-3
20. Karlupia N, Schalek RL, Wu Y, Meirovitch Y, Wei D, Charney AW, Kopell BH, Lichtman JW (2023) Immersion fixation and staining of multicubic millimeter volumes for electron microscopy–based connectomics of human brain biopsies. Biol Psychiatry 94:352–360. https://doi.org/10.1016/j.biopsych.2023.01.025
21. Palade GE (1952) A study of fixation for electron microscopy. J Exp Med 95:285–298. https://doi.org/10.1084/jem.95.3.285
22. Edgar JM, Smith RS, Duncan ID (2020). Transmission Electron Microscopy and Morphometry of the CNSWhite Matter. In: Babetto, E. (eds) Axon Degeneration. Methods in Molecular Biology, vol 2143. Humana, NewYork, NY. https://doi.org/10.1007/978-1-0716-0585-1_18
23. Bakken TE, Jorstad NL, Hu Q, Lake BB, Tian W, Kalmbach BE, Crow M, Hodge RD, Krienen FM, Sorensen SA, Eggermont J, Yao Z, Aevermann BD, Aldridge AI, Bartlett A, Bertagnolli D, Casper T, Castanon RG,

Crichton K, Daigle TL, Dalley R, Dee N, Dembrow N, Diep D, Ding S-L, Dong W, Fang R, Fischer S, Goldman M, Goldy J, Graybuck LT, Herb BR, Hou X, Kancherla J, Kroll M, Lathia K, Lew B, van, Li YE, Liu CS, Liu H, Lucero JD, Mahurkar A, McMillen D, Miller JA, Moussa M, Nery JR, Nicovich PR, Niu S-Y, Orvis J, Osteen JK, Owen S, Palmer CR, Pham T, Plongthongkum N, Poirion O, Reed NM, Rimorin C, Rivkin A, Romanow WJ, Sedeño-Cortés AE, Siletti K, Somasundaram S, Sulc J, Tieu M, Torkelson A, Tung H, Wang X, Xie F, Yanny AM, Zhang R, Ament SA, Behrens MM, Bravo HC, Chun J, Dobin A, Gillis J, Hertzano R, Hof PR, Höllt T, Horwitz GD, Keene CD, Kharchenko PV, Ko AL, Lelieveldt BP, Luo C, Mukamel EA, Pinto-Duarte A, Preissl S, Regev A, Ren B, Scheuermann RH, Smith K, Spain WJ, White OR, Koch C, Hawrylycz M, Tasic B, Macosko EZ, McCarroll SA, Ting JT, Zeng H, Zhang K, Feng G, Ecker JR, Linnarsson S, Lein ES (2021) Comparative cellular analysis of motor cortex in human, marmoset and mouse. Nature 598:111–119. https://doi.org/10.1038/s41586-021-03465-8

24. Kalmbach BE, Hodge RD, Jorstad NL, Owen S, Frates R, de, Yanny AM, Dalley R, Mallory M, Graybuck LT, Radaelli C, Keene CD, Gwinn RP, Silbergeld DL, Cobbs C, Ojemann JG, Ko AL, Patel AP, Ellenbogen RG, Bakken TE, Daigle TL, Dee N, Lee BR, McGraw M, Nicovich PR, Smith K, Sorensen SA, Tasic B, Zeng H, Koch C, Lein ES, Ting JT (2021) Signature morpho-electric, transcriptomic, and dendritic properties of human layer 5 neocortical pyramidal neurons. Neuron 109:2914–2927.e5. https://doi.org/10.1016/j.neuron.2021.08.030

25. Braak H, Braak E (1976) The pyramidal cells of Betz within the cingulate and precentral gigantopyramidal field in the human brain. A Golgi and pigmentarchitectonic study. Cell Tissue Res 172:103–119

26. Mohan H, Verhoog MB, Doreswamy KK, Eyal G, Aardse R, Lodder BN, Goriounova NA, Asamoah B, Brakspear ABCB, Groot C, van der Sluis S, Testa-Silva G, Obermayer J, Boudewijns ZSRM, Narayanan RT, Baayen JC, Segev I, Mansvelder HD, de Kock CPJ (2015) Dendritic and axonal architecture of individual pyramidal neurons across layers of adult human neocortex. Cereb Cortex 25:4839–4853. https://doi.org/10.1093/cercor/bhv188

GPSR Compliance

The European Union's (EU) General Product Safety Regulation (GPSR) is a set of rules that requires consumer products to be safe and our obligations to ensure this.

If you have any concerns about our products, you can contact us on ProductSafety@springernature.com

In case Publisher is established outside the EU, the EU authorized representative is:

Springer Nature Customer Service Center GmbH
Europaplatz 3
69115 Heidelberg, Germany

Batch number: 10770324

Printed by Printforce, the Netherlands